ADVANCES IN CHEMICAL ENGINEERING

Volume 10

CONTRIBUTORS TO THIS VOLUME

P. C. Kapur
Richard S. H. Mah
G. E. O'Connor
T. W. F. Russell
J. Robert Selman
Mordechai Shacham
Charles W. Tobias

ADVANCES IN
CHEMICAL ENGINEERING

Edited by

THOMAS B. DREW

Department of Chemical Engineering
Massachusetts Institute of Technology
Cambridge, Massachusetts

GILES R. COKELET

Department of Chemical Engineering
University of Rochester
Rochester, New York

JOHN W. HOOPES, Jr.

Imperial Chemical Industries United States, Inc.
Wilmington, Delaware

THEODORE VERMEULEN

Department of Chemical Engineering
University of California
Berkeley, California

Volume 10

Academic Press • New York • London 1978

A Subsidiary of Harcourt Brace Jovanovich, Publishers

COPYRIGHT © 1978, BY ACADEMIC PRESS, INC.
ALL RIGHTS RESERVED.
NO PART OF THIS PUBLICATION MAY BE REPRODUCED OR
TRANSMITTED IN ANY FORM OR BY ANY MEANS, ELECTRONIC
OR MECHANICAL, INCLUDING PHOTOCOPY, RECORDING, OR ANY
INFORMATION STORAGE AND RETRIEVAL SYSTEM, WITHOUT
PERMISSION IN WRITING FROM THE PUBLISHER.

ACADEMIC PRESS, INC.
111 Fifth Avenue, New York, New York 10003

United Kingdom Edition published by
ACADEMIC PRESS, INC. (LONDON) LTD.
24/28 Oval Road, London NW1 7DX

LIBRARY OF CONGRESS CATALOG CARD NUMBER: 56-6600

ISBN 0-12-008510-0

PRINTED IN THE UNITED STATES OF AMERICA

CONTENTS

LIST OF CONTRIBUTORS vii
PREFACE ix
CONTENTS OF PREVIOUS VOLUMES xi

Heat Transfer in Tubular Fluid-Fluid Systems
G. E. O'CONNOR AND T. W. F. RUSSELL

I.	Introduction: The Flow Behavior	1
II.	Heat Transfer without Phase Change	9
III.	Heat Transfer with Phase Change	28
IV.	Concluding Remarks	48
	References	51

Balling and Granulation
P. C. KAPUR

I.	Introduction	56
II.	Balling and Granulation Equipment	57
III.	Bonding Mechanisms in Agglomerates	62
IV.	Compaction and Growth Mechanisms	75
V.	Kinetics of Balling and Granulation	84
VI.	Bonding Liquid and Additives	100
VII.	Granulation of Fertilizers	105
VIII.	Miscellaneous Topics	112
	References	120

Pipeline Network Design and Synthesis
RICHARD S. H. MAH AND MORDECHAI SHACHAM

I.	Introduction	126
II.	Steady-State Pipeline Network Problems: Formulation . . .	127
III.	Steady-State Pipeline Network Problems: Methods of Solution	148
IV.	Design Optimization and Synthesis	170
V.	Transient and Compressible Flows in Pipeline Networks . .	190
VI.	Concluding Remarks	198
	References	205

Mass-Transfer Measurements by the Limiting-Current Technique

J. Robert Selman and Charles W. Tobias

I.	Introduction	212
II.	Basic Theory	213
III.	Limiting-Current Measurement	217
IV.	Interpretation of Results	229
V.	Conditions for Valid Measurement and Interpretation of Limiting Currents	252
VI.	Review of Applications	253
VII.	Concluding Remarks	279
	References	310

Author Index	319
Subject Index	331

CONTRIBUTORS TO VOLUME 10

P. C. KAPUR, *Department of Metallurgical Engineering, Indian Institute of Technology, Kanpur—208016, India*

RICHARD S. H. MAH, *Department of Chemical Engineering, The Technological Institute, Northwestern University, Evanston, Illinois 60201*

G. E. O'CONNOR*, *Department of Chemical Engineering, University of Delaware, Newark, Delaware*

T. W. F. RUSSELL, *Department of Chemical Engineering, University of Delaware, Newark, Delaware*

J. ROBERT SELMAN, *Department of Chemical Engineering, Illinois Institute of Technology, Chicago Illinois 60616*

MORDECHAI SHACHAM, *Department of Chemical Engineering, Ben Gurion University of the Negev, Beersheva, Israel*

CHARLES W. TOBIAS, *Materials and Molecular Research Division, Lawrence Berkeley Laboratory, and Department of Chemical Engineering, University of California, Berkeley, California 94720*

---

* Present address: Monsanto Company–Rubber Chemicals, 260 Springside Road, Akron, Ohio 44313.

PREFACE

The aim of Volume 10, as with each of its predecessors in this serial publication, is perhaps best expressed by quoting the first two paragraphs of our instructions to our authors:

> Ideally, a chapter in *Advances in Chemical Engineering* is a short monograph in which the author summarizes the current state of knowledge of his topic for the benefit of professional colleagues in engineering who by reason of their normal duties have not been able to make a study of the subject in depth. They want an authoritative account not cloaked in unintelligible specialized terminology. They will read it, not only for general information, but also because as sophisticated engineers they know that major progress in science and engineering is made by those who see connections between matters others have imagined unrelated: they may spot in the author's specialty a method or an idea with an analog useful in theirs. Many readers will not have ready access to large university libraries and many with such access are by hypothesis too inexpert to assess the validity of journal articles on the author's topic—they expect such assessment in the chapter.
>
> Typically, one expects a chapter to be a critical review and an evaluation of the results and opinions which various workers have presented in journal articles or books. The author is expected to point out discrepancies in previous work and, if he cannot resolve them, to suggest the nature of further studies needed for that purpose. Except where it may be necessary to introduce them to justify his evaluations and conclusions, an article in *Advances in Chemical Engineering* is not ordinarily the appropriate place of first publication for new experimental or theoretical results of the author. In exceptional cases, especially when the space required for intelligible presentation would exceed that normally available in a journal article, the Editors will consider chapters that are essentially reports of previously unpublished work by the author.

The Editors and their assisting reviewers feel that the chapters herein satisfy adequately the criteria set forth above. However, statements and opinions in these chapters represent judgments of the respective authors who by submitting the chapter may be presumed to aver that the references listed have been personally studied and compared, and that the formulas and data quoted have been personally verified unless the contrary is stated.

Thomas B. Drew
Giles R. Cokelet
John W. Hoopes, Jr.
Theodore Vermeulen

CONTENTS OF PREVIOUS VOLUMES

Volume 1

Boiling of Liquids
  J. W. Westwater
Non-Newtonian Technology: Fluid Mechanics, Mixing, and Heat Transfer
  A. B. Metzner
Theory of Diffusion
  R. Byron Bird
Turbulence in Thermal and Material Transport
  J. B. Opfell and B. H. Sage
Mechanically Aided Liquid Extraction
  Robert E. Treybal
The Automatic Computer in the Control and Planning of Manufacturing Operations
  Robert W. Schrage
Ionizing Radiation Applied to Chemical Processes and to Food and Drug Processing
  Ernest J. Henley and Nathaniel F. Barr
AUTHOR INDEX—SUBJECT INDEX

Volume 2

Boiling of Liquids
  J. W. Westwater
Automatic Process Control
  Ernest F. Johnson
Treatment and Disposal of Wastes in Nuclear Chemical Technology
  Bernard Manowitz
High Vacuum Technology
  George A. Sofer and Harold C. Weingartner
Separation by Adsorption Methods
  Theodore Vermeulen

Mixing of Solids
  Sherman S. Weidenbaum
AUTHOR INDEX—SUBJECT INDEX

Volume 3

Crystallization from Solution
  C. S. Grove, Jr., Robert V. Jelinek, and Herbert M. Schoen
High Temperature Technology
  F. Alan Ferguson and Russell C. Phillips
Mixing and Agitation
  Daniel Hyman
Design of Packed Catalytic Reactors
  John Beek
Optimization Methods
  Douglass J. Wilde
AUTHOR INDEX—SUBJECT INDEX

Volume 4

Mass-Transfer and Interfacial Phenomena
  J. T. Davies
Drop Phenomena Affecting Liquid Extraction
  R. C. Kintner
Patterns of Flow in Chemical Process Vessels
  Octave Levenspiel and Kenneth B. Bischoff
Properties of Cocurrent Gas–Liquid Flow
  Donald S. Scott
A General Program for Computing Multistage Vapor–Liquid Processes
  D. N. Hanson and G. F. Somerville
AUTHOR INDEX—SUBJECT INDEX

Volume 5

Flame Processes—Theoretical and Experimental
  J. F. Wehner
Bifunctional Catalysts
  J. H. Sinfelt
Heat Conduction or Diffusion with Change of Phase
  S. G. Bankoff

The Flow of Liquids in Thin Films
  George D. Fulford
Segregation in Liquid–Liquid Dispersions and Its Effect on Chemical Reactions
  K. Rietema
AUTHOR INDEX—SUBJECT INDEX

Volume 6

Diffusion-Controlled Bubble Growth
  S. G. Bankoff
Evaporative Convection
  John C. Berg, Andreas Acrivos, and Michel Boudart
Dynamics of Microbial Cell Populations
  H. M. Tsuchiya, A. G. Fredrickson, and R. Aris
Direct Contact Heat Transfer between Immiscible Liquids
  Samuel Sideman
Hydrodynamic Resistance of Particles at Small Reynolds Numbers
  Howard Brenner
AUTHOR INDEX—SUBJECT INDEX

Volume 7

Ignition and Combustion of Solid Rocket Propellants
  Robert S. Brown, Ralph Anderson, and Larry J. Shannon
Gas-Liquid-Particle Operations in Chemical Reaction Engineering
  Knud Østergaard
Thermodynamics of Fluid-Phase Equilibria at High Pressures
  J. M. Prausnitz
The Burn-Out Phenomenon in Forced-Convection Boiling
  Robert V. Macbeth
Gas-Liquid Dispersions
  William Resnick and Benjamin Gal-Or
AUTHOR INDEX—SUBJECT INDEX

Volume 8

Electrostatic Phenomena with Particulates
  C. E. Lapple
Mathematical Modeling of Chemical Reactions
  J. R. Kittrell

Decomposition Procedures for the Solving of Large Scale Systems
  W. P. Ledet and D. M. Himmelblau
The Formation of Bubbles and Drops
  R. Kumar and N. R. Kuloor
AUTHOR INDEX—SUBJECT INDEX

Volume 9

Hydrometallurgy
  Renato G. Bautista
Dynamics of Spouted Beds
  Kishan B. Mathur and Norman Epstein
Recent Advances in the Computation of Turbulent Flows
  W. C. Reynolds
Drying of Solid Particles and Sheets
  R. E. Peck and D. T. Wasan
AUTHOR INDEX—SUBJECT INDEX

ADVANCES IN CHEMICAL ENGINEERING

Volume 10

HEAT TRANSFER
IN TUBULAR FLUID–FLUID SYSTEMS

G. E. O'Connor[1] and T. W. F. Russell

Department of Chemical Engineering
University of Delaware
Newark, Delaware

I.	Introduction: The Flow Behavior	1
	A. Flow Patterns	2
	B. Two-Phase Hydrodynamics	6
II.	Heat Transfer without Phase Change	9
	A. Basic Model Equations	11
	B. Parameter Evaluation	19
III.	Heat Transfer with Phase Change	28
	A. Vaporization Phenomena	28
	B. The Rate of Phase Change	31
	C. Basic Model Equations	37
	D. Parameter Evaluation	41
	E. Model Behavior	46
IV.	Concluding Remarks	48
	Nomenclature	49
	References	51

I. Introduction: The Flow Behavior

This chapter has two goals, to provide a critical review of the current state of the art in the field of two-phase flow with heat transfer and to provide procedures which can be used for the design of tubular fluid–fluid systems. We hope that this work will help point out areas in which further theoretical and experimental research is critically needed, and that it will motivate design engineers to test out our procedures (in combination with details from the original references) in solving pragmatic problems.

---

[1] Present address: Monsanto Company–Rubber Chemicals, 260 Springside Road, Akron, Ohio.

A systematic, rational analysis of both isothermal and nonisothermal tubular systems in which two fluids are flowing must be carried out, if optimal design and economic operation of these pipeline devices is to be achieved. The design of all two-phase contactors must be based on a firm knowledge of two-phase hydrodynamics. In addition, a mathematical description is needed of the heat and mass transfer and of the chemical reaction occurring within a particular system.

To predict the heat transfer effects, the engineer must have an adequate quantitative description of heat transfer between the tube wall and the fluid phases, heat transfer between the tube wall and the fluid phases, heat transfer between the two phases, the rate of phase change within the system, and the rate of heat transfer resulting from phase change. Unfortunately, present design procedures only provide estimates of the system performance. Many procedures have not been formulated in a systematic manner, and therefore it is difficult to pinpoint areas where the present understanding of the design process is weakest.

Most of the research studies on heat transfer in nonisothermal two-phase systems have been conducted without a firm understanding of the hydrodynamics, and consequently useful mathematical expressions for the conservation of mass and energy have not been properly derived. Empirical correlations for the wall and the interfacial heat transfer coefficients exist, but at the present time these correlations cannot be used with great confidence because frequently the lack of knowledge of the gross fluid motions is incorporated into the correlations. Little attention has been given to developing an analytical expression for the rate of phase change at the gas–liquid interface in nonisothermal systems, and this must be known before systems in which phase change is important can be analyzed.

A. Flow Patterns

The analysis of two-phase tubular contactors and pipelines is complicated because of the variety of configurations that the two-phase mixture may assume in these systems. The design engineer must have knowledge of the flow pattern that results from a given set of operating conditions if the in situ quantities such as pressure drop, holdup of each phase, phase Reynolds numbers, and interfacial area are to be determined. These in situ quantities must be known if the rate of heat transfer is to be predicted.

Through visual identification, Alves (A2) has defined the different flow patterns that occur in horizontal gas–liquid systems; Nicklin and Davidson (N2) have defined the different visual flow patterns appearing in vertical gas–liquid systems. These flow patterns are depicted in Figs. 1 and 2. The

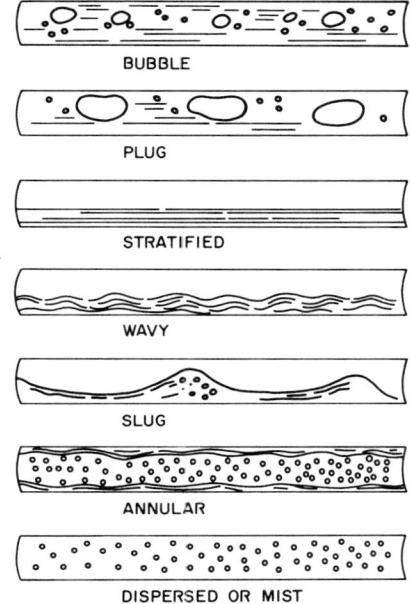

FIG. 1. Horizontal gas–liquid flow patterns.

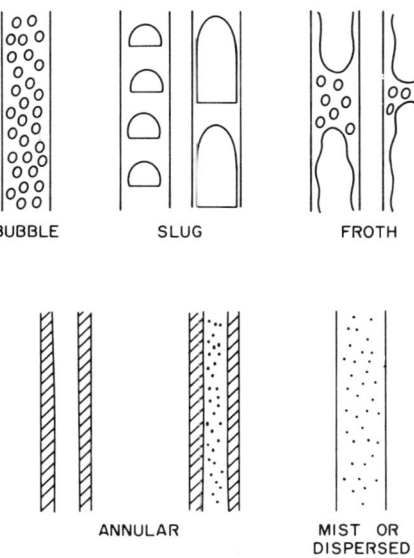

FIG. 2. Vertical gas–liquid flow patterns.

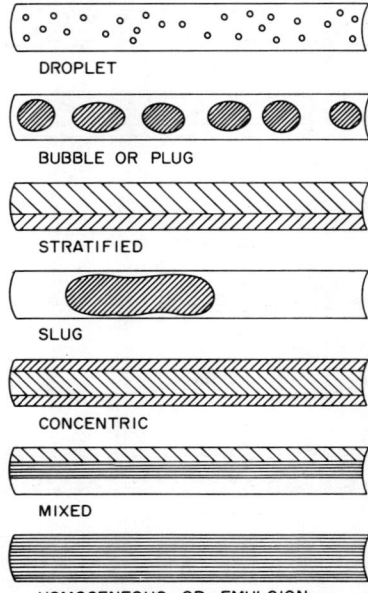

FIG. 3. Horizontal liquid–liquid flow patterns.

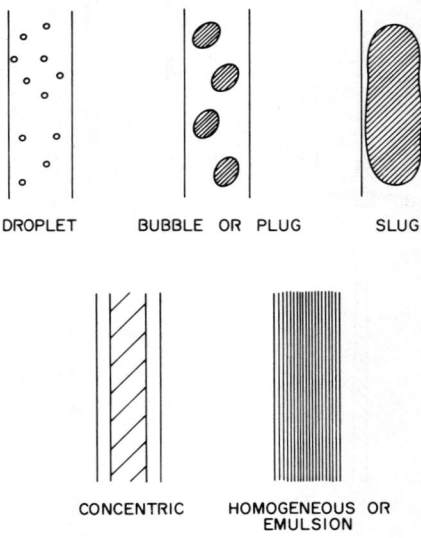

FIG. 4. Vertical liquid–liquid flow patterns.

descriptive definitions of the flow patterns have been given in a good detailed review of gas–liquid flows by Scott (S4), in a recent book by Govier and Kaziz (G1), and in papers by Cichy et al. (C5), and Anderson and Russell (A3). The flow patterns in liquid–liquid flows have been identified by Charles et al. (C3), are further discussed in Govier and Short (G2), and are shown in Figs. 3 and 4. It must be noted that, for liquid–liquid systems, either phase can exist as the continuous phase in the central core in concentric flow, or in the form of drops, bubbles, plugs or slugs in the dispersed-flow patterns.

Baker (B1) developed a flow pattern map for horizontal gas–liquid systems that is shown in Fig. 5. The coordinates are functions of gas and liquid mass flow rates, phase densities, liquid viscosity, and surface tension. Using the same coordinates, Cichy et al. (C5) have presented a modification of the flow-pattern maps of Govier and co-workers (B6, G2, G3) for vertical gas liquid systems.

Etchells (E1) has pointed out that the Baker chart has four major shortcomings: (1) the data used to define the flow patterns are based upon the independent visual observations of many researchers, each having his own description for a particular flow pattern; (2) air–water measurements in 1- and 2-in. pipes represent a major portion of the data; (3) the chart is prepared from a limited number of data, not all taken at the transition points; and (4) many experiments were performed in short pipes or pipes with unusual inlets, causing entrance and transition effects that may not have died out in the region of observation. Similar comments can be made about the other flow pattern charts.

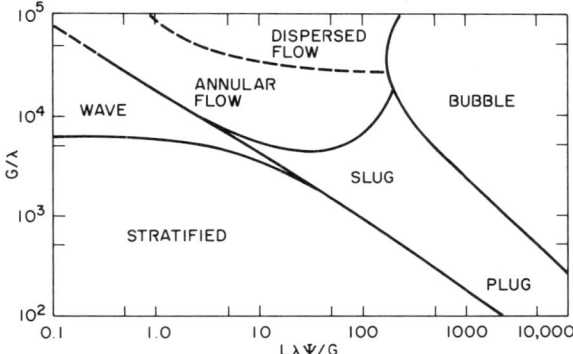

FIG. 5. The Baker chart, $\lambda = [(\rho_G/0.075)(\rho_L/62.3)]^{1/2}$ and $\psi = (73/\gamma_L)[\mu_L(62.3/\rho_L)]^{1/3}$. The units are: ρ_G, lb_{mass} ft^{-3}; γ_L, dyn cm^{-1}; ρ_L, lb_{mass} ft^{-3}; μ_{L_1}, centipoise; G, lb_{mass} ft^{-2} hr^{-1}; L, lb_{mass} ft^{-2} hr^{-1}.

Additional research on the prediction of flow patterns is a necessity, for until detailed stability criteria are developed for the transition from one flow pattern to another, there is no alternative to the empirical flow pattern charts. Some progress in theoretically defining the transition from stratified to wavy or slug flow has been made by Russell and Etchells (R3). Inaccuracy and uncertainty in flow pattern prediction makes estimation of the in situ hydrodynamic quantities and the rate of heat transfer a difficult task.

The flow patterns in liquid–liquid systems have not been as extensively studied as those in gas–liquid systems. However, Russell *et al.* (R6), and Charles *et al.* (C3) have studied the flow of oils and water in horizontal pipes and have presented flow-pattern charts for the various oil–water systems. It is very difficult to predict the flow pattern for a liquid–liquid system, unless the liquids have physical properties similar to those of water and the oils used by Govier and co-workers. The Baker chart might be used to give a first estimate of the flow pattern for a liquid–liquid system, but the viscosity of the less-dense phase is not included in the coordinate parameters, and the feasibility of such an approach has never been investigated.

B. Two-Phase Hydrodynamics

The design of two-phase heat transfer devices requires a knowledge of the hydrodynamics so that the pressure drop, holdup, phase velocities, and interfacial areas can be determined. In this discussion Phase I will generally refer to the less-dense phase, and Phase II to the more-dense phase. The holdup of Phase I has been conventionally defined as

$$R_I = \frac{\text{cross-sectional area of the pipe occupied by Phase I}}{\text{cross-sectional area of the pipe}} \quad (1)$$

and similarly for Phase II. Thus

$$R_I + R_{II} = 1.0 \quad (2)$$

The principles of conservation of mass and momentum must be applied to each phase to determine the pressure drop and holdup in two phase systems. The differential equations used to model these principles have been solved only for laminar flows of incompressible, Newtonian fluids, with constant holdups. For this case, the momentum equations become

$$\partial^2 v_I/\partial x^2 + \partial^2 v_I/\partial y^2 = -(1/\mu_I)(\Delta P/L) \quad (3a)$$

$$\partial^2 v_{II}/\partial x^2 + \partial^2 v_{II}/\partial y^2 = -(1/\mu_{II})(\Delta P/L) \quad (3b)$$

where y is distance measured perpendicular to the interface in the plane of the cross section and the x axis lies in this plane and extends perpendicular to the y axis. The boundary conditions are

$$v_\text{I} = 0 \quad \text{at the duct wall} \quad (4a)$$

$$v_\text{II} = 0 \quad \text{at the duct wall} \quad (4b)$$

$$\mu_\text{I}\, \partial v_\text{I}/\partial y = \mu_\text{II}\, \partial v_\text{II}/\partial y \quad \text{at} \quad y = h \quad (4c)$$

$$v_\text{I} = v_\text{II} \quad \text{at} \quad y = h \quad (4d)$$

All solutions of Eqs. (3) such as that by Yu and Sparrow (Y1) yield the velocity profiles in each phase as a function of the interfacial position h and the pressure drop. The volumetric flow rates Q_I and Q_II are obtained by integrating each velocity profile over the respective phase cross-sectional area. The ratio of the flow rates can then be determined as a function of only the interfacial position, and since the volumetric flow rates are known, this yields an implicit fourth order equation for the interfacial position h. The holdups R_I and R_II can be calculated once the interfacial position is known. Since each equation for the volumetric flow rates is linear with respect to the pressure drop, once the interfacial position is known the pressure drop may be easily computed. An analytical procedure for determining pressure drop and holdup for turbulent gas–laminar liquid flows has been developed by Etchells (E1) and verified by comparison with experimental data in horizontal systems (A7).

When both phases are in turbulent flow, or when one phase is discontinuous as in bubble flow, it is not presently possible to formulate the proper boundary conditions and to solve the equations of motion. Therefore, numerous experimental studies have been conducted where the holdups and/or the pressure drop were measured and then correlated as a function of the operating conditions and system parameters. One of the most widely used correlations is that of Lockhart and Martinelli (L12), who assumed that the pressure drop in each phase could be calculated from the equations

$$(\Delta P/\Delta z)_{TP_\text{G}} = 2f_\text{G}(\rho_\text{G} v_\text{G}^2/D_\text{G} g_\text{c}) \quad (5a)$$

$$(\Delta P/\Delta z)_{TP_\text{L}} = 2f_\text{L}(\rho_\text{L} v_\text{L}^2/D_\text{L} g_\text{c}) \quad (5b)$$

where the friction factors f_G and f_L are given by empirical relations of the form

$$f_i = C\, \text{Re}_i^n$$

and D_L and D_G are the unknown hydraulic diameters of each phase. Rather than correlating the hydraulic diameters and the holdups as functions of the operating conditions, they defined two general parameters as follows:

$$\chi = [(\Delta P/\Delta z)_L/(\Delta P/\Delta z)_G]^{1/2} \tag{6a}$$

$$\phi_G = [(\Delta P/\Delta z)_{TP}/(\Delta P/\Delta z)_G]^{1/2} \tag{6b}$$

The terms $(\Delta P/\Delta z)_L$ and $(\Delta P/\Delta z)_G$ are easily evaluated since they represent the pressure drop which would exist if the liquid and gas flowed alone in the pipe.

Lockhart and Martinelli divided gas–liquid flows into four cases: (1) laminar gas–laminar liquid; (2) turbulent gas–laminar liquid; (3) laminar gas–turbulent liquid; and (4) turbulent gas–turbulent liquid. They measured two-phase pressure drops and correlated the value of ϕ_g with parameter χ for each case. The authors presented a plot of ϕ_g versus χ for each case, and one plot of R_G and R_L versus χ that is applicable to all cases. The assumptions applied by Lockhart and Martinelli are: (1) constant holdups, R_G and R_L; (2) no acceleration effects, incompressible flow; (3) no interaction at the interface; and (4) the pressure drop in the gas phase equals the pressure drop in the liquid phase.

Numerous other correlations for pressure drop and holdup have been developed, but none has been accepted by the practicing engineer to the extent that Lockhart and Martinelli's has. Charles and Lilliheht (C2) have developed a correlation, which is analogous to that of Lockhart and Martinelli, for pressure drop in stratified laminar liquid–turbulent liquid systems. Unfortunately they did not include a holdup correlation. Anderson and Russell (A4) and Dukler et al. (D4) have reviewed the applicability and accuracy of the more useful correlations. A designer must be aware that, while a correlation is supposedly applicable to a specific flow pattern, it can yield greatly inaccurate results in some cases.

Martinelli and Nelson (M7) developed a procedure for calculating the pressure drop in tubular systems with forced-circulation boiling. The procedure, which includes the accelerative effects due to phase change while assuming each phase is an incompressible fluid, is an extrapolation of the Lockhart and Martinelli χ parameter correlation. Other pressure drop calculation procedures have been proposed for forced-circulation phase-change systems; however, these suffer severe shortcomings, and have not proved more accurate than the Martinelli and Nelson method.

At the present time most of what we know about the fluid mechanics of two-phase flow in pipes is thoroughly discussed in books by Govier and Aziz (G1), Wallis (W2), Brauer (B5), and Hewitt and Hac-Taylor (H1).

While it is apparent that the study of mass transfer and heat transfer in two-phase tubular systems requires a thorough understanding of the hydrodynamics of two-phase flow, it is equally apparent that most real problems in two-phase contactors involve heat and mass transfer. The solutions to these problems cannot be delayed until the hydrodynamics is completely understood. They must be attempted realizing that the lack of knowledge about the hydrodynamics presents severe limitations on the degree of agreement one can expect between theory and physical reality. With these limitations in mind, the remainder of this chapter is devoted to the analysis of heat transfer in two-phase flow.

II. Heat Transfer without Phase Change

Heat transfer in two-phase flow can be studied most conveniently by adopting a two-part classification: heat transfer without phase change, and heat transfer with phase change. In this section, the non-phase-change problems will be discussed; the phase-change problem will be dealt with in Section III.

There are a number of two-phase systems for which the design engineer must predict either the rate of heat transfer between the phases or between each fluid and the tube wall. Such operations may occur as a result of heat effects due to reaction or absorption, or they may simply arise because it is necessary to change the mixture temperature. Common adiabatic operations in which heat transfer through the tube wall is not of prime importance are liquid–liquid, direct-contact heat transfer, and mass-contacting or reacting systems in which there is no need to add or remove heat. Nonadiabatic operations that require the addition or removal of heat are encountered in the heating of two-phase mixtures, or as a result of strong thermal effects in mass transfer contactors or reacting systems.

At present there is not a general systematic approach for solving any non-phase-change design problem. There are only a few studies in the literature, and, in most cases, the results presented are not very useful for design. Most investigators fail to recognize that the flow pattern in which the experiments are performed must be properly defined, and too frequently this lack of knowledge of the gross fluid mechanics is reflected in the values of the heat-transfer coefficients reported.

Two-phase mass transfer and heat transfer without phase change are analogous, and the results of mass-transfer studies can be used to help clarify the heat-transfer problems. Cichy *et al.* (C5) have formulated basic design equations for isothermal gas–liquid tubular reactors. The authors arranged the common visually defined flow patterns into five basic flow regimes, each regime being characterized by a specific set of hydrodynamic properties. A set of model equations for the conservation of mass was developed for each regime. Cichy and Russell (C4) have discussed the evaluation of the parameters in the model equations from a design viewpoint. The flow configurations for liquid–liquid systems can be defined in an analogous manner, based on the works of Charles *et al.* (C3), and Russell *et al.* (R6):

1. Stratified, smooth interface, laminar liquid, laminar liquid.
2. Stratified, smooth interface, less dense liquid: turbulent, more dense liquid: laminar.
3. Stratified, smooth interface, turbulent liquid, turbulent liquid.
4. Stratified flow, wavy interface.
5. Stratified flow, mixed phase separating two pure liquid phases.
6. Horizontal, concentric flow.
7. Vertical, concentric flow.
8. Horizontal, slugs.
9. Vertical, slugs.
10. Horizontal, noninteracting drops.
11. Horizontal, interacting plugs.
12. Vertical, noninteracting drops.
13. Vertical, interacting plugs.
14. Horizontal, emulsions.
15. Vertical, emulsions.

These flow configurations are grouped into the five basic flow regimes defined by Cichy *et al.* (the numbers refer to the above list):

I. Continuous fluid phases with a well-defined interface: 1, 2, 3, 4, 5, 6, 7.
II. Continuous fluid phases with complex interfaces and fluid phase interchange: none.
III. Alternating discrete fluid phases: 8, 9.
IV. One continuous fluid phase and one discrete fluid phase: 10, 11, 12, 13.
V. Homogeneous two-phase flow: 14, 15.

The importance of the gross fluid mechanics is emphasized in this chapter by developing a set of model equations for the following flow regimes. Regime I is characterized by the presence of two continuous fluid phases. Regime II includes the fluid–fluid flows which maintain two continuous phases plus the entrainment of one phase as drops or bubbles in the other. Regime IV represents those systems with one continuous phase and one discrete phase. Regimes III and V are not considered in this chapter. Regime III includes those flows characterized by the periodic transition from principally Phase I flow to principally Phase II flow. This generates a relatively small interfacial area and large pipeline vibrations; both conditions are generally undesirable in pragmatic situations. In Regime V the two-phase system is considered to be a homogeneous mixture. This approach can be useful in analyzing the hydrodynamics, but the interphase heat and mass transfer must be described by applying a Regime IV or possibly a Regime I model.

For each regime considered, the system variables are assumed to be functions of axial position, and a general set of steady state mass and energy balances are proposed for each regime, since the focal point of any design is the steady-state system performance. The unsteady-state model equations have been developed by O'Connor (O1). Simplified assumptions applicable to both gas–liquid and liquid–liquid systems are specified and the equations reduced accordingly. The specific properties of the liquid–liquid and the gas–liquid systems are discussed separately, and the evaluation of the necessary parameters is described in detail. With these equations and methods for evaluating the parameters, an effective design can be performed for nonisothermal two-phase systems when phase change can be neglected.

A. Basic Model Equations

1. Regime I

This regime is characterized by the presence of two continuous fluid phases and an interface which can easily be described. The term *separated flows* is frequently employed to describe these situations in both horizontal and vertical systems. Some flow patterns in Regime I are advantageous for transferring heat between the tube wall and the fluid mixture or for carrying out two-phase reactions. The special case of laminar–laminar flow is included in this regime, and two studies seem to be of interest, Byers and King (B7) and Bentwich and Sideman (B3).

In general, only turbulent–turbulent flows are of pragmatic interest. The basic Regime I mass balances for the steady state turbulent-turbulent case are

$$d(\rho_I R_I v_I)/dz = dW_I/dz = 0 \tag{7a}$$

$$d(\rho_{II} R_{II} v_{II})/dz = dW_{II}/dz = 0 \tag{7b}$$

and the energy balances are

$$W_I \, dH_I/dz = (d/dz[R_I k_I \, dT_I/dz]) + Ua(T_{II} - T_I) + q_{wI} \tag{8a}$$

$$W_{II} \, dH_{II}/dz = (d/dz[R_{II} k_{II} \, dT_{II}/dz]) - Ua(T_{II} - T_I) + q_{wII} \tag{8b}$$

In formulating Eqs. (7)–(8), the cross-sectional area of the contactor was assumed to be constant, changes in kinetic and potential energy were assumed to be negligible, and cocurrent flow was assumed. Countercurrent flow can only be established in vertical systems, and only over a rather narrow range of flow conditions.

a. *Liquid–Liquid Systems.* Heat transfer in liquid–liquid systems is encountered in the design of adiabatic units for direct contact heat transfer and nonisothermal solvent extraction, and of nonadiabatic units for heating two immiscible liquids. In analyzing these systems, it is often reasonable to assume that the thermal conductivities are constant, that the holdup of each phase is constant, and that the enthalpies are solely functions of temperature. Equations (8) can be rearranged using the following dimensionless groups:

$$Pe_I = W_I C_{PI} L / (R_I k_{I,\,eff}), \qquad Pe_{II} = W_{II} C_{PII} L / (R_{II} k_{II,\,eff})$$
$$N_I = UaL / (W_I C_{PI}), \qquad M = W_I C_{PI} / (W_{II} C_{PII})$$

Equations (8) are based on the assumption of plug flow in each phase but one may take account of any axial mixing in each liquid phase by replacing the molecular thermal conductivities k_I and k_{II} with the effective thermal conductivities $k_{I,\,eff}$ and $k_{II,\,eff}$ in the definition of the Peclet numbers. The evaluation of these conductivity terms is discussed in Section II,B,1. The wall heat-transfer terms may be defined as

$$N_{WI} = q_{WI} L / (W_I C_{PI}), \qquad N_{WII} = q_{WII} L / (W_{II} C_{PII})$$

and Eqs. (8) can be expressed as

$$(1/Pe_I) \, d^2 T_I / d\zeta^2 - dT_I / d\zeta = -N_I (T_{II} - T_I) - N_{WI} \tag{9a}$$

$$(1/Pe_{II}) \, d^2 T_{II} / d\zeta^2 - dT_{II} / d\zeta = N_I M (T_{II} - T_I) - N_{WII} \tag{9b}$$

The solution of Eqs. (9) is straightforward if the six parameters are known and the boundary conditions are specified. Two boundary conditions are necessary for each equation. Pavlica and Olson (P1) have discussed the applicability of the Wehner–Wilhelm boundary conditions (W3) to two-phase mass-transfer model equations, and have described a numerical method for solving these equations. In many cases this is not necessary, for the second-order differentials can be neglected. Methods for evaluating the dimensionless groups in Eqs. (9) are given in Section II,B,1.

b. *Gas–Liquid Systems.* Heat transfer without significant phase change in gas–liquid systems is most commonly encountered because of the design considerations involved in nonisothermal mass-transfer or reaction processes. The processes are either carried out adiabatically with interphase transfer or nonadiabatically with both interphase transfer and transfer between the fluids and the tube wall. In both cases, the general fluid–fluid equations of Section II,A,1 may be applied.

As in Section II,A,1,a, the thermal conductivities are assumed to be constant, and the liquid enthalpy is assumed to be only a function of temperature. The gas phase enthalpy term can be separated into two parts:

$$dH_I/dz = C_{PI}\, dT_I/dz + (\partial H_I/\partial P)_{T_I}\, dP/dz \qquad (10)$$

and through the use of known thermodynamic relationships, Eq. (10) becomes

$$dH_I/dz = C_{PI}\, dT_I/dz + \{1/\rho_I + (T_I/\rho_I^2)(\partial \rho_I/\partial T_I)_P\}\, dP/dz \qquad (11)$$

For an ideal gas, Eq. (11) simplifies to yield

$$dH_I/dz = C_{PI}\, dT_I/dz \qquad (12)$$

By assuming constant holdups and by using Eq. (12), Eqs. (8) become

$$R_I k_I\, d^2 T_I/dz^2 - W_I C_{PI}\, dT_I/dz = -Ua(T_{II} - T_I) - q_{wI} \qquad (13a)$$

$$R_{II} k_{II}\, d^2 T_{II}/dz^2 - W_{II} C_{PII}\, dT_{II}/dz = Ua(T_{II} - T_I) - q_{wII} \qquad (13b)$$

Equations (13) may be transformed into a form analogous to Eqs. (9) with the dimensionless groups defined in the same manner as in Section II,A,1,a.

2. Regime II

This regime is defined to include the fluid–fluid flows that maintain two continuous phases, with a portion of one phase leaving the continuous flow stream and entering the second phase as discrete drops or bubbles. The phase exchange process causes the interface to be highly disturbed and thus more difficult to describe than Regime I interfaces. There are no liquid–liquid flow configurations in this regime, and therefore the discussion is limited to gas–liquid systems. Heat transfer without phase change is encountered in the design of nonisothermal mass transfer or reaction processes. These systems may be either adiabatic or nonadiabatic, depending upon the specific process.

The gas phase exists as a continuum in the central core of the conduit, and the liquid phase exists as a continuous film along the tube wall and as droplets in the gas phase. New droplets are continually being formed at the gas–liquid interface. The motion of the gas phase accelerates the droplets to a velocity approaching that of the gas phase. Droplets in the central core continually impinge on the film. This phenomenon is known as *interchange*. For the purposes of modeling, the droplets are referred to as the entrained phase. The following assumptions are made with respect to the droplets at each axial position; the droplets can be characterized by (1) an average size, (2) an average interfacial area, (3) an average velocity, and (4) an average enthalpy. A steady-state balance on the number of droplets is expressed as

$$d(\alpha v_e)/dz = Q_g - Q_d \tag{14}$$

where v_e is the characteristic velocity of the entrained phase, Q_g the rate of droplet generation and Q_d the rate of droplet deposition on the liquid film. The Regime II mass balances are

$$d(\rho_l R_l v_l)/dz = d(W_l)/dz = 0 \tag{15a}$$

$$d(\rho_f R_e v_e)/dz = (Q_g - Q_d)\rho_f V_d \tag{15b}$$

$$d(\rho_f R_f v_f)/dz = -(Q_g - Q_d)\rho_f V_d \tag{15c}$$

where $R_f = A_f/A_c$ and

$$R_e = \alpha A_c V_d / A_c \tag{16}$$

If the assumption of constant entrainment is made, then from Eq. (15b), $Q_g = Q_d$. This is equivalent to assuming that the mass flow rates of the entrained phase and the liquid film are constant.

By applying these assumptions, the energy balances can be written as

$$W_l \, dH_l/dz = (d/dz(k_l R_l \, dT_l/dz)) + Ua(T_f - T_l)$$
$$+ U'a'V_d \alpha (T_e - T_l) \tag{17a}$$

$$W_e \, dH_e/dz = -U'a'V_d \alpha (T_e - T_l) + Q_g \rho_f V_d (H_f - H_e) \tag{17b}$$

$$W_f \, dH_f/dz = (d/dz(k_f R_f \, dT_f/dz)) - Ua(T_f - T_l)$$
$$- Q_g \rho_f V_d (H_f - H_e) + q_{wf} \tag{17c}$$

The application of Eqs. (17) requires a detailed knowledge of the two-phase hydrodynamics so that the parameters R_l, W_e, U, a, U', a', V_d, α, and Q_g can be evaluated. The fluid dynamics of annular flows have been investigated experimentally. Russell and Lamb (R4) have studied the flow mechanism of horizontal annular flow; Cousins *et al.* (C8) have dealt with droplet movement in vertical annular flow. Anderson and Russell (A6) have analyzed the interchange of droplets in horizontal systems, and numerous other hydrodynamic studies have been reported in Hewitt *et al.* (H1). Cichy *et al.* (C4) have reviewed the methods for evaluating the same hydrodynamic parameters that are also used in mass transfer studies. It is difficult to estimate the values of these parameters. Because of the necessity to describe the heat-transfer process in terms of parameters that can be measured easily or correlated from experimental data, three simplified cases are developed.

a. *Case I.* For this case, the assumption is made that the droplets do not contribute to the heat-transfer processes, and Eqs. (17) reduce to yield

$$W_I\, dH_I/dz = (d/dz(k_I R_I(dT_I/dz))) + Ua(T_{II} - T_I) \tag{18a}$$

$$W_{II}\, dH_{II}/dz = (d/dz(k_{II} R_{II}(dT_{II}/dz)))$$
$$- Ua(T_{II} - T_I) + q_{WII} \tag{18b}$$

Equations (18) are identical with Eqs. (8), which were used to describe Regime I. The evaluation of the parameters in Eqs. (18) is described in Section II,B,2,a.

b. *Case II.* Case II is based on the assumption of a large rate of droplet generation, and therefore the droplets and the liquid are always in thermal equilibrium. By applying this assumption, Eqs. (17) become

$$W_I\, dH_I/dz = (d/dz(k_I R_I(dT_I/dz))) + (Ua + U'a_e')(T_{II} - T_I) \tag{19a}$$

$$(W_f + W_e)\, dH_{II}/dz = (d/dz(k_{II} R_f(dT_{II}/dz)))$$
$$-(Ua + U'a_e')(T_{II} - T_I) + q_{Wf} \tag{19b}$$

where the dimensionless groups are defined as before, except for N_I and Pe_{II}, Eqs. (19) may be rearranged to yield

$$(1/Pe_I)\, d^2T_I/d\zeta^2 - dT_I/d\zeta = -N_I(T_{II} - T_I) \tag{20a}$$

$$(1/Pe_{II})\, d^2T_{II}/d\zeta^2 - dT_{II}/d\zeta = +N_I M(T_{II} - T_I) - N_{Wf} \tag{20b}$$

where the dimensionless groups are defined as before, except for N_I and Pe_{II}, which are defined as

$$N_I = (Ua + U'a_e')L/(W_I C_{PI}), \qquad Pe_{II} = W_{II} C_{PII} L/(R_f k_{II,\,eff})$$

and the heat-transfer term is defined as

$$N_{Wf} = q_{Wf} L/(W_{II} C_{PII})$$

Methods for evaluating the parameters which are in the dimensionless groups are described in Section II,B,2,b.

c. *Case III.* This case is based on two assumptions: (1) the interfacial heat transfer coefficient U' is small, and (2) the temperature of the entrained phase does not equal the temperature of the film at a given axial position.

The latter assumption has been proposed because the droplets are known to have a velocity of approximately $80\% \, v_g$ and a gas-phase residence time on the order of 0.01 sec, according to Russell and Rogers (R5). Equations (17) become

$$W_l \, dH_l/dz = (d/dz(k_l R_l \, dT_l/dz)) + Ua(T_f - T_l) \tag{21a}$$

$$W_f \, dH_f/dz = (d/dz(k_f R_f \, dT_f/dz)) - Ua(T_f - T_l)$$
$$\qquad - R_g \rho_{II} V_d (H_f - H_e) + q_{wf} \tag{21b}$$

$$W_e \, dH_e/dz = R_g \rho_{II} V_d (H_f - H_e) \tag{21c}$$

By applying the standard assumptions of constant thermal conductivities, constant holdup, and an ideal gas phase, Eqs. (21) can be rearranged to yield

$$k_l R_l/(W_l C_{Pl}) \, d^2 T_l/dz^2 - dT_l/dz = -(Ua/W_l C_{Pl})(T_f - T_l) \tag{22a}$$

$$(1/W_f C_{PII})(d/dz(-W_e H_e + k_f R_f \, dT_f/dz)) - dT_f/dz$$
$$= Ua/(W_f C_{PII})(T_f - T_l) - q_{wf}/(W_f C_{PII}) \tag{22b}$$

Now, by defining

$$k_{II, \, eff} R_f \, d^2 T_f/dz^2 = (d/dz(-W_e H_e)) + k_{f, \, eff} R_f \, d^2 T_f/dz^2 \tag{23}$$

and by using $k_{I, \, eff}$ in place of k_l, Eqs. (22) can be rearranged using the dimensionless groups as defined in Section II,A,1,a, with N_1 and M changed to

$$N_1 = UaL/(W_l C_{Pl}), \qquad M = W_l C_{Pl}/(W_f C_{PII}) \tag{24a}$$

and the wall heat transfer term defined as

$$N_{Wf} = q_{wf} L/W_f C_{PII} \tag{24b}$$

The evaluation of the parameters in the dimensionless groups is discussed in Section II,B,2,c.

3. *Regime IV*

This regime is characterized by the presence of one continuous fluid phase and one discrete fluid phase in tubular systems. The existence of the discrete phase generates a large interfacial area per unit tube volume for all flow configurations included in this regime. For that reason, Regime IV is of pragmatic interest when interphase heat and mass transfer are of key importance.

Most of the published studies of heat transfer without phase change in two-phase flow are for flow configurations included in Regime IV. Adiabatic operations of pragmatic interest include direct contact heat transfer between two liquids, nonisothermal solvent extraction operations, and two-phase reacting systems. Nonadiabatic processes of importance include the heating or cooling of two-phase mixtures and mass-transfer and reacting systems with strong thermal effects.

It is assumed that the discrete phase exists as drops (or bubbles) that can be characterized by an average velocity, an average size, and an average enthalpy at each axial position. A steady-state balance on the number of drops (or bubbles) present at each axial position is given by

$$d(\alpha v_d)/dz = Q_g \tag{25}$$

where Q_g is the net rate of drop generation per unit volume of conduit. Coalescence is simply a negative generation. By defining the holdup of Phase I as

$$R_I = \alpha A_c V_d/A_c \tag{26}$$

the basic mass balances become

$$d(R_I \rho_I v_d)/dz = d(W_I)/dz$$
$$= 0 \tag{27a}$$
$$d(R_{II} \rho_{II} v_{II})/dz = d(W_{II})/dz$$
$$= 0 \tag{27b}$$

The kinetic and potential energy terms are assumed to be negligible, and therefore the energy balances are

$$W_I \, dH_d/dz = (d/dz(R_I k_I \, dT_I/dz)) + U'a'R_I(T_{II} - T_I) \tag{28a}$$

$$W_{II} \, dH_{II}/dz = (d/dz(R_{II} k_{II} \, dT_{II}/dz))$$
$$- U'a'R_I(T_{II} - T_I) + q_{wII} \tag{28b}$$

In the rest of this development the rate of drop (bubble) generation will be assumed to be zero. This assumption could easily be removed when analyzing systems in which coalescence exists or a reaction generates a gaseous product.

a. *Liquid–Liquid Systems.* Heat transfer in liquid–liquid systems is encountered in the design of both vertical countercurrent flow spray columns or towers for direct contact heat transfer, and cocurrent tubular contactors for direct contact heat transfer and for heating or cooling a liquid–liquid system. The fluid–fluid equations for this regime can be simplified by assuming constant holdups, constant thermal conductivities, and that the enthalpy of each liquid is solely a function of temperature. Equations (28) become

$$W_I C_{PI}\, dT_I/dz = R_I k_I\, d^2T_I/dz^2 + U'a'R_I(T_{II} - T_I) \tag{29a}$$

$$W_{II} C_{PII}\, dT_{II}/dz = R_{II} k_{II}\, d^2T_{II}/dz^2$$
$$\qquad - U'a'R_I(T_{II} - T_I) + q_{WII} \tag{29b}$$

In this section subscript I refers to the discrete phase and subscript II to the continuous phase.

By following the same procedures as before, Eqs. (29) can be transformed with the aid of dimensionless groups to

$$(1/Pe_I)\, d^2T_I/d\zeta^2 - dT_I/d\zeta = -N_I(T_{II} - T_I) \tag{30a}$$

$$(1/Pe_{II})\, d^2T_{II}/d\zeta^2 - dT_{II}/d\zeta = +N_I M(T_{II} - T_I) - N_{WII} \tag{30b}$$

The dimensionless groups are defined as in Section II,A,1,a, with the exception of N_I, which is defined as

$$N_I = U'a'R_I/(W_I C_{PI})$$

and the wall heat transfer term is defined as

$$N_{WII} = q_{WII} L/(W_{II} C_{PII})$$

b. *Gas–Liquid Systems.* Heat transfer without significant phase change in gas–liquid systems is encountered, because of the design considerations involved, in nonisothermal mass-transfer or reaction processes. These processes are either carried out adiabatically with interphase heat transfer, or nonadiabatically with both interphase transfer and heat transfer between the continuous phase and the tube wall. In both cases the general fluid–fluid equations of Section III,A,3 may be applied to the design of these systems.

As in the previous sections the thermal conductivities and holdups are assumed to be solely functions of temperature. Therefore, Eqs. (28) can be simplified to yield

$$W_I C_{PI}\, dT_I/dz = R_I k_I\, d^2T_I/dz^2 + U'a'R_I(T_{II} - T_I) \tag{31a}$$

$$W_{II} C_{PII}\, dT_{II}/dz = R_{II} k_{II}\, d^2T_{II}/dz^2$$
$$\qquad - U'a'R_I(T_{II} - T_I) + q_{WII} \tag{31b}$$

Using the dimensionless groups as defined in Section II,A,1,a, except for N_I, which is defined as

$$N_I = U'a'R_I/(W_I C_{PI})$$

Eqs. (31) can be transformed into a form analogous to Eqs. (30). The methods of evaluating the parameters for gas–liquid systems in Regime IV are described in Section II,B,3,b.

B. Parameter Evaluation

The model equations in Section II,A have been formulated to describe the energy and mass transfer processes occurring in two-phase tubular systems. The accuracy of these model equations in representing the physical processes depends on the parameters of the equations being correctly evaluated. Constitutive equations that relate each of the parameters to the physical properties, system properties, and dependent variables of the system are discussed in the following sections.

1. Regime I

The evaluation of the parameters for this flow regime requires the calculation of the Reynolds number and hydraulic diameter for each continuous phase. The hydraulic diameter can be determined only if the holdup of each phase is known. This again illustrates the importance of understanding the fluid mechanics of two phase systems. Once the hydraulic diameter is known, the Reynolds number can be evaluated with the knowledge of the in situ phase velocity, and the parameters of the model equations can be evaluated.

a. *Liquid–Liquid Systems.* The basic model equations for liquid–liquid system are Eqs. (9), and the dimensionless parameters used in these equations are defined in Section II,A,1,a. The dimensionless group M is evaluated easily from the operating conditions and the physical properties of the liquids. The wall heat-transfer terms N_{WI} and N_{WII} can be evaluated from a knowledge of the mass flow rates of each phase, once the wall heat fluxes are evaluated. It is common practice to assume that the wall heat flux is given by

$$q_{WI} = h_{WI} A_{WI}(T_W - T_I) \tag{32a}$$

$$q_{WII} = h_{WII} A_{WI}(T_W - T_{II}) \tag{32b}$$

where T_W is known. The wall area wetted by each phase can be determined from a knowledge of the flow configuration, the tube geometry, and the holdup of each phase. In the absence of experimental evidence to the contrary, it seems reasonable to assume that the wall heat-transfer coefficients can be evaluated by using the following constitutive equation:

$$h_{Wi} D_i / k_i = 0.027 (\text{Re}_i)^{0.8} (\text{Pr}_i)^{1/3} \tag{33}$$

This requires the calculation of the phase Reynolds number and Prandtl number. This method of evaluating the wall heat flux is exactly the same as that commonly used for evaluating the heat flux for single-phase flow in an irregularly shaped duct.

The effective thermal conductivities $k_{I,\text{eff}}$ and $k_{II,\text{eff}}$ were introduced to compensate for many of the simplifying assumptions made in the development of the model equations. Ideal plug flow does not exist in either phase, and the axial mixing that occurs in the phase causes a corresponding energy transfer in both the axial and radial direction. These effects, which cannot be easily measured, are lumped together in the concept of an effective thermal conductivity. It is known that the magnitude of these effects is directly dependent on the hydrodynamics of the fluid phase, and without experimental results to show otherwise the effective conductivities of the Peclet numbers should be evaluated as though each phase were flowing in single-phase flow in an irregularly shaped duct. Levenspiel (L10) presents a graph from which the Peclet number can be evaluated if the phase Reynolds number is known.

The final dimensionless group to be evaluated is the interfacial heat-transfer number, and therefore the interfacial heat-transfer coefficient and the interfacial area must be determined. The interface is easily described for this regime, and, with a knowledge of the holdup and the tube geometry, the interfacial area can be calculated. The interfacial heat trasfer coefficient is not readily evaluated, since experimental values for U are not available. A conservative estimate for U is found by treating the interface as a stationary wall and calculating U from the relationship

$$1/U = 1/h_{wI} + 1/h_{wII} \tag{34}$$

If the hydrodynamics of the two-phase systems are understood, and the holdups and phase Reynolds numbers known, the rate of heat transfer can be estimated with about the same confidence as that for single-phase systems.

For turbulent flow in single-phase systems, the predicted temperature profile is not changed significantly if the Peclet number is assumed to be infinite. Therefore, in turbulent two-phase systems the second-order terms in Eqs. (9) probably do not have a significant effect on the resulting temperature profiles. In view of the uncertainties in the present state of the art for determining the holdups and the heat-transfer coefficients, the inclusion of these second-order terms is probably not justified, and the resulting first-order equations should adequately model the process.

Using the coupled first-order Eqs. (9a) and (9b) to describe the problem of heating benzene water mixture from 70 to 200°F in a 1-in. tube with 40-psig

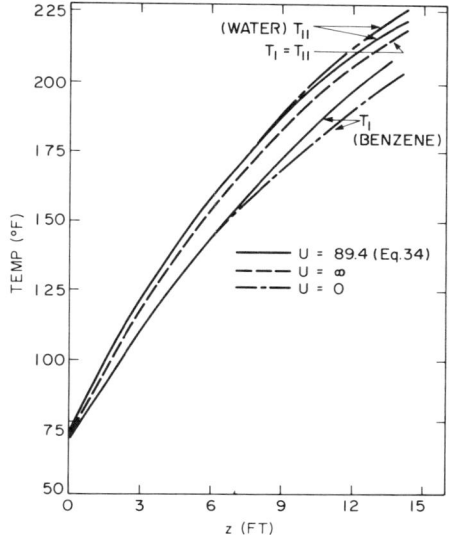

Fig. 6. Temperature profiles for Regime I liquid–liquid systems: $W_I = 254$ lb_{mass} hr^{-1}; $W_{II} = 254$ lb_{mass} hr^{-1}; $\rho_I = 53.8$ lb_{mass} ft^{-3}; $\rho_{II} = 61.2$ lb_{mass} ft^{-3}; C_{PI} 0.446 BTU lb_{mass}^{-1} $°F^{-1}$; C_{PII} BTU lb_{mass}^{-1} $°F^{-1}$; $\mu_I = 0.445$ cP; $\mu_{II} = 0.640$ cP; $k_I = 0.087$ BTU hr^{-1} ft^{-1} $°F^{-1}$; $k_{II} = 0.373$ BTU hr^{-1} ft^{-1} $°F^{-1}$

steam in a 2-in. jacketing pipe, O'Connor (O1) obtained the results shown in Fig. 6. As can be seen, the results are not sensitive to the value of U, the interfacial heat transfer coefficient, for this example.

b. *Gas–Liquid Systems.* It was shown in Section II,A,1,b that the basic model equations for Regime I gas–liquid flows are represented by Eqs. (9). The dimensionless groups are defined in Section II,A,1,b. The evaluation of the parameters in these groups depends heavily on a firm understanding of the gross fluid mechanics.

For gas–liquid flows in Regime I, the Lockhart and Martinelli analysis described in Section I,B can be used to calculate the pressure drop, phase holdups, hydraulic diameters, and phase Reynolds numbers. Once these quantities are known, the liquid phase may be treated as a single-phase fluid flowing in an open channel, and the liquid-phase wall heat-transfer coefficient and Peclet number may be calculated in the same manner as in Section II,B,1,a. The gas-phase Reynolds number is always larger than the liquid-phase Reynolds number, and it is probable that the gas phase is well mixed at any axial position; therefore, Pe_I is assumed to be infinite. The dimensionless group M is easily evaluated from the operating conditions and physical properties.

The gas-phase wall heat-transfer coefficient can be evaluated by using the gas-phase Reynolds number and Prandtl number in Eq. (33). The thermal conductivities of liquids are usually two orders of magnitude larger than the thermal conductivities of gases; therefore, the liquid-phase wall heat-transfer coefficient should be much larger than the gas-phase wall heat-transfer coefficient, and Eq. (34) simplifies to

$$U = h_{WI} \tag{35}$$

The interfacial area can be calculated from the description of the interface, the knowledge of the holdup and the tube geometry. The methods for evaluating all of the parameters in Eqs. (9) have been discussed, and these equations can be solved for the temperature profiles.

If the no-phase-change restriction does not rigorously apply, a simple design procedure can be formulated based on the results discussed in Section III, where it is shown that thermal equilibrium is quickly achieved in gas–liquid systems because of the large heat effects associated with evaporation or condensation. Although the total mass transfer between the phases may be small, it is not unrealistic to assume that the gas and liquid phases have the same temperature at each axial position.

Since the gas-phase wall heat flux q_{WI} is much smaller than the liquid-phase wall heat flux, the energy transferred from the liquid phase to the gas phase is approximately equal to the increase in energy in the gas phase. Therefore, Eqs. (9a) and (9b) can be combined to yield

$$1/\text{Pe}_{II}\, d^2 T_{II}/d\zeta^2 - (1 + M)\, dT_{II}/d\zeta = -N_{W_{II}} \tag{36}$$

and only the three parameters Pe_{II}, M, and $N_{W_{II}}$ must be evaluated.

The results obtained by O'Connor (O1) when Eq. (36) is solved for an air–water system in a 1-in. pipe being heated by saturated steam in a jacket pipe are shown in Fig. 7.

2. Regime II

The evaluation of the parameters for Regime II flows is difficult, due to the complexity of the fluid mechanics in this regime. It was because of this complexity that the model equations, Eqs. (15) and (17), were simplified by considering three special cases for these gas–liquid systems.

a. *Case I.* The model equations for this case are Eqs. (18), which were shown to be identical in form with Eqs. (7), and can be transformed into Eqs. (9) with the dimensionless groups defined in Section II,A,1,b. For an annular flow pattern, the liquid-phase Reynolds number is given by

$$\text{Re}_{II} = 4(\text{flow area})\rho_{II} v_{II}/[(\text{wetted perimeter})\mu_{II}] \tag{37a}$$

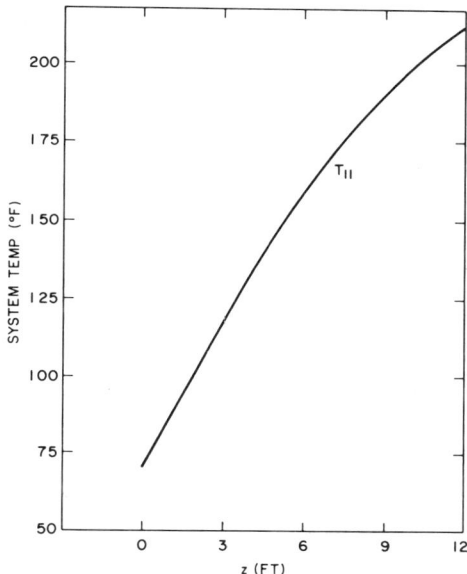

FIG. 7. Temperature profile for Regime I gas–liquid.

or

$$\text{Re}_{\text{II}} = (\pi D^2) R_{\text{II}} \rho_{\text{II}} v_{\text{II}} / (\pi D \mu_{\text{II}}) = D W_{\text{II}} / \mu_{\text{II}} \tag{37b}$$

and the gas-phase Reynolds number is calculated by using the film interfacial area per foot of pipe as the wetted perimeter. By using these definitions of the phase Reynolds numbers and by applying the discussion of Section II,B,1,b, all of the parameters of Eqs. (9) can be evaluated for this case.

b. *Case II.* The model equations for this case are Eqs. (20). The holdups R_{I} and R_{II} can be determined by applying the Lockhart and Martinelli χ parameter correlation (L12) or the Hughmark correlation (H3). Wicks and Dukler (W6) and Magiros and Dukler (M6) have developed correlations for calculating the entrainment. By using one of these correlations, W_e can be calculated from W_{II}. The film Reynolds number is defined as

$$\text{Re}_f = (\pi D^2 R_f) \rho_{\text{II}} v_f / (\pi D \mu_{\text{II}}) = D(W_{\text{II}} - W_e) / \mu_{\text{II}} \tag{38}$$

and the wall heat-transfer coefficient is evaluated by using Eqs. (33) with the film Reynolds number and the liquid Prandtl number.

As in Section II,B,1,b, the gas-phase Peclet number is assumed to be infinite, and the dimensionless group M is easily evaluated. The interfacial area a can be calculated with a knowledge of the holdup of the film phase and the tube geometry. The interfacial area a'_e can be evaluated if the droplet shape, characteristic size, and number density are known, since

$$a'_e = a' V_d \alpha \qquad (39)$$

The droplets are usually assumed to be spherical with a diameter in the range 0.005–0.05 in. The number density can be found by rearranging Eq. (15c) to yield

$$\alpha = W_e/(V_d \rho_{II} V_e) \qquad (40)$$

The velocity v_e must be determined before α can be evaluated. It is usually assumed that v_e is equal to approximately 80% of the gas phase velocity v_I. The value of v_I is known from Eq. (15a).

The heat-transfer coefficient U can be evaluated by using the gas-phase Reynolds number and the gas-phase Prandtl number in Eq. (33). The term U' can also be estimated by using Eq. (33), but the gas-phase Reynolds number used for evaluating U' should probably be based on the relative velocity v_r, which is defined as

$$v_r = v_g - v_e \qquad (41)$$

The final parameter to be evaluated is the liquid-phase Peclet number, and the graph given by Levenspiel (L10) can be used for this purpose. It must be remembered that the film Reynolds number should be used in estimating the Peclet number.

The methods proposed for evaluating the parameters in this case are based on engineering judgments; further experimental studies will provide better means of evaluation.

c. *Case III.* Equations (22) are the model equations for this case, and the dimensionless groups are defined in Section II,A,2,c. The holdups R_l and R_f, pressure drop, and entrainment are determined in exactly the same manner as in Case II. The film Reynolds number, defined by Eq. (38), and the wall heat-transfer coefficient, given in Eq. (33), are also evaluated in the same manner as for Case II. The dimensionless group M can be evaluated, since the film mass flow rate has been determined, and the gas-phase Peclet number is again assumed to be infinite. The film interfacial area a and the heat-transfer coefficient U are evaluated exactly as in Case II.

The Peclet number defined for this case is unique to Regime II flows, due to the definition of $k_{II, eff}$ given in Eq. (23). Therefore, Pe_{II} must be correlated with experimental data from Regime II flows.

The effectiveness of Cases I, II, and III in modeling the heat-transfer process has not been determined, due to the lack of available experimental data. In a recent study of mass transfer in Regime II flows, Ostermaier (O2) has followed an approach analogous to that presented here, and has shown that Case I is best for low values of the entrainment $W_e/(W_e + W_f)$. At high values of the entrainment, the droplets contribute to the transfer processes, and either equations similar to Eq. (20) or Eq. (22) must be used to describe the experimental data of Ostermaier.

Again as in Section II,B,1,b, if the no-phase-change restriction does not rigorously apply, a simpler design procedure can be formulated. Based on the results of Section III, the gas and liquid are assumed to be in thermal equilibrium. For Case II, Eqs. (20a) and (20b) are combined to yield

$$1/\text{Pe}_{II}\, d^2 T_{II}/dz^2 - (1 + M)\, dT_{II}/dz = -N_{wf} \qquad (42)$$

Under the same conditions, Eq. (42) also applies to Case III. As in Section II,B,1,b, the number of parameters that must be evaluated is reduced to three, N_{wf}, M, and Pe_{II}, when the gas and liquid are assumed to be in thermal equilibrium.

3. *Regime IV*

Regime-IV flow patterns are of pragmatic interest when interphase heat and mass transfer are of key importance because the existence of the discrete phase generates a large interfacial area per unit tube volume. Evaluation of the interfacial area is made difficult because the bubbles or drops of the discrete phase are usually not of uniform size or shape. By assuming a characteristic size and shape for the drops or bubbles, the interfacial area and the other parameters can be estimated with reasonable accuracy for many situations.

    a. *Liquid–Liquid Systems.* The model equations for these systems are given in Eqs. (30). Due to the current interest in the desalinization of seawater, the results of numerous experimental studies of direct-contact heat transfer have recently been published. Many of these studies and some industrial processes have been carried out in vertical adiabatic countercurrent-flowing spray columns or towers. For these systems $N_{WII} = 0$. Letan and Kehat (L3–L9) have performed extensive experimental investigations on direct-contact heat transfer for the water–kerosene system. These studies show that a severe temperature change occurs in each phase at the inlets. It is most difficult to predict these inlet temperature jumps, since very little useful quantitative information has been published on these effects. However, it is known that these temperature changes can correspond to at least 65% of the total temperature change in each phase, thereby greatly decreasing the efficiency of these units.

Equations (30) can be applied to these systems when the end effects are understood and properly evaluated. Pavlica and Olson (P1), reviewing the use of Peclet numbers for direct-contact heat transfer in spray towers, have presented a graph of the Peclet numbers versus the superficial liquid velocities. Mixon *et al.* (M8) describe the evaluation of the interfacial heat coefficient U'; the review of direct-contact heat transfer by Sideman (S5) is also useful in evaluating U'. Mixon *et al.* have discussed the limiting case in which U' is assumed to be infinite, and therefore the two phases are in equilibrium at all points within the column. The interfacial area a' can be evaluated by assuming spherical drops and knowing the average drop size. The flow rates within these systems are governed by gravitational effects, and therefore the range of possible operating conditions is severely limited.

The design of devices to promote cocurrent drop flows for heating or cooling a two-phase system, or for direct-contact heat transfer between two liquids, is difficult. The study by Wilke *et al.* (W7) is typical of the approach frequently used to analyze these processes. Wilke *et al.* described the direct-contact heat transfer between Aroclor (a heavy organic liquid) and seawater in a 3-in. pipe. No attempt was made to describe the flow pattern that existed in the system. The interfacial heat-transfer coefficient was defined by

$$(Ua/V) = Q/(\Delta T_{1m} V) \tag{43}$$

The interfacial area a was not measured, and the quantity (Ua/V) was correlated versus the total mass flow rate. From results of this type, the general parameters given in Eqs. (30) cannot be evaluated. Therefore, the only design that can be safely performed is for the Aroclor–water system in a 3-in. pipe.

The proper analysis of liquid–liquid systems requires an understanding of the system. When the model equations are derived so that the heat-transfer coefficient does not include the lack of knowledge of the gross fluid mechanics, a correlation for the coefficient, formulated from experiment analysis, is most useful. The correlation can be employed in physical situations where the gross fluid mechanics is entirely different, as in a different-diameter pipe.

As shown in Section I, very little is known about predicting liquid–liquid flow patterns and calculating pressure drops, holdups, and interfacial areas; however, some estimates can be made by assuming no slip between the phases, using

$$R_{\mathrm{I}} = Q_{\mathrm{I}}/(Q_{\mathrm{I}} + Q_{\mathrm{II}}) \tag{44}$$

Hinze (H2) has developed a correlation for predicting the average value of the maximum drop diameter that can exist in a turbulent continuous fluid phase:

$$D_d = 0.75[D(W_I + W_{II})\rho_{II}/(A_c \mu_{II})]^{0.08}[DA_c^3/(W_I + W_{II})^3]^{0.4}[\sigma/\rho_{II}]^{0.6} \quad (45)$$

Collins and Knudsen (C6) recently reported drop-size distribution produced by two immiscible liquids in turbulent flow, and the average drop size can be calculated from these distributions. From a knowledge of the average drop size, the interfacial area per drop a' and the drop volume can be calculated. The number of drops per unit volume is given by

$$\alpha = R_I/V_d \quad (46)$$

Once α is known, the total interfacial area can be evaluated.

The interfacial heat transfer coefficient can be evaluated by using the correlations described by Sideman (S5), and then the dimensionless parameter N_I can be calculated. If the Peclet numbers are assumed to be infinite, Eqs. (30) can be applied to the design of adiabatic cocurrent systems. For nonadiabatic systems, the wall heat flux must also be evaluated. The wall heat flux is described by Eqs. (32) and the wall heat-transfer coefficient can be estimated by Eq. (33) with

$$\mathrm{Re}_H = D(W_I + W_{II})/(R_I \mu_I + R_{II} \mu_{II}) \quad (47a)$$

$$\mathrm{Pr}_H = \mu_{II} C_{PII}/k_{II} \quad (47b)$$

Now Eq. (30) can be applied to the design of nonadiabatic systems.

From this discussion of parameter evaluation, it can be seen that more research must be done on the prediction of the flow patterns in liquid–liquid systems and on the development of methods for calculating the resulting holdups, pressure drop, interfacial area, and drop size. Future heat-transfer studies must be based on an understanding of the fluid mechanics so that more accurate correlations can be formulated for evaluating the interfacial and wall heat-transfer coefficients and the Peclet numbers. Equations (30) should provide a basis for analyzing the heat-transfer processes in Regime IV.

b. *Gas–Liquid Systems.* The model equations for gas–liquid systems with a flow configuration in Regime IV are Eqs. (31). By using either the Lockhart and Martinelli χ parameter correlation (L12) or the Dukler *et al.* (D4) and Hughmark (H4) correlations, the pressure drop and holdups can be evaluated. The continuous-phase Reynolds number can be calculated from Eqs. (37a), (33) can be used to evaluate the wall heat-transfer coefficient. Therefore, N_{WI} can be evaluated. The value of M is easily calculated from the knowledge of the flow rates and physical properties.

The average bubble size is usually in the range of 2–4 mm, and by assuming spherical bubbles the interfacial area a' can be determined. The interfacial heat-transfer coefficient can be evaluated by one of the correlations of Sideman (S5). Pavlica and Olson (P1) describe methods for evaluating the Peclet numbers. As in the previous sections, if the nonphase-change restriction does not apply rigorously, the gas and liquid can be assumed to be in thermal equilibrium. Therefore, Eqs. (31) can be combined to yield

$$1/P_{eII} \, d^2 T_{II}/d\zeta^2 - (1 + M) \, dT_{II}/d\zeta = -N_{WII} \qquad (48)$$

and only the three parameters M, N_{WII}, and P_{eII} must be evaluated.

III. Heat Transfer with Phase Change

A. Vaporization Phenomena

The design of a large number of two-phase processing operations requires a quantitative understanding of heat transfer with phase change. Steam boilers and thermosiphon reboilers are but two examples of two-phase processing units designed specifically to produce a vapor. Large heat effects and phase change can also be produced in nonisothermal two-phase reactors. The essence of phase-change problems in nonisothermal two-phase flow can be illustrated by looking at the process of completely vaporizing a liquid flowing in a tube.

In most situations the liquid enters the tube below its saturation temperature, and since heat is transferred into the system through the tube wall, the liquid is heated by the well-understood process of single-phase forced convection. As the liquid temperature increases, the fluid elements closest to the wall are heated above their boiling point; with a sufficient degree of superheating, small bubbles appear on the tube wall. The bulk liquid motion causes these bubbles to be stripped off the wall and mixed with the liquid phase. The bubbles initially generated condense when they come in contact with the subcooled bulk liquid. Farther down the tube, the bulk liquid temperature reaches the saturation temperature, and from this point onward bubbles continue to grow after they are removed from the tube wall. The velocity of the mixture increases as vapor is formed, with the gas bubbles having a velocity greater than the liquid. This increase in the mixture velocity results in a larger pressure drop in the axial direction than if the mixture were single phase.

As the pressure in the pipe decreases in the direction of the flow, the saturation temperature of the liquid decreases, promoting an increase in vaporization. More bubbles are formed; these interact to form a continuous vapor phase, and an annular flow pattern is produced. After annular flow is achieved, the liquid flows as a film, wetting the tube wall, with a velocity greater than the initial liquid velocity. This greater velocity tends to prevent further bubble formation at the wall. Continued addition of heat causes the liquid film to decrease in thickness as a result of both vaporization at the gas–liquid interface and further fluid acceleration. Eventually dry spots begin to appear on the tube wall.

As dry spots appear, the wall temperature increases and the tensile strength of the tube decreases. If the heating medium is at a sufficiently high temperature, the wall temperature will increase until the tube ruptures. This phenomenon is termed *burn-out*. If the heating medium is not at a temperature high enough to cause burn-out, the tube wall will become completely dry with the liquid flowing entirely as droplets in the vapor phase. With continued heating, all of the droplets will vaporize, and only a saturated or superheated vapor will be present. A graphical depiction of this phase change process is given in Figs. 8 and 9 for horizontal and vertical systems.

Three main flow patterns exist at various points within the tube: bubble, annular, and dispersed flow. In Section I, the importance of knowing the flow pattern and the difficulties involved in predicting the proper flow pattern for a given system were described for isothermal processes. Nonisothermal systems may have the added complication that the same flow pattern does not exist over the entire tube length. The point of transition from one flow pattern to another must be known if the pressure drop, the holdups, and the interfacial area are to be predicted. In nonisothermal systems, the heat-transfer mechanism is dependent on the flow pattern. Further research on predicting flow patterns in isothermal systems needs to be undertaken

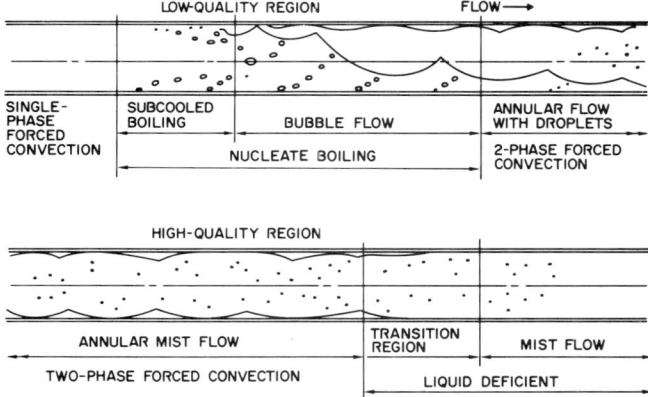

Fig. 8. The process of phase change inside horizontal tubes.

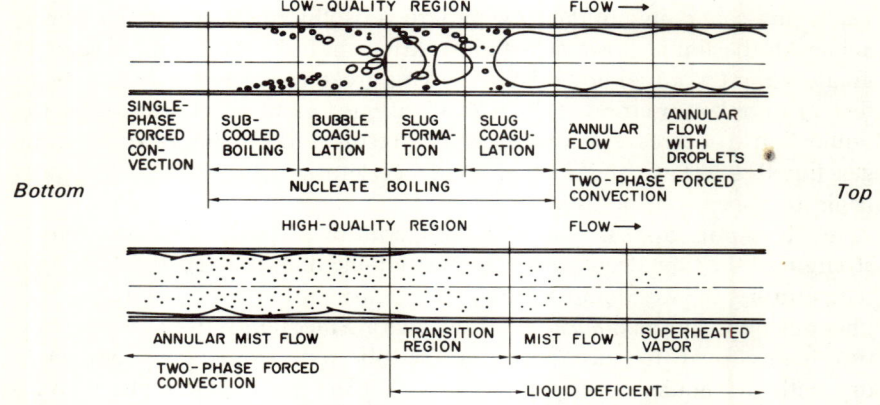

FIG. 9. The process of phase change inside vertical tubes.

with an awareness of the problems in designing nonisothermal two-phase systems.

For the total vaporization process, it is generally assumed that four basic heat-transfer regions exist within the tube, each representing a distinct heat-transfer mechanism with the transition from one region to another being gradual:

Region I: single-phase convection.
Transition I: bubble formation begins.
Region II: nucleate boiling.
Transition II: suppression of nucleation.
Region III: Two-phase forced convection.
Transition III: dry spots appear on the wall.
Region IV: liquid deficient heat transfer

The approximate correspondence between the heat-transfer regions and the flow patterns is shown in Figs. 8 and 9.

In many design problems, the determination of a wall heat-transfer coefficient or the heat flux between the tube wall and the fluid mixture is only part of the required information. The pressure drop within the system, the rate of phase change at the gas–liquid interface, the point at which the tube walls become dry, and the holdup of the fluids at each point in the pipe must all be determined.

In a vaporization process, the pressure at each axial position must be known so that the saturation temperature can be calculated and the rate of phase change predicted. In Section I, the present state of the art for calculating pressure drop and holdups in isothermal incompressible two-phase systems was reviewed, and the need for a better understanding of the hydrodynamics was demonstrated. In nonisothermal systems, the density of the vapor can change significantly, and the accelerative effects resulting from phase change can be appreciable. Therefore, the correlations based on the four Lockhart–Martinelli assumptions are useful only in providing a first estimate of the true values in nonisothermal systems. The Martinelli–Nelson correlation (M7), although described in Section I,B as the most widely used method for predicting pressure drops in two-phase systems with phase change, has not been quantitatively tested over a wide range of experimental conditions, and should therefore be used with an awareness of its limitations. Further research into the prediction of pressure drop and into configurations in two-phase systems needs to be undertaken, with an understanding of the difficulties in designing both isothermal and nonisothermal systems.

The hydrodynamics and the heat-transfer processes for Regions II, III, and IV are described in the rest of this section. Region I is widely discussed in the chemical engineering literature; Bird *et al.* (B4) and Kern (K1) provide detailed summaries of single-phase forced convection. General introductions into the subject of heat transfer in two-phase flow are given by Anderson and Russell (A5), DeGance and Atherton (D2), and Scott (S4). Tong (T1) has written a book on the general subject of heat transfer and two-phase flow that provides a fairly extensive set of references.

B. The Rate of Phase Change

Before the phase change process can be described in two-phase tubular systems, it is necessary to understand the molecular motions that result in mass transfer due to phase change, and to determine an analytical expression for the rate of phase change. In most engineering applications, the rate of phase change is generally ignored, and the design procedures are carried out under the assumption of local liquid–vapor equilibrium.

1. *Molecular Interchange Process*

This section describes the phase change process for a single component on a molecular level, with both vaporization and condensation occurring simultaneously. Molecules escape from the liquid surface and enter the bulk vapor phase, whereas other molecules leave the bulk vapor phase by becoming attached to the liquid surface. Analytical expressions are developed for the absolute rates of condensation and vaporization in one-component systems. The net rate of phase change, which is defined as the difference between the absolute rates of vaporization and condensation, represents the rate of mass transfer between phases, and is the net result of the molecular interchange process.

An expression for the absolute rate of condensation can be developed readily if the simple kinetic theory of gases and the ideal gas law are applied (S2):

$$r_c = E_c P_g (M/(2\pi RT))^{1/2} \tag{49}$$

E_c is a condensation coefficient. Schrage (S2) has shown that the absolute rate of vaporization is given by

$$r_v = E_v P_L^* (M/(2\pi RT_L))^{1/2} \tag{50}$$

where P_L^* is the vapor pressure of the liquid at temperature T_L. The net rate of phase change is obtained by combining Eqs. (49) and (50). The recent work of Nabavian and Bromley (N1) and Maa (M1) has shown that the evaporation and condensation coefficients can be assumed to be unity. The absolute rate expressions can be combined to yield an equation for the net rate of phase change:

$$r_{net} = (M/2\pi R)^{1/2} (P_L^*/T_L^{1/2} - P_g/T_g^{1/2}) \tag{51}$$

This equation is commonly designated the Hertz–Knudsen equation.

The assumptions inherent in the derivation of the Hertz–Knudsen equation are: (1) the vapor phase does not have a net motion; (2) the bulk liquid temperature and corresponding vapor pressure determine the absolute rate of vaporization; (3) the bulk vapor phase temperature and pressure determine the absolute rate of condensation; (4) the gas–liquid interface is stationary; and (5) the vapor phase acts as an ideal gas. The first assumption is rigorously valid only at equilibrium. For nonequilibrium conditions there will be a net motion of the vapor phase due to mass transfer across the vapor–liquid interface. The derivation of the expression for the absolute rate of condensation has been modified by Schrage (S2) to account for net motion in the vapor phase. The modified expression is

$$r_c = P_g (M/(2\pi RT_g))^{1/2} \Gamma \tag{52}$$

where

$$\Gamma = \exp[(-\beta^2 U_0^2) - \pi^{1/2} \beta U_0 (1 - \text{erf}(\beta U_0))] \tag{53}$$

If the system is not under any external forces, the velocity U_0 is defined by

$$U_0 = r_{net}/P_{net} \tag{54}$$

where P_{net} is the vapor-phase density of the material-changing phase. The absolute rate of vaporization is not affected by a net motion in the vapor phase.

The validity of the second assumption has been examined by Maa (M2), who postulated that the liquid surface temperature and not the bulk liquid temperature determines the absolute rate of vaporization. The rate expression using the surface temperature is

$$r_v = (M/(2\pi R))^{1/2}(P_S^*/T_S^{1/2}) \tag{55}$$

The vapor pressure of the liquid at the surface P_S^* can be evaluated from an integrated from of the Clausius–Clapeyron equation if the surface temperature T_S is known.

Maa (M2) developed a procedure for calculating the liquid surface temperature as a function of the time each liquid element is in contact with the vapor. He assumed that the latent heat of vaporization is transferred from the interior of the liquid to the interface by pure conduction. Consequently, the sole source of energy for vaporization is the sensible heat made available by a change in the liquid temperature. If exposure time is short, only the liquid near the surface will undergo a temperature change. The heat transfer within the liquid is modeled by

$$\partial T/\partial t = \alpha_L \partial^2 T/\partial x^2 \tag{56}$$

where $\alpha_L = k_L/(\rho_L C_{PL})$, and the boundary conditions are

$$T = T_L \quad \text{at} \quad t = 0 \quad \text{for all} \quad x \tag{57a}$$

$$T = T_L \quad \text{at} \quad x \to \infty \quad \text{for all} \quad t \tag{57b}$$

$$k \, \partial T/\partial X \bigg|_{X=0} = \lambda_g r_{net} \tag{57c}$$

Equation (57c) implies that the latent heat transferred from the surface by vaporization must be transported to the surface by conduction. For condensation, a development completely analogous to this would apply, but the heat must be transferred from the surface to the interior of the liquid.

In evaluating Eq. (57), Maa (M2) calculated the net rate of phase change by combining Eqs. (52) and (55) to yield

$$r_{net} = (M/(2\pi R))^{1/2}(P_S^*/T_S^{1/2} - P_g \Gamma/T_g^{1/2}) \tag{58}$$

Since the liquid surface temperature changes with time, the quantity of interest is not the instantaneous value of r_{net} but the average value of the net rate of phase change defined by

$$\bar{r}_{net} = (1/\theta) \int_0^\theta r_{net} \, dt \tag{59}$$

where θ is the exposure time period for each surface element. In terms of the absolute rate expressions, the result is

$$\bar{r}_{\text{net}} = (M/(2\pi R))^{1/2} \left[(1/\theta) \int_0^\theta (P_S/T_S^{1/2}) \, dt - P_g \Gamma/T_g^{1/2} \right] = \bar{r}_v - r_c \quad (60)$$

2. *Verification of the Absolute Rate Expressions*

The third, fourth, and fifth assumptions inherent in the Hertz–Knudsen equation are also used in the derivation of Eq. (60). Maa (M3) has shown by experimental studies that Eq. (60) provides a good model for predicting the net rate of phase change. Maa (M4) and O'Connor (O1) have shown that if these assumptions are replaced by significantly less restrictive ones, and the derivation repeated, the calculated values of the net rate of phase change differ by less than 10% under the most extreme conditions.

Numerous experiments have been conducted to determine the validity of the rate expressions in Eqs. (52) and (55). Langmuir (L1) used Eq. (50) with $E_v = 1.0$ to calculate the vapor pressure of nonvolatile metals such as tungsten, molybdenum, platinum, nickel, and iron. Equation (50) was applied to calculate the vapor pressure of each metal, and the results were found to be in good agreement with previously known values. Volmer and Estermann (V1) measured the absolute rate of vaporization of liquid mercury at temperatures up to 140°F, and found good agreement between the experimental results and those calculated from Eq. (50) with E_v equal to unity. Tschudin (T2) also used Eq. (50) with E_v equal to unity to compare the theoretical absolute rate of vaporization to measured values for ice; agreement was good.

Waldman and Houghton (W1) have analyzed the growth of a vapor bubble in a superheated liquid by coupling Eq. (58) with a modified form of the Rayleigh equation and with the thermal diffusion equation. The Rayleigh equation describes the hydrodynamic phenomena in the liquid surrounding the bubble. The thermal diffusion equation describes the heat transport from the bulk liquid to the liquid interface. These three coupled equations were solved numerically, and the calculated growth rates were shown to be in good agreement with the experimental results of Dergarabedian (D1), who photographed the growth of steam bubbles in superheated water at 1-atm pressure. The degree of superheat ranged from 1 to 9°F. From the photographs Dergarabedian was able to plot the bubble radius as a function of time. The initial bubble radii were on the order of 0.1 mm and the final radii about 1.0 mm. The time period of observation was about 15 msec.

Maa (M4) determined experimental values of $\bar{r}_{\text{net}}$ from measurements of the rate of phase change for water, toluene, carbon tetrachloride, isopropyl alcohol, and isoamyl acetate for temperatures up to 68°F. The agreement between the measured values of $\bar{r}_{net}$ and those calculated from Eqs. (52)–(60) was within 7%. Rates of phase change as large as 120 lb$_{mass}$ ft^{-2} hr^{-1} were measured.

The numerical procedure used to solve these equations has been completely described by Maa (M4). The procedure has been improved by O'Connor (O1) to require significantly less computation time.

3. Penetration Theory: A Net Rate Expression

An expression for the net rate of phase change can also be derived by assuming that the phase-change process is controlled solely by the rate at which heat can be transferred between the bulk liquid and the liquid surface. In this penetration theory approach, the liquid surface temperature is assumed to equal the gas-phase temperature. The heat transfer within a liquid element is assumed to occur by pure conduction, and therefore Eq. (56) is applicable:

$$\partial T/\partial t = \alpha_L \, \partial^2 T/\partial X^2 \qquad (61)$$

but for this case, the boundary conditions are

$$T = T_L \quad \text{at} \quad t = 0 \quad \text{for all} \quad X \qquad (62a)$$

$$T = T_L \quad \text{as} \quad X \to \infty \quad \text{for all} \quad t \qquad (62b)$$

$$T = T_g \quad \text{at} \quad X = 0 \quad \text{for all} \quad t \qquad (62c)$$

The solution of Eqs. (61) and (62) is

$$T = T_g + (T_L - T_g)\,\text{erf}(X/(2(\alpha_L t)^{1/2})) \qquad (63)$$

For the phase-change process, it is assumed that the latent heat transferred to the surface must be removed from the surface by pure conduction within the liquid elements. Therefore

$$k\,\partial T/\partial X \Big|_{X=0} = \lambda_g r_{net}$$

By using Eq. (63) to evaluate the derivative, an expression for $\bar{r}_{net}$ is obtained:

$$\bar{r}_{net} = (1/\lambda_g)(k_L C_{PL} P_L/\pi t)^{1/2}(T_L - T_g) \qquad (64)$$

and by using Eq. (59)

$$\bar{r}_{net} = -(2/\lambda_g)(k_L C_{PL} P_L/\pi\theta)^{1/2}(T_g - T_L) \qquad (65)$$

where θ is the exposure time period.

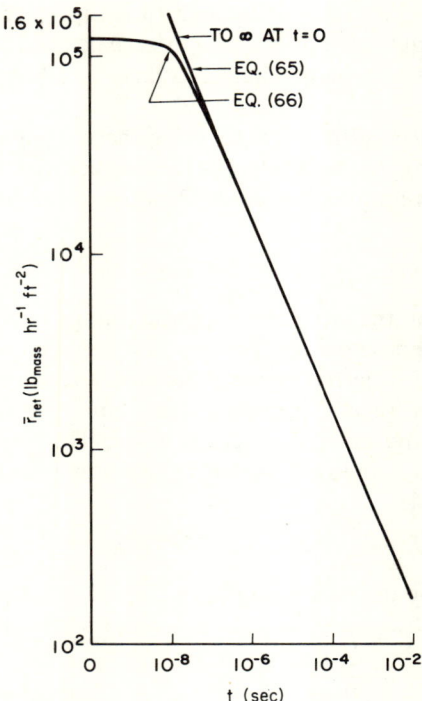

FIG. 10. Comparison of penetration and kinetic theory rate expressions.

In Fig. 10, the values of $\bar{r}_{net}$ calculated from Eq. (65), for various exposure times, are compared to the values calculated using the equations

$$r_v = (M/(2\pi R))^{1/2}(P_S^*/T_S^{1/2}) \tag{55}$$

$$r_c = (M/(2\pi R))^{1/2}(P_g \Gamma/T_g^{1/2}) \tag{52}$$

$$\bar{r}_{net} = (1/\theta) \int_0^\theta (r_v - r_c)\, dt \tag{66}$$

To compute the value of $\bar{r}_{net}$ from Eq. (66), Eqs. (53), (54), and (56) must be used to evaluate Γ and T_S^*. The values of $\bar{r}_{net}$ calculated from Eqs. (64) and (66) are within 2% for exposure times greater than 10^{-7} sec, and therefore they are plotted on the same curve in Fig. 10. The two values of $\bar{r}_{net}$ are given in Table I; they agree within 1% over the entire range of T_L.

The agreement between these two developments is not simply a coincidence of having chosen the proper numerical example. O'Connor (O1) has shown that these two developments always yield an identical expression for

TABLE I

VALUES OF $\bar{r}_{net}$ FROM PENETRATION THEORY AND KINETIC THEORY RATE EXPRESSIONS[a]

T_L (°F)	$\bar{r}_{net}$[b] ($lb_{mass}\ hr^{-1}\ ft^{-1}$)	$\bar{r}_{net}$[c] ($lb_{mass}\ hr^{-1}\ ft^{-1}$)
200	127	128
190	233	234
180	339	340
170	445	447
160	551	553
150	657	660
140	763	766
130	869	872
120	975	980
110	1,080	1,084
100	1,186	1,190
90	1,291	1,297
80	1,398	1,407
70	1,503	1,510

[a] $\theta = 0.001$ sec.; $T_g = 212°F$.
[b] From Eq. (66) using Eq. (60).
[c] From Eq. (65).

$\bar{r}_{net}$ when the exposure time is greater than 10^{-5} sec, and the gas phase is saturated. From this discussion it can be concluded that the net rate of phase change, that is, the rate of mass transfer, can be calculated from either Eq. (65) or Eq. (66) with an equal degree of accuracy. However, the absolute rates of vaporization and condensation can only be calculated from the rate expressions based on the molecular interchange process. It is shown later in this chapter that the absolute rates must be determined to describe accurately the energy transfer that accompanies phase change.

C. BASIC MODEL EQUATIONS

In Section III,A, the basic process of vaporization in a tubular two-phase contacting device was described, and four basic heat-transfer regions were shown to exist in the general case. The first region is that of single-phase forced convection, where the methods for modeling the heat transfer and hydrodynamics are well known. The remaining three regions are regions of heat transfer with two-phase flow, and the model equations to describe the transport processes are developed in this section. Methods for evaluating the parameters of these model equations are presented in Section III,D. Model behavior is discussed in Section III,E.

1. Nucleate Boiling

When heat is transferred to a pure liquid near its boiling point, vaporization begins with the formation of microscopic bubbles on the tube wall in a process identified as nucleation. The analysis of heat transfer in nucleate boiling can be separated into two parts, the formation and growth of bubbles on a surface, and the subsequent growth of these bubbles after they leave the surface. Both processes are complex and in an attempt to better understand the basic mechanisms they are most often studied in nonflowing batch systems.

Westwater (W4, W5) has written a detailed review of boiling in liquids with emphasis on nucleation at surfaces. Although written in 1956, this is still very useful and it provides a detailed description of the factors affecting nucleation. In a more recent review, Leppert and Pitts (L2) have described the important factors in nucleate boiling and bubble growth, and Bankoff (B2) has reviewed the field of diffusion-controlled bubble growth in nonflowing batch systems.

The information presented in these three reviews provides a qualitative understanding of the factors involved in pool boiling and bubble growth, but does not provide sufficient insight to make the quantitative statements necessary for the design of flowing systems. By formulating a set of steady-state mass and energy balances for the nucleate heat transfer process in flowing systems, the parameters which must be evaluated can be pinpointed. Nucleate boiling produces a bubble flow pattern and therefore, the design equations are formulated in a manner analogous to that used in Section II,A,3. A balance on the number of bubbles is

$$\frac{d(\alpha A_c v_d)}{dz} = \pi D f n - R_{c1} A_c - R_{c2} A_c \qquad (67)$$

where n represents the number of nucleation centers per unit wall area, f the bubble departure frequency, and R_{c1} the rate of bubble collapse. The rate of bubble coalescence R_{c2} is the negative of the rate of bubble generation. The phase mass balances are

$$d(\alpha A_c \rho_I V_b v_b)/dz$$
$$= \pi D f n \rho_I V_b - R_{c1} A_c \rho_I V_b + \alpha A_c a' V_b (\bar{r}_{vb} - r_{cb}) \qquad (68a)$$

$$d(R_{II} A_c \rho_{II} v_{II})/dz$$
$$= -\pi D f n \rho_I V_b + R_{c1} A_c \rho_I V_b - \alpha A_c a' V_b (\bar{r}_{vb} - r_{cb}) \qquad (68b)$$

where $\bar{r}_{vb}$ is the average absolute rate of vaporization at the vapor–liquid interface of a bubble and r_{cb} the absolute rate of condensation. The phase energy balances are

$$d(\alpha A_c \rho_I V_b v_b H_I)/dz$$
$$= \pi D f n \rho_I V_b H_I$$
$$+ (d/dz\{\alpha V_b A_c k_I dT_b dz\}) - R_{c1} A_c \rho_I V_b H_I$$
$$+ \alpha A_c a' V_b (\bar{r}_{vb} \bar{H}_{vb} - r_{cb} H_{cb})$$
$$+ \alpha A_c a' V_b U'(T_{II} - T_I) \tag{69a}$$

$$d(R_{II} A_c \rho_{II} H_{II})/dz$$
$$= -\pi D f n \rho_I V_b H_I$$
$$+ R_{c1} A_c \rho_I V_b H_I - \alpha A_c a' V_b(\bar{r}_{vb}\bar{H}_{vb} - r_{cb}H_{cb})$$
$$+ (d/dz(R_{II} k_{II} A_c\, dT_{II}/dz)) - \alpha A_c a' V_b U'(T_{II} - T_I) + q_{wII} \tag{69b}$$

where $\bar{H}_{vb}$ is the average enthalpy of the molecules vaporizing and $\bar{H}_{cb}$ is the enthalpy of the vapor phase per unit mass. The value of $\bar{H}_{vb}$ is determined from

$$\bar{r}_{vb}\bar{H}_{vb} = (1/\theta)\int_0^\theta r_{vb} H_{vb}\, dt \tag{70}$$

where θ is the exposure time of a surface element.

It has been assumed that all bubbles in the dispersed phase can be represented by one characteristic bubble size and velocity. This assumption is not as valid for nucleate boiling as it is for the non-phase-change Regime IV flows, and in fact stochastic equations are probably necessary to adequately describe this process. Due to the limited knowledge of the hydrodynamics associated with forced nucleation, however, such added complexity cannot be justified at present. By comparing the phase-change equations to the non-phase-change equations in Section II,A,3, it is readily seen that five additional parameters R_{c1}, f, n, r_{cb}, and r_{vb} must be determined for the phase change case. Constitutive equations relating each of these parameters to the physical properties and the dependent variables of the system must be formulated before Eqs. (68) and (69) can be used to design nucleate boiling systems. The evaluation of these parameters is discussed in detail in Section III,D,1.

2. *Two-Phase Forced Convection*

In the vaporization process, the formation of a continuous vapor phase causes a transition from bubble to annular flow. This flow-pattern transition is accompanied by a gradual change in the heat-transfer mechanism. Both the resulting increase in liquid velocity and the decrease in the degree of superheat in the liquid suppress bubble formation, and the heat-transfer rate again becomes dependent on the fluid velocity. The transition zone cannot be predicted accurately due to present difficulties in estimating the rate of nucleation and bubble growth in the nucleate boiling region and to uncertainties in predicting the conditions under which annular and bubble flow occur.

As in Section II,A, a set of steady-state mass and energy balances are formulated so that the parameters that must be evaluated can be identified. The annular flow patterns are included in Regime II, and the general equations formulated in Section II,A,2,a, require a detailed knowledge of the hydrodynamics of both continuous phases and droplet interactions. Three simplified cases were formulated, and the discussion in this section is based on Case I. The steady-state mass balances are

$$d(\rho_I R_I v_I)/dz = a(\bar{r}_v - r_c) \tag{71a}$$

$$d(\rho_{II} R_{II} v_{II})/dz = -a(\bar{r}_v - r_c) \tag{71b}$$

and the energy balances are

$$d(\rho_I R_I v_I H_I)/dz$$
$$= (d/dz(R_I k_I \, dT_I/dz))$$
$$+ Ua(T_{II} - T_I) + a(\bar{r}_v \bar{H}_{IIv} - r_c H_I) \tag{72a}$$

$$d(\rho_{II} R_{II} v_{II} H_{II})/dz$$
$$= (d/dz(R_{II} k_{II} \, dT_{II}/dz))$$
$$- Ua(T_{II} - T_I) - a(\bar{r}_v \bar{H}_{IIv} - r_c H_I) + q_{wII} \tag{72b}$$

By comparing Eqs. (71) and (72) to the non-phase-change equations in Section II,A,2, it can be seen that the only additional parameters to be evaluated are r_v and r_{c1}, the absolute rates of vaporization and condensation at the gas–liquid interface. The methods for evaluating all parameters in these model equations are given in Section III,D,2.

3. *Liquid-Deficient Heat Transfer*

As 100% vaporization is approached, there is not sufficient liquid flowing in the system to continuously wet the entire tube wall, and thus dry spots appear. Transition Zone III is characterized by the initial appearance of these dry spots, and Region IV is characterized by a completely dry tube wall and a dispersed flow pattern. Due to the effects of gravity, dry spots appear at higher values of the liquid holdup for horizontal systems than for vertical systems.

Transition Zone III is of utmost importance, since the formation of dry spots is accompanied by a dramatic change in the heat transfer mechanism. In such units as gas-fired boilers, the dry spots may cause the tube wall temperature to approach the temperature of the heating gas. However, before the tube wall temperature reaches a steady-state value, the tensile strength of the tube wall is reduced, and rupture may occur. This phenomenon, called burn-out, may also occur at any point along the tube wall if the wall heat flux $q_{w_{II}}$ is large enough so that a vapor film forms between the tube wall and the liquid surface.

Macbeth (M5) has recently written a detailed review on the subject of burn-out. The review contains a number of correlations for predicting the maximum heat flux before burn-out occurs. These correlations include a dependence upon the tube geometry, the fluid being heated, the liquid velocity, and numerous other properties, as well as the method of heating. Silvestri (S6) has reviewed the fluid mechanics and heat transfer of two-phase annular dispersed flows with particular emphasis on the critical heat flux that leads to burn-out. Silvestri has stated that phenomena responsible for burn-out, due to the formation of a vapor film between the wall and the liquid, are believed to be substantially different from phenomena causing burn-out due to the formation of dry spots that produce the liquid-deficient heat transfer region. It is known that the value of the liquid holdup at which dry spots first appear is dependent on the heat flux $q_{w_{II}}$. The correlations presented by Silvestri and Macbeth (S6, M5) can be used to estimate the burn-out conditions.

D. Parameter Evaluation

The model equations in Section III,C, have been formulated to describe those energy and mass-transfer processes in two-phase tubular systems for which one cannot neglect phase change. Constitutive equations for the parameters in these model equations are discussed in this section.

1. *Nucleate Boiling*

The rate of bubble collapse R_{c1} is primarily important in the first transition zone where the bulk liquid is subcooled. A number of studies have been published on subcooled boiling as well as the prediction of the point of net vapor generation, characteristics defining Transition Zone I, and the onset of nucleation. These studies all result in empirical correlations, and have not led to quantitative conclusions which can be generalized. The radial velocity and temperature profiles in the liquid must be known before the point of net vaporization can be predicted and the value of R_{c1} determined. The hydrodynamics of bubble flow are not understood well enough to predict these radial profiles. The most comprehensive studies on subcooled boiling have been reported by Levy (L11), Staub (S8), and Zuber *et al.* (Z1).

Both the bubble departure frequency f and the number of nucleation centers n are difficult to evaluate. These quantities are known to be dependent on the magnitude of the heat flux, material of construction of the tube, roughness of the inside wall, liquid velocity, and degree of superheat in the liquid elements closest to the tube wall. Koumoutsos *et al.* (K2) have studied bubble departure in forced-convection boiling, and have formulated an equation for calculating bubble departure size as a function of liquid velocity.

The absolute rates of vaporization and condensation are evaluated by using the rate expressions discussed in Section III,B. The net rate of phase change at the bubble interface or equivalently the rate of bubble growth, has been widely studied for single bubbles in stationary systems. Bankoff (B2) has reviewed the results of these studies. Ruckenstein (R2) has analyzed bubble growth in flowing systems.

The other parameters in Eqs. (67) and (68) are evaluated in the same manner as in Section II,B,3, with the single exception of $q_{W_{II}}$. The evaluation of the wall heat flux for nucleate boiling in flowing systems has been described briefly in the reviews by Westwater (W4, W5) and Leppert and Pitts (L2). Less quantitative information is available for this case than for pool boiling. According to these authors, the heat flux in forced-convection nucleation is independent of the fluid velocity. Westwater (W5) and Rohsenow (R1) have presented a number of correlations for calculating the total heat flux in forced-convection nucleation; the general form of these correlations is

$$q/A = g(T_w - T_{BP})^{a_1} \tag{73}$$

where g is a constant, T_{BP} is the boiling point of the liquid at the pressure in the tube, and a is a constant usually greater than 3.

At present, Eq. (68) only provides a simple estimate of the mass and energy transfer processes in forced-flow nucleation. The methods for evaluating the parameters must be improved by further detailed research on forced-flow nucleation. In particular, the calculation of the rate of nucleation n and of the bubble departure frequency f are the weakest points in the analysis of this heat-transfer region. Obviously, accurate prediction of the pressure drop and holdups are also needed.

2. Two-Phase Forced Convection

This heat-transfer region has been more widely studied than either Region II or IV. However, the methods for evaluating the parameters have not been tested over a wide range of experimental conditions. Equations (71) and (72) must be coupled with a knowledge of the pressure drop, holdups, and other system parameters if the temperature and mass flow-rate profiles are to be determined. As mentioned earlier in this chapter, the pressure drop and holdups in phase change systems can be estimated by using the Martinelli–Nelson Correlation.

All of the parameters in Eqs. (71) and (72) except q_{WII}, r_{cb}, and $\bar{r}_{vb}$ can be evaluated as in Section II,B,2, but with the added complication that the holdups are not constant.

The wall heat flux q_{WII} cannot be evaluated as in Section II,B. Numerous experimental studies on heat transfer in this two-phase forced-convection region have been carried out, and the results of these investigations are usually presented in the form of a correlation for the wall heat-transfer coefficient h_{WII}, which is defined as in Eq. (32b). Most of these correlations fit one of two generalized forms. The first is

$$h_{WII} = a_2 h_L (1/\chi)^{b_1} \tag{74}$$

where a_2 and b_1 are constants determined by fitting Eq. (74) with the results of the experimental data. The symbol h_L represents the heat-transfer coefficient which would exist if only liquid flowed in the tube, and it is generally calculated by using Eq. (33). In some studies, the total flow rate $W_I + W_{II}$ was used in evaluating h_L; in others an average value of W_{II} was used. The parameter χ is the Lockhart–Martinelli parameter discussed in Section I. The values of a_2 and b_1 determined by various researchers are shown in Table II, and in many cases data from the nucleate boiling region are included in the formulation of the correlation. Dengler and Addams (D3) and Guerrieri and Talty (G4) both present a method for calculating a correction factor used to multiply h_{WII} to yield the heat-transfer coefficient during flow nucleation. In each case, flow nucleation is assumed to occur when the correction factor has a value greater than unity.

A second generalized form for the heat-transfer coefficient correlations is

$$h_{WII} = h_L a_3 [B_0 \times 10^4 + a_4(1/\chi)^{b_2}]^{b_3} \tag{75}$$

where χ and h_L have the same definition as in Eq. (74); a_3, a_4, b_2, and b_3 are constants determined by fitting Eq. (75) to the experimental results, and the boiling number B_0 is defined as

$$B_0 = \frac{q_{WII}/A_{WII}}{(W_I + W_{II})\lambda_g} \tag{76}$$

TABLE II

Heat-Transfer Studies Fitting General Form I

Reference	a	b	Fluid	Flow direction	Inlet quality (%)	Exit quality (%)	h_L basis
Collier et al. (C7)	2.167	0.699	Water	Upward	—	66	W_{II}
Dengler and Addams (D3)	3.5	0.5	Water	Upward	0	70	$W_I + W_{II}$
Guerrieri and Talty (G4)	3.4	0.45	Pentane Heptane	Upward	0	12	W_{II}
Pujol and Stenning (P2)	4.0	0.37	Refrigerant 113	Upward Downward	0	70	$W_I + W_{II}$
Schrock and Grossman (S3)	2.5	0.75	Water	Upward	—	59	$W_I + W_{II}$
Somerville (S7)	7.55	0.328	n-Butanol	Downward	—	31	W_{II}
Wright (W8)	2.721	0.581	Water	Downward	0	19	W_{II}

The values of the constants that have been determined by various researchers are given in Table III. In all cases Eq. (76) is based on data from both the nucleate boiling and the two-phase forced-convection regions.

Most of the studies listed in Table II and III have the same shortcomings. In all cases mathematical models for the conservation of mass and energy were not formulated for each phase, and therefore, the definition of the heat-transfer coefficient is not explicitly given. The authors did not state how they calculated the rate of phase change at the interface, and the vapor-phase temperature was not measured. The pressure drop and holdups were reported only in the study by Dengler and Addams, but no correlations for these quantities were reported. Because of these shortcomings the design engineer cannot use any of these correlations with complete confidence. However, until experimental investigations are undertaken by researchers who have an understanding of the gross fluid mechanics, and who have described the systems with mathematical descriptions for the conservation of mass and energy, similar to those presented here, design engineers have no choice but to use the correlation that was developed for conditions similar to those of their proposed designs.

The absolute rates of vaporization and condensation are evaluated from the rate expressions given in Section III,B. In the past, the rate of mass transfer (which is the net rate of phase change) has not been calculated from an understanding of the physics of the phase-change process at the interface. The rate is generally evaluated by applying some simplifying assumptions to the process, rather than from an expression in terms of the dependent variables of the model equations.

If Eqs. (72a) and (72b) are combined, and if the interfacial heat transfer

TABLE III

Heat Transfer Studies Fitting General Form II

Reference	a_1	a_2	b_1	b_2	Fluid	Flow direction	Inlet quality (%)	Exit quality (%)	h_L basis
Chaddock and Brunemann (C1)	1.91	1.5	2/3	0.6	Refrigerant 12 Refrigerant 22	Horizontal	0	75	$W_l + W_{ll}$
Pujol and Stenning (P2)	0.90	4.45	0.37	1	Refrigerant 113	Upward	0		$W_l + W_{ll}$
	0.53	7.55	0.37	1		Downward	0	70	$W_l + W_{ll}$
Sani (S1)	1.48	1.5	2/3	1	Water	Downward	0	14	W_{ll}
Schrock and Grossman (S3)	0.739	1.5	2/3	1	Water	Upward	0	59	$W_l + W_{ll}$
Somerville (S7)	2.45	1.5	2/3	1	n-Butanol	Downward	0	31	W_{ll}
Wright (W8)	1.39	1.5	2/3	1	Water	Downward	0	19	W_{ll}

and the axial conduction terms are assumed to be negligible, then the rate term is given by

$$\bar{r}_{\text{net}} a = \frac{(q_{W\text{II}} + W_{\text{II}} \, dH_{\text{II}}/dz)}{(H_{\text{I}} - H_{\text{II}})} \tag{77}$$

The evaluation of $\bar{r}_{\text{net}} a$ in this case requires an a priori description of the phase temperature profiles. In general, investigators have assumed that gas and liquid are in thermal equilibrium. Therefore, Eq. (77) can be simplified to yield

$$\bar{r}_{\text{net}} a = (q_{W\text{II}} + W_{\text{II}} C_{P\text{II}} \, dT_{\text{II}}/dz)/\lambda_g \tag{78}$$

The studies listed in Tables II and III do not explicitly state how the mass flow rate profiles were evaluated, but they do state the amount of vapor formed. From the discussions in these papers, it can be assumed that one of two methods was followed. In the first, the liquid is assumed to be at a constant temperature and the rate term is evaluated from

$$\bar{r}_{\text{net}} a = q_{W\text{II}}/\lambda_g \tag{79}$$

Equation (79) implies that all of the heat entering the system is used to produce a vapor. The second method requires a knowledge of the liquid temperature profile, since the rate term is given by

$$\bar{r}_{\text{net}} a = q_{W\text{II}} + W_{\text{II}} C_{P\text{II}} (\Delta T/\Delta z) \lambda_g \tag{80}$$

In either case, either Eq. (79) or (80) can be coupled with Eq. (71) to determine the mass flow-rate profiles.

From the design viewpoint, Eq. (78) could be coupled with Eq. (71) to obtain an approximation of the system performance and if the liquid temperature profile can be estimated, the same procedure can be followed with Eq. (80). However, in general the design engineer needs to use analytical expressions for the absolute rates of vaporization and condensation, so that with a knowledge of the rate terms and the other parameters, Eqs. (71) and (72) could be solved for the temperature and mass flow-rate profiles.

E. Model Behavior

By examining the model equations in Section III,C it can be seen that the net rate of phase change

$$r_{\text{net}} = r_v - r_c$$

can be used to eliminate the absolute rate terms for the mass-transfer equations. The absolute rate terms cannot be eliminated from the energy transfer equations unless the average enthalpy of the vaporizing molecules $\bar{H}$ is equal to the enthalpy of the condensing molecules. This condition is only rigorously true when both phases are in equilibrium. O'Connor (O1) has shown that equivalence of these terms is a reasonable assumption when dealing with the liquid-phase energy equation under most conditions of pragmatic interest. The assumption is not equally as attractive when applied to the vapor-phase energy equation.

1. *Saturated Vapor Phase*

It is possible, however, to simplify the calculation of the energy transfer by assuming that the vapor phase is always a saturated vapor. O'Connor (O1) has shown that the rate of approach of a superheated vapor to saturated conditions is extremely rapid when the superheated vapor is in direct contact with its liquid phase. If the vapor phase is assumed to be saturated, the temperature of the phase can be calculated from an integrated form of the Clausius–Clapeyron equation instead of from the vapor-phase energy-transfer equation.

The assumption of a saturated vapor phase greatly simplifies the mathematical descriptions. For example, the equations given in Section III,C,2 can be written as

$$(d/dz(\rho_1 R_1 v_1)) = a\bar{r}_{net} \tag{81a}$$

$$(d/dz(\rho_{II} R_{II} v_{II})) = -a\bar{r}_{net} \tag{81b}$$

$$(d/dz(\rho_{II} R_{II} v_{II} H_{II})) = (d/dz(R_{II} k_{II} dT_{II}/dz)) \tag{82a}$$

$$- V_a(T_{II} - T_I) - aH_I\bar{r}_{net} + q_{wII} \tag{82b}$$

These equations can be evaluated by using the rate expression for $\bar{r}_{net}$ given in Section III,B,3. As shown in that section, the results are equivalent to evaluating the absolute rate terms but the computation procedure is greatly simplified.

If the vapor-phase temperature is to be evaluated from the Clausius–Clapeyron equation, the pressure in the two-phase tubular contactor must be known at each axial position. This need once again illustrates the necessity of obtaining an understanding of the hydrodynamics of two-phase systems in order to carry out the design of heat-transfer contactors.

2. *Condensation*

In discussing the phase-change problem in this chapter, the discussion has been limited to the process of vaporizing a liquid. Equally important is the process of condensing a vapor. The equations developed in Section III,C can be applied directly to this process.

In the condensation process the vapor, which has a higher velocity than the liquid, releases kinetic energy as it decelerates when changing phase. This reduction in the kinetic energy of the mixture tends to counterbalance the energy dissipated by friction at the wall. Therefore, the pressure drop in a two-phase tubular condensor should be less than the calculated valve based on the gas volume fraction at the entrance conditions. As a first approximation, the pressure in the contactor can be assumed to be constant. This assumption coupled with the assumption of a saturated vapor permits the design engineer to obtain a first estimate for the condensation process by solving the mass-transfer equations for each phase and the liquid-phase energy equation.

IV. Concluding Remarks

The two goals of this chapter were to provide a critical review of the current state of the art in the field of two-phase flow with heat transfer and to provide procedures which can be used for the design of tubular fluid–fluid systems. Both heat transfer without phase change and with phase change were discussed in detail. In each case the analysis was based on an understanding of the flow patterns and the hydrodynamics of the system.

In heat transfer without phase change, the hydrodynamics of these systems have been classified into five basic flow regimes, three of which were shown to be of pragmatic interest. For each regime the analysis was based on an understanding of the flow patterns included in that regime, a recognition of the similarities and differences between single- and two-phase heat transfer, and a realization that the ultimate goal of this analysis was the development of methods for designing two-phase processes.

It was shown that in heat transfer with phase change it is necessary to understand the phase-change phenomenon on the molecular level to model effectively the mass- and heat-transfer processes. An analytical expression for the rates of vaporization and condensation was developed. It was also shown that the assumption of a saturated vapor phase greatly simplified the calculation without a significant loss in accuracy for given examples. However, experimental verification of this simplified assumption is currently lacking.

The analysis of tubular contactors for heat transfer with phase changes in fluid–fluid systems was shown to be heavily dependent on a proper understanding of two-phase hydrodynamics. It was shown that three basic flow patterns exist within a tube, each with a different heat-transfer mechanism. The formulation of the proper mass and energy models pinpointed three key areas in which further research must be performed before the total phase change process in two phase tubular contactors can be described: (1) the calculation of the rate of nucleation and bubble departure frequency; (2) the accurate prediction of the flow patterns, pressure drop and holdups; and (3) additional experimental verification of the analytical model for the rate of phase change at the gas–liquid interface.

The design of two-phase contactors with heat transfer requires a firm understanding of two-phase hydrodynamics in order to model effectively the heat- and mass-transfer processes. In this chapter we have pointed out areas where further theoretical and experimental research is critically needed. It is hoped that design engineers will be motivated to test the procedures presented, in combination with their use of the details from the original references, in the solution of pragmatic problems.

Nomenclature[2]

a	Interfacial area between two continuous phases per unit volume of tube ($L^2\ L^{-3}$)	D	Diameter of the pipe (L)
		D_i	Hydraulic diameter for phase i (L)
a'	Interfacial area of a drop per unit drop volume ($L^2\ L^{-3}$)	E_c	Condensation coefficient
		E_v	Vaporization coefficient
a'_e	Interfacial area of entrained phase per unit volume of tube ($L^2\ L^{-3}$)	f	Bubble departure frequency (bubble time^{-1})
		f_i	Friction factor for Phase i
a_1	Constant (Eq. (73))	g	Constant (Eq. (73))
a_2	Constant (Eq. (74))	g_c	32.2 lb$_{mass}$ ft (lb$_f$ sec^2)$^{-2}$
a_3	Constant (Eq. (75))	G	Gas mass flux (lb$_{mass}$ ft^{-2} hr^{-1}) (Fig. 5)
a_4	Constant (Eq. (75))		
a_c	Cross-sectional area of the pipe (L^2)	h	Interfacial position in the y direction (L)
A_i	Cross-sectional area of the pipe filled with phase i (L^2)	h_{w_i}	Wall heat transfer coefficient for Phase i (F Lθ^{-1} degree^{-1})
A_{w_i}	Wall area in contact with Phase i per unit volume of tube ($L^2\ L^{-3}$)	H_i	Enthalpy of phase i ($FL\ M^{-1}$)
		$\bar{H}_v$	Average enthalpy of the material vaporizing (FL M^{-1})
b_1	Constant (Eq. (74))		
b_2	Constant (Eq. (75))	$\bar{H}_c$	Average enthalpy of the material condensing (FL M^{-1})
b_3	Constant (Eq. (75))		
B_0	Boiling number, defined in Eq. (76)	k_i	Thermal conductivity of Phase i (F θ^{-1} degree^{-1})
C_{p_i}	Heat capacity at constant pressure for Phase i (FL degree^{-1})	$k_{i_{eff}}$	Effective thermal conductivity of Phase i (F θ^{-1} degree^{-1})
		L	Characteristic length (L)

[2] F = force dimension; L = length dimension; θ = time dimension; M = mass dimension.

L	Liquid mass flux (lb_{mass} ft^{-2} hr^{-1}) (Fig. 5)	U	Interfacial heat transfer coefficient between two continuous phases, (F L^{-1} θ^{-1} degrees^{-1})
M	Molecular weight (M)		
n	Number of nucleation centers per unit area (L^{-2})	U_0	Net velocity in the gas phase (L θ^{-1})
N_1	Interfacial heat-transfer number	U'	Interfacial heat transfer coefficient between droplets and a gas (F L^{-1} θ^{-1} degrees^{-1})
N_{w_i}	Wall heat-transfer term for Phase i		
Pe_i	Peclet number for Phase i		
Pr_i	Prandtl number of Phase i	v_i	Average velocity of Phase i (L θ^{-1})
P_i	Pressure of Phase i (F L^{-2})		
P_i^*	Vapor pressure of liquid at temperature T_i (F L^{-2})	v_r	$v_g - v_e$, relative velocity of gas phase to entrained liquid phase (L θ^{-1})
$(\Delta P/L)$	Pressure drop per unit length (F L^{-3})	V_d	Volume of a drop or bubble (L^3)
$(\Delta P/\Delta z)_{TP_i}$	Pressure drop in Phase i of a two-phase system (F L^{-3})	W_i	Mass flow rate of phase i per unit area of pipe (M L^{-2} θ^{-1})
$(\Delta P/\Delta z)_i$	Pressure drop in Phase i assuming only Phase i flows in the tube (F L^{-2})		
q_{w_i}	Wall heat flux to Phase i per unit volume of tube (F L^{-2})	GREEK LETTERS	
Q	Heat flux (FL)	α	Number of bubbles or drops per unit volume of tube (L^{-3})
Q_d	Rate of droplet deposition per unit volume of tube (L^{-3})		
Q_g	Rate of droplet generation per unit volume of tube (L^{-3})	α_i	Thermal diffusivity of phase i (L^2 θ^{-1})
Q_i	Volumetric flow rate of Phase i (L^3 θ^{-1})	Γ	Defined by Eq. (53)
		γ	Surface tension (dyne cm^{-1})
r_c	Absolute rate of condensation (M L^{-2} θ^{-1})	ζ	Dimensionless length
		θ	Surface renewal time period (θ)
$\bar{r}_{net}$	Net rate of phase change (M L^{-2} θ^{-1})	λ	Baker chart parameter, $[(\rho_G/0.075)(\rho_L/62.3)]^{1/2}$ (Fig. 5)
$\bar{r}_v$	Absolute rate of vaporization (M L^{-2} θ^{-1})		
		λ_i	Latent heat of phase change at temperature T_1 (FL θ^{-1})
R_i	Hold up of Phase i		
R_{c1}	Rate of bubble condensation per unit volume of tube, bubbles (L^{-3} θ^{-1})	μ_i	Viscosity of Phase i (centipoise in Fig. 5) (F L^{-2} θ^{-1})
		π	3.14159
R_{c2}	Rate of bubble coalescence per unit volume of tube, bubbles (L^{-3} θ^{-1})	ρ_i	Density of Phase i (lb_m ft^{-3}) (Fig. 5) (M L^{-3})
		ϕ_G	Lockhart and Martinelli parameter
Re_i	Reynolds number of Phase i		
t	Surface exposure time (θ)	χ	Lockhart and Martinelli parameter
T_i	Temperature of Phase i (degrees)		
		Ψ	Baker chart parameter, $(73/\gamma_L)[\mu_L(62.3/\rho_L)]^{1/3}$ (Fig. 5)
T_W	Temperature of the tube wall (degrees)		

SUBSCRIPTS

d	Drop or bubble	i	Component or phase
e	Entrained phase	L	Bulk liquid phase
f	Film phase	I	Less dense phase
G	Bulk gas phase	II	More dense phase

References

A1. Al-Sheikh, J. N., Saunders, D. E., and Brodkey, R. S., *Can. J. Chem. Eng.* **48**, 21 (1970).
A2. Alves, G. E., *Chem. Eng. Prog.* **50**, No. 9, 444 (1954).
A3. Anderson, R. J., and Russell, T. W. F., *Chem. Eng.* **72**, No. 25, 134 (1965).
A4. Anderson, R. J., and Russell, T. W. F., *Chem. Eng.* **72**, No. 26, 99 (1965).
A5. Anderson, R. J., and Russell, T. W. F., *Chem. Eng.* **73**, No. 1, 87 (1966).
A6. Anderson, R. J., and Russell, T. W. F., *Ind. Eng. Chem. Fundam.* **9**, 130 (1970).
A7. Arruda, P. J., M.Ch.E. thesis, University of Delaware, Newark, Delaware, 1970.
B1. Baker, O., *Oil Gas J.* **53**, No. 12, 185 (1954).
B2. Bankoff, S. G., *Adv. Chem. Eng.* **6**, 1 (1966).
B3. Bentwich, M., and Sideman, S., *Can. J. Chem. Eng.* **42**, 9 (1964).
B4. Bird, R. B., Stewart, W. E., and Lightfoot, E. N., "Transport Phenomena." Wiley, New York, 1960.
B5. Brauer, H., "Grundlagen der Einphasen-und Mehrphasen Strömungen." Verlag Sauerländer, Aabrau und Frankfort am Main, 1971.
B6. Brown, R. A. S., Sullivan, G. A., and Govier, G. W., *Can. J. Chem. Eng.* **38**, 62 (1960).
B7. Byers, C. H., and King, C. J., *AIChE J.* **13**, 628 (1967).
C1. Chaddock, J. B., and Brunemann, H., Report HL-113, School of Engineering, Duke University, July, 1967.
C2. Charles, M. E., and Lilliheht, L. U., *Can. J. Chem. Eng.* **44**, 47 (1966).
C3. Charles, M. E., Govier, G. W., and Hodgson, G. W., *Can. J. Chem. Eng.* **39**, 27 (1961).
C4. Cichy, P. T., and Russell, T. W. F., *Ind. Eng. Chem.* **61**, No. 8, 15 (1969).
C5. Cichy, P. T., Ultman, J. S., and Russell, T. W. F., *Ind. Eng. Chem.* **61**, No. 8, 6 (1969).
C6. Collins, S. B., and Knudsen, J. G., *AIChE J.* **16**, 1072 (1970).
C7. Collier, J. G., Lacey, P. M. C., and Pulling, D. J., *Trans. Inst. Chem. Eng.* **42**, 127 (1964).
C8. Cousins, L. B., Denton, W. H., and Hewitt, G. F., UKAEA Research Group Report, AERE-R4926 (1965).
D1. Dergarabedian, P., *J. Appl. Mech.* **75**, 537 (1953).
D2. DeGance, A. E., and Atherton, R. W., *Chem. Eng.* **77**, No. 13, 95 (1970).
D3. Dengler, C. E., and Addams, J. N., *Chem. Eng. Prog. Sym. Ser.* **52**, No. 18, 95 (1956).
D4. Dukler, A. E., Wicks, M., and Cleveland, R. G., *AIChE J.* **10**, 44 (1964).
E1. Etchells, A. W., Ph.D. thesis, University of Delaware, Newark, Delaware, 1970.
G1. Govier, G. W., and Aziz, K., "The Flow of Complex Mixtures in Pipes." Van Nostrand-Reinhold, Princeton, New Jersey, 1972.
G2. Govier, G. W., and Short, W. L. *Can. J. Chem. Eng.* **36**, 195 (1958).
G3. Govier, G. W., Radford, B. A., and Dunn, J. S. C., *Can. J. Chem. Eng.* **35**, 58 (1957).
G4. Guerrieri, S. A., and Talty, R. D., *Chem. Eng. Prog. Sym. Ser.* **52**, No. 18, 69 (1956).
H1. Hewitt, G. F., and Hac-Taylor, N. S., "Annular Two-Phase Flow." Pergamon, Oxford, 1970.
H2. Hinze, J. O., *AIChE J.* **1**, 289 (1955).

H3. Hughmark, G. A., *Chem. Eng. Prog.* **58**, No. 4, 62 (1962).
H4. Hughmark, G. A., *Chem. Eng. Sci.* **20**, 1007 (1965).
K1. Kern, D. Q., "Process Heat Transfer." McGraw-Hill, New York, 1950.
K2. Koumoutsos, N., Moissis, R., and Spyridonos, A., *J. Heat Transfer* **90**, 223 (1968).
L1. Langmuir, I., *J. Am. Chem. Soc.* **54**, 2798 (1932).
L2. Leppert, G., and Pitts, C. C., *Adv. Heat Transfer* **1**, 185 (1964).
L3. Letan, R. and Kehat, E., *AIChE J.* **11**, 804 (1965).
L4. Letan, R., and Kehat, E., *Chem. Eng. Sci.* **20**, 856 (1965).
L5. Letan, R., and Kehat, E., *AIChE J.* **13**, 443 (1967).
L6. Letan, R., and Kehat, E., *AIChE J.* **14**, 398 (1968).
L7. Letan, R., and Kehat, E., *AIChE J.* **14**, 831 (1968).
L8. Letan, R., and Kehat, E., *AIChE J.* **15**, 4 (1969).
L9. Letan, R., and Kehat, E., *AIChE J.* **16**, 955 (1970).
L10. Levenspiel, O., "Chemical Reaction Engineering," p. 276. Wiley, New York, 1962.
L11. Levy, S., *Int. J. Heat Mass Transfer* **10**, 951 (1967).
L12. Lockhart, R. W., and Martinelli, R. C., *Chem. Eng. Prog.* **45**, No. 1, 39 (1949).
M1. Maa, J. R., *Ind. Eng. Chem. Fundam.* **6**, 504 (1967).
M2. Maa, J. R., *Ind. Eng. Chem. Fundam.* **8**, 564 (1969).
M3. Maa, J. R., *Ind. Eng. Chem. Fundam.* **8**, 560 (1969).
M4. Maa, J. R., *Ind. Eng. Chem. Fundam.* **9**, 283 (1970).
M5. Macbeth, R. V., *Adv. Chem. Eng.* **7**, 208 (1968).
M6. Magiros, P. G., and Dukler, A. E., *Dev. Mech.* **1**, 532 (1961).
M7. Martinelli, R. C., and Nelson, D. B., *Trans. Am. Soc. Mech. Eng.* **70**, 695 (1948).
M8. Mixon, F. O., Whitaker, D. R., and Orcutt, J. C., *AIChE J.* **13**, 21 (1967).
N1. Nabavian, K., and Bromley, L. A., *Chem. Eng. Sci.* **18**, 651 (1963).
N2. Nicklin, D. J., and Davidson, J. F., Symp. Two-Phase Flow, Inst. Mech. Eng., London, Feb., 1962, No. 4.
O1. O'Connor, G. E., Ph.D. thesis, University of Delaware, Newark, Delaware, 1971.
O2. Ostermaier, J., M.Ch.E. thesis, University of Delaware, Newark, Delaware, 1971.
P1. Pavlica, R. T., and Olson, J. H., *Ind. Eng. Chem.* **62**, No. 12, 45 (1970).
P2. Pujol, L., and Stenning, A. H., Int. Sym. Res. Cocurrent Gas–Liquid Flow, University of Waterloo, Sept., 1968.
R1. Rohsenow, W. M., "Developments in Heat Transfer." MIT Press, Cambridge, Mass., 1964.
R2. Ruckenstein, E., *Chem. Eng. Sci.* **10**, 22 (1959).
R3. Russell, T. W. F., and Etchells, A. W., Paper presented at Int. Symp. Two-Phase Systems, Haifa, Israel, August, 1971.
R4. Russell, T. W. F., and Lamb, D. E., *Can. J. Chem. Eng.* **43**, 234 (1965).
R5. Russell, T. W. F., and Rogers, R. W., *AIChE Symp. Ser.* **69**, No. 127, 60 (1971).
R6. Russell, T. W. F., Hodgson, G. W., and Govier, G. W., *Can. J. Chem. Eng.* **37**, 9 (1959).
S1. Sani, R. L., Lawrence Radiation Lab., UCRL-9023, Jan., 1960.
S2. Schrage, R. W., "A Theoretical Study of Interphase Mass Transfer." Columbia Univ. Press, New York, 1953.
S3. Schrock, V. E., and Grossman, L. M., USACE Rep. TID-14632 (1959).
S4. Scott, D. S., *Adv. Chem. Eng.* **4**, 199 (1963).
S5. Sideman, S., *Adv. Chem. Eng.* **4**, 207 (1966).
S6. Silvestri, M., *Adv. Heat Transfer* **1**, 355 (1964).
S7. Somerville, G. F., Lawrence Radiation Lab., UCRL-10527, Oct., 1962.
S8. Staub, F. W., *J. Heat Transfer* **90**, 151 (1968).

T1. Tong, L. S., "Boiling Heat Transfer and Two-Phase Flow." Wiley, New York, 1965.
T2. Tschudin, K., *Helv. Phys. Acta* **19**, 91 (1946).
V1. Volmer, M., and Estermann, I., *Z. Phys.* **7**, 1 (1921).
W1. Waldman, L. A. and Houghton, G., *Chem. Eng. Sci.* **20**, 625 (1965).
W2. Wallis, G. B., "One-Dimensional Two-Phase Flow." McGraw-Hill, New York, 1967.
W3. Wehner, J. F., and Wilhelm, R. H., *Chem. Eng. Sci.* **6**, 89 (1956).
W4. Westwater, J. W., *Adv. Chem. Eng.* **1**, 1 (1958).
W5. Westwater, J. W., *Adv. Chem. Eng.* **2**, 1 (1958).
W6. Wicks, M., and Dukler, A. E., *AIChE J.* **6**, 463 (1960).
W7. Wilke, C. R., Cheng, C. T., Ladesma, V. L., and Porter, J. W., *Chem. Eng. Prog.* **59**, No. 12, 69 (1963).
W8. Wright, R. M., USAEC Rep. 9744 (1961).
Y1. Yu, H. S., and Sparrow, E. M., *AIChE J.* **13**, 10 (1962).
Z1. Zuber, N., Staub, F. W. and Bijwaard, G., *Proc. Int. Heat Transfer Conf.*, 3rd, **5**, 24 (1966).

BALLING AND GRANULATION

P. C. Kapur

Department of Metallurgical Engineering
Indian Institute of Technology
Kanpur, India

I.	Introduction	56
II.	Balling and Granulation Equipment	57
	A. Drums and Disks	58
	B. Miscellaneous Devices	61
III.	Bonding Mechanisms in Agglomerates	62
	A. Tensile Strength of Agglomerates	63
	B. Capillary Bonds	66
	C. Solid Bridges	72
	D. Attractive Forces in the Absence of Material Bridges	73
	E. Deformation of Agglomerates	74
IV.	Compaction and Growth Mechanisms	75
	A. Compaction	75
	B. Growth Mechanisms	77
	C. Regions of Agglomerate Growth	81
V.	Kinetics of Balling and Granulation	84
	A. Snowballing Kinetics	85
	B. Crushing and Layering Kinetics	86
	C. Random Coalescence Kinetics	90
	D. Nonrandom Coalescence Kinetics	93
	E. Empirical Kinetic Models	99
VI.	Bonding Liquid and Additives	100
	A. Water Content	100
	B. Bentonite Additive in Iron Ore Balling	102
VII.	Granulation of Fertilizers	105
	A. Granulation in Coalescence Mode	106
	B. Granulation in Snowballing Mode	109
VIII.	Miscellaneous Topics	112
	A. Dry Pelletization	113
	B. Spherical Agglomeration in Liquid Suspension	115
	C. Inadvertent Agglomeration	117
	Nomenclature	118
	References	120
	Supplementary References	123

I. Introduction

The agglomeration of moist particulate solids into nearly spherical pellets by tumbling, rolling, or some other systematic agitation of the mass is called *granulation* in the fertilizer industry and *balling* or *wet* (or *green*) *pelletization* in the iron ore industry. Dry pelletization refers to the agglomeration of finely divided moisture-free powder into dense free-flowing spheroids (M6). The formation of granules from finely divided solids in liquid suspension in the presence of a small amount of a second immiscible liquid, which preferentially wets the solid, is known as *spherical agglomeration* (F1). In this review the terms *granulation, balling, pelletization,* and *agglomeration* are used interchangeably in the appropriate context. The spheroidal ensemble of particles is called a *granule, ball, pellet,* or an *agglomerate*. It is the balling and granulation of moist or wet particulate masses with which the following is primarily concerned.

Nucleation, compaction, size enlargement, and spheroidization of the pellets take place in the course of balling and granulation and related agglomeration processes. Depending on the nature of the pelletizing system, the sequence of the last three actions may occur in different orders and with extensive overlapping, as indeed happens in most instances. A characteristic feature is that, unlike in briquetting, tableting, and extrusion (B8, P2), no external pressure is applied for the densification and shaping of the particulate masses into larger bodies. In yet another class of the size enlargement processes, which includes sintering and nodulizing (B8, P2), heat treatment of some sort results in relatively strong bondings between the constituent particles by controlled fusion at their points of contact. In general, elevated temperatures are not essential to the formation of pellets in balling and granulation. However, the pellets are significantly porous, and the interparticle bond is invariably weaker than the chemical bond within the individual particles. For these reasons, particle enlargement by balling and granulation is only superficially the reverse of comminution in which the size reduction occurs by breakage of the chemical bonds in the solids.

Many chemical, ceramic, metallurgical, nuclear, and pharmaceutical industries, to mention a few, are concerned with the processing of finely divided solids and find it advantageous, or even necessary, to carry out size enlargement by some method of agglomeration. The properties desired are densification, homogenization, strength, flowability, and uniformity. Balling and granulation has proved to be a highly versatile technique for large-scale production of particulate spheroids at relatively low capital and operational costs. Moreover, it is possible to achieve a wide range of strengths, sizes, and porosities of the agglomerates.

In tonnage, by far the main application of balling is in the agglomeration of the run-of-mine iron ore dust and of the concentrates from low-grade iron ores. It is estimated that in 1976 the amounts involved are likely to exceed 200 million tons (R2). The green pellets are indurated or hardened in some manner and then charged to the blast furnace. The relatively uniform size of the balls ranging from about 10 to 12 mm in diameter results, first, in an even distribution of the burden in the furnace and, second, in greater bed permeability to the ascending hot gases with virtual elimination of channeling. The porosity of the sintered pellets is normally 20–30%, which again improves solid–gas contact, resulting in an enhanced rate of reduction of the iron oxides with marked economy of operation (E3).

Currently in the chemical industry, most solid fertilizers are granulated together with other fertilizer constituents in the form of slurry, solution, or melt. Granulation permits the production of straight or mixed high-analysis products of higher bulk density, which results in lower cost of handling and transportation. In addition, granulated fertilizers are dust free and easy to handle. They present fewer problems of setting up and caking during storage, and can be applied conveniently to the soil in a uniform manner (H7).

The feed to glass-melting furnaces has been pelletized advantageously. Handling and transportation of the charge is facilitated, feeding is more uniform, and dust losses from the furnace are minimized. The intimate and uniform mixture of the reactants in the pellets results in a faster rate of melting and increased uniformity and superior quality of the product (E2). In particular, the number of bubbles trapped in the frozen melt may be reduced substantially when a granulated feed charge is employed (I1). Similar advantages are gained with the agglomeration of the feed to portland cement kilns (T4).

Many minerals, industrial chemicals, and waste products are now either routinely pelletized in large tonnage or have been investigated for agglomeration on the laboratory or pilot-plant scale. The list includes fluorspar (A1), fly ash (V1), detergents and food products (E1), pharmaceuticals (P9), flue dust from steel plant furnaces (C1), manganese ore fines (M7), ceramic and catalysts (E2), nuclear fuels (G1), and many more. A large patent literature exists on the varied and diverse applications of balling and granulation.

II. Balling and Granulation Equipment

Rotating drums, disks, and to a lesser extent, cones are used in continuous large-scale balling of iron ores (E3). A pelletizing disk of industrial type is shown in Fig. 1. Fertilizers are granulated in twin-shaft pug mills, drums,

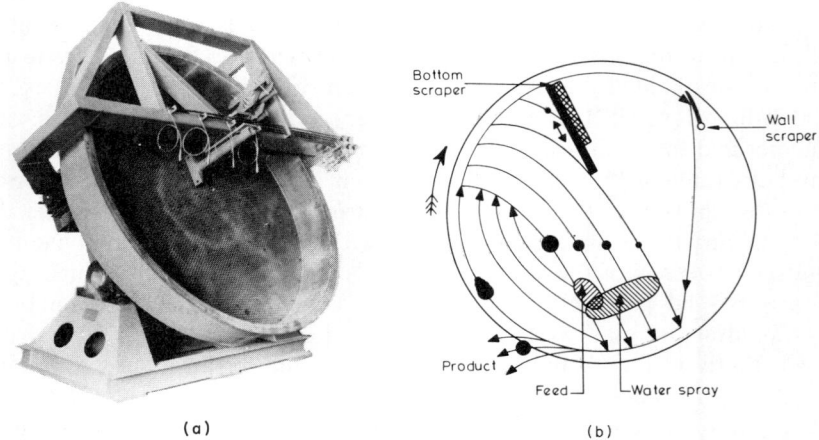

FIG. 1. (a) Industrial disk pelletizer (diameter 8 ft) (courtesy of Ferro-Tech, Inc.). (b) Schematic front view of a disk.

disks, spherodizers, and by prilling (C10). Fluidized- and spouted-bed units (B5) are also used for this purpose. Ceramic materials and catalysts are pelletized in ribbon blenders, mix mullers, spray driers, drums, and disks (E2). Disks are also standard equipment for balling feeds to the cement kilns and glass furnaces. High-speed blenders, colloid mills, pans, drums, and reciprocating shakers are reported to be suitable for spherical agglomeration (S11). Dry pelletization has been carried out in drums and in a ro-tap sieve shaker (F3). Clearly, drums and disks are by far the most widely used equipment for balling and granulation.

A. Drums and Disks

The published data on design, scale-up, and performance characteristics of the balling drum are substantially fewer than those available for the disk. This presumably reflects the fact that the latter is a more versatile agglomerator; it is, however, more difficult to control for a stable operation.

In general, the underlying principles of agglomeration in the drum and in the disk are similar (B2). The pelletizer is rotated at a fixed speed, and dry or wet particulate material is fed into it continuously. If required, a spray of water, solution, or slurry is introduced over the rotating charge. The rolling action imparted to the agglomerates by the rotation gives rise to small spherical pellets that grow in size until ejected from the machine. The pelletizer is arranged, if necessary, in a closed circuit with a screen or a deck of screens for the separation of the undersize and oversize pellets. The undersize and, after crushing, the oversize pellets are recycled to the device as seed pellets. The off size material may range up to 400% of the production.

Normally, the length of an iron ore balling drum is two to three times its diameter and the inclination 2–5 degrees. A drum of 2.5-m diameter and 7.5-m length, operating at 10.75 rpm or 40% critical speed, should produce about 1000 tons of 19-mm pellets per day. This is equivalent to a specific production rate of approximately 13 kg/min^{-1} m^{-2} of the drum surface area (M2). In order to prevent a buildup of moist charge on the surface, the drum is equipped with a cutter bar, which scrapes off the excess material and maintains a uniform lining. Such a lining has been found to be essential for the production of good-quality balls (E3).

The difference between the balling drum and the pelletizing disk is noteworthy on two counts. First, the area of the disk used in pelletizing, (i.e., covered under the agglomerating charge at any given time) is roughly twice that in the drum, typically 70% as against 40% (M2). Second, size classification of the agglomerates, at least in case of the large balls, is inherent in the working of the disk (this is illustrated schematically in Fig. 1b). The path of the nuclei or seed pellets lies near the center and the bottom of the disk. The nucleation zone is most likely to occur in the localized regions under water sprays where water droplets impinging on the feed particles quickly form clumps of nuclei or seeds. These small pellets, as well as the recycled pellets, grow in size by coalescing with other nuclei or by picking up incoming feed particles, that is, by snowballing. As the nuclei acquire mass, centrifugal forces overcome the friction, and the pellets gradually move outward and rise to the top of the granulating charge, where they continue to grow. Eventually they flip out of the disk as a narrow size fraction product (B6). Because of this classifying action, the disk may be operated in an open circuit without the screen, unless of course, the size specifications are quite stringent. In the case of fertilizer granulation however, the situation is somewhat more complex. The product size seldom exceeds 6 mm, and the frictional forces between the granules of relatively large surface areas impose some constraint on the classifying action of the machine. Moreover, the size and size distribution of the granules formed in the disk may be determined, in the first instance, by the size of the feed particles and the liquid content of the charge, as is discussed in Section VII.

The performance of the pelletizing disk depends on its size, speed, angle to the horizontal, collar height, positions of the feed chutes, water sprays and scrapers, and the feed throughput. The critical speed at which the balls no longer roll down but stick to the collar wall under centrifugal force is given by (P3)

$$n_c = 42.3(\sin \beta_d/D_d)^{1/2} \tag{1}$$

where n_c is the critical speed (rpm), β_d the inclination angle of the disk to the horizontal, and D_d the disk diameter (m). The disks are usually operated at speeds of 0.6–0.75 times the critical speed (K11, P1, P3). The collar height is about one-fifth of the disk diameter (K11). As a rule of thumb, the output varies as the square of the diameter (C13, K11). The inclination angle of the disk ranges from about 45 to 55 degrees (P1); a change from lower to steeper angle is accompanied by up to 50% reduction in the average pellet size in the product. Steeper inclination causes the granules to roll down from points of greater height, but the holdup and the mean residence time of the material are reduced. With increasing throughput the average size decreases almost linearly (B1), and concurrently the size dispersion increases (K9). Bhrany *et al.* (B6) and Kayatz (K9) have presented data for the residence time as a function of the throughput. The average granule size has been found to increase linearly with the moisture content (P1). The driving power varies directly with D_d^2, where the constant of proportionality lies between 1 and 1.2 kW m^{-2} (B1, K11). The design and scale-up of disks have been discussed by Bazilevich (B3) and Macavei (M1).

The position of the feed chute, of the water sprays, and of the bottom scraper are critical for stable operation of the disk and for an optimal production rate. However, the "best" configuration is highly dependent on the feed material and the product size desired, and usually must be determined by trial and error. Figure 2 is an illustration of a possible arrangement in which the granule growth occurs primarily by snowballing and layering. Pellets, soaked under a fine spray of water, roll through the feed zone where they snowball by picking up loose powder or are layered by

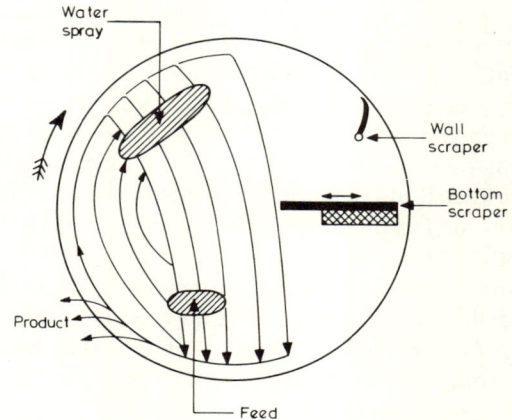

FIG. 2. Disk operating in the snowballing mode of pellet growth.

small, porous, easily deformable nuclei. When the feed zone and the spray are in the vicinity of each other (Fig. 1b), coalescence between the granules may become the predominant growth mechanism.

B. Miscellaneous Devices

Drums and disks are not without some disadvantages:

1. Since densification of the granules in the drums and disks is brought on by gravitational force, there is a definite upper limit on both the rate and the extent of compaction that is attainable. This is specially true in the case of fine solids. In general, compaction of the granules occurs concurrently with their growth, and it is seldom possible to exercise independent control over these two actions of the pelletizer.

2. Drums and disks are not suitable for the agglomeration of sticky and highly plastic materials, for example, clays or solids containing colloids and very fine components.

3. The proper functioning of the disks and drums is possible only over a narrow range of the liquid content of the agglomerating charge, and very little flexibility is available in this respect. This limitation may lead to operational problems as, for instance, when the balling feed is iron ore concentrates from a filter press and is excessively moist.

A number of new agglomeration devices meant for improvement in the conventional drums and disks or designed for a specific size enlargement task at hand, have been reported. Disks constructed with steps are capable of producing pellets of highly uniform size (B1, K11). The residence time of the granules is shortened substantially, but the porosity is increased (P3). Sterling (S12) has described a multiple-cone drum that combines the beneficial rolling action of the drum with the classifying action of the disk. Scottish Agricultural Industries (SAI) have designed a double drum for granulation of ammonium nitrate and phosphate fertilizers in which the various steps of reaction, mixing, granulation, and drying are combined into a single operation (C10, S10). Wet filter cake has been granulated in a machine that comprises a female cone and a rotating male cone, with a 12-degree angle between the surfaces (C8). The cake is fed to the center and the granules ejected at the periphery are agitated in dry powder in order to absorb the excess moisture.

In the conventional drums and disks, the compressive forces due to the weight of the rotating charge mass and the rolling action of the pellets are insufficient for a very high degree of compaction and rapid thinning of the water film separating the particles. Stoev and Watson (S13) have employed a

Russell vibrator for pelletizing on a flat surface using three-dimensional ellipsoidal vibrations. The vibrator imparts very little bulk translational and rotational motions to the charge. The energy input is expended mostly into pellet–pellet and pellet–plate collisions, which, supplemented by the rolling action, lead to better compaction and faster growth of the agglomerates. Very effective compaction and good spheroidization are obtained when the granules are subjected to a gyratory or swirling motion in a planetary mill type device (G1). The charge is placed in a gyrating pot, which moves in a circle but does not rotate by more than a few degrees, and is subjected to both a centrifugal force ($9-18g$) and a tumbling action. Uranium oxide powders have been granulated in sizes ranging from 0.2 to 3 mm. In general, it is not possible to make dense, well-rounded granules in this size range in the conventional balling equipment. A somewhat simpler gyrating device for dry pelletization of uranium oxide powders and its admixtures with carbon was constructed by Ford and Shennan (F3), in which the container is mounted on a ro-tap sieve shaker.

A single or twin-shaft paddle mixer or pug mill is widely used in the fertilizer granulation processes. Advantages claimed are intimate mixing of fertilizer slurry and recycle fines, production of harder and uniform-size granules, and flexibility in operation (C10). Sherrington (S9) has given details of a laboratory paddle mixer. A peg granulator, suitable for sticky and plastic materials, has been described by Brociner (B7). It consists of a cylinder with a rotating shaft carrying a number of pegs arranged in a helix. Feed is introduced at one end of the machine and emerges in the form of granules at the other end.

III. Bonding Mechanisms in Agglomerates

The formation of viable agglomerates in the balling and granulation processes requires interparticle bonds of finite strength to maintain necessary structural integrity of the pellets. The strength of the particulate ensemble is a function of its bulk density, or porosity, of the size, size distribution, shape, and packing arrangement of the constituent particles, and of the particle–particle bonding mechanisms. Rumpf (R3, R4) has provided a systematic classification of the various bonding mechanisms:

1. Capillary bonds due to negative capillary pressure (suction) and interfacial forces:
   (a) Liquid bridges or pendular bonds
   (b) Funicular bonds

(c) Capillary pressure bonds
(d) Liquid envelope (droplet) bonds
2. Solid bridges between the particles formed by:
   (a) Inorganic bonding agents (cementitious bonds)
   (b) Chemical reaction (chemical bonds)
   (c) Crystallization of dissolved material
   (d) Melting at points of contact between particles by friction and pressure
   (e) Sintering
3. Bridges with limited mobility:
   (a) Viscous binders
   (b) Adsorbed layers
4. Attractive forces between particles in the absence of liquid and solid bridges:
   (a) Molecular forces, including valence and van der Waals forces
   (b) Electrostatic forces
   (c) Magnetic forces
5. Mechanical bonds due to interlocking, microcontacts, friction, and arching of particles

The strength of the moist or liquid–wet granules is mainly due to capillary bonds. On an industrial scale the agglomerating charge is invariably wet; hence, capillary bonding plays a central role in the nucleation and growth of the pellets. Dry pelletization of finely divided powders is brought about by nonspecific electrostatic (N1) and van der Waals forces (M4), perhaps aided when present by adsorbed layers (R3). Capillary bonds are replaced by solid bridges when fertilizer granules are dried. Such bridges may arise by the crystallization of the dissolved salts when the liquid phase evaporates, or upon the freezing of the original liquid, or by chemical reaction. (B9). When relatively small amounts of bentonite or a suitable organic polymer are incorporated in the feed as an additive, viscous bonds contribute significantly to the strength of the wet iron ore balls, in particular to the impact strength (R2, W1). Even in the absence of nonmechanical bonds, a well-compacted pellet does exhibit some cohesive strength on account of friction and interlocking of particles (T5).

A. Tensile Strength of Agglomerates

Although measurement of the compressive strength of agglomerates is certainly more convenient for routine checking and quality control, the tensile strength is a more fundamental property, since in theory it can be directly related to the interparticle bonding force. Schubert (S7) has summarized a number of empirical and theoretical models of the tensile strength of pellets. All the models explicitly recognize the very strong dependence of strength on the degree of compaction, that is, porosity. In the model of Ashton *et al.* (A2), for example, the tensile strength σ is related to the volume void fraction ϵ by a power function

$$\sigma \propto (1 - \epsilon)^m \tag{2}$$

A majority of the models incorporate what are essentially curve fitting parameters or functions. Some (C11, K12) are more pertinent to the pressed, briquetted, or tableted beds of particles rather than to granulated ensembles of particles, even though the distinction between the two kinds of pellets is necessarily somewhat arbitrary.

Rumpf (R4) has derived an explicit relationship for the tensile strength as a function of porosity, coordination number, particle size, and bonding forces between the individual particles. The model is based on the following assumptions: (1) particles are monosize spheres; (2) fracture occurs through the particle–particle bonds only and their number in the cross section under stress is high; (3) bonds are statistically distributed across the cross section and over all directions in space; (4) particles are statistically distributed in the ensemble and hence in the cross section; and (5) bond strength between the individual particles is normally distributed and a mean value can be used to represent each one. Rumpf's basic equation for the tensile strength is

$$\sigma = \tfrac{9}{8}[(1 - \epsilon)/\pi D_p^{\,2}]kH(D_p) \tag{3}$$

where D_p is the diameter of the particle, k is coordination number, and H is interparticle bonding force. In a regular and irregular closed-packed system of monosize spheres, the coordination number is essentially a function of the void fraction ϵ only (D1, M5) and the two parameters can be related by the following expression:

$$k\epsilon = \pi \tag{3a}$$

Equation (3) then becomes

$$\sigma = \tfrac{9}{8}[(1 - \epsilon)/\epsilon D_p^{\,2}]H(D_p) \tag{4}$$

For a medium porosity of $\epsilon = 0.35$, the criterion for failure in tension reduces to

$$\sigma = 2H(D_p)/D_p^{\,2} \tag{5}$$

In normal practice the constituent particles of a pellet invariably exhibit a size dispersion, and the fines play a dominant role in determining the strength (R4). Unfortunately, no completely satisfactory modification of the Rumpf's model for strength of an assemblage of particles of varying size is available (R5). A rough approximation is based on the replacement of D_p by the surface-equivalent diameter.

According to Cheng (C11), the two major factors in determining the tensile strength are the particle-size distribution and the strong dependence of the interparticle force on surface separation between the particles. The latter is manifested in large variations of the tensile strength with small changes in porosity. The attractive force between fine particles is the sum total of van der Waals, electrostatic, and other forces that are known to act over short distances of separation (K14). In addition microcontacts between asperities on particle surfaces contribute significantly to the interparticle bonds, more so, when the pellet is compacted under high external pressure, for example, in briquetting, tableting, and pressing. In Cheng's model, the tensile strength is given by (C11)

$$\sigma = (a'b'k/2)(\bar{s}/\bar{v})(\rho/\rho_S)F_c(l) \qquad (6)$$

where a' is the ratio of number of particle pairs per unit area to the number per unit volume, b' the ratio of overall area of contact per particle pair to the surface area of the smaller particle of the pair, $\bar{s}$ the mean effective surface area per particle, $\bar{v}$ the mean volume per particle, ρ the bulk density of the pellet, ρ_S the density of particles, and $F_c(l)$ the attractive force per unit contact area, which is a function of the mean separation distance l between the particle surfaces. The mean separation distance is given by

$$l = l_0 - (\bar{D}_p/3)(\rho/\rho_0 - 1) \qquad (7)$$

where l_0 is the effective range of the interparticle forces, ρ_0 the bulk density when the strength vanishes, and $\bar{D}_p$ the mean effective diameter of the particle. Kočova and Pilpel (K12) have extended this model to binary and three-component mixtures of powders. The force function $F_c(l)$ is a composite quantity that includes the unspecified bonding forces and the effect of the shape and surface geometry of the particles on these forces as l is varied. At present it is not possible to derive the form of this function from theoretical assumptions alone. However, Cheng (C11) has established a law of corresponding states (H2) in the following form:

$$(l_0^3/E)F_c(l) = \psi(l/l_0) \qquad (8)$$

in terms of one energy parameter E, which may be taken as a measure of the "strength" of the interparticle force, and one length parameter l_0. The function ψ, defined as a reduced form of interparticle force per unit overall area of contact, may be determined from experimental data. Thus, an estimate of the strength and length parameters that characterize the interparticle bond and a knowledge of the form of the function ψ would enable the tensile strength to be predicted from a particle size distribution alone without the need for further measurements. Capillary bonds are of principal interest in the balling and granulation of the moist particulate solids, and it remains to be seen whether the approach taken by Cheng can be extended to these bonding mechanisms.

B. Capillary Bonds

In the course of balling and granulation, the agglomerates undergo a continuous gradual compaction, at least in the initial period of growth. As a consequence, the void spaces become increasingly filled with liquid, as shown in Fig. 3. Newitt and Conway-Jones (N2) and Rumpf (R4) have formulated the wet pellet strength in terms of three-phase air–liquid–solid and two-phase liquid–solid regimes of the particulate ensemble. The former regime gives rise to the pendular and funicular bonds and the latter to the capillary pressure and liquid-envelope (droplet) bonds.

In the pendular state, the particles are held together by discrete bridges or lens-shaped rings of liquid at the point of contact or the point of close approach, air being the continuous phase. Bonding by capillary pressure prevails when the granules are saturated; the pore spaces, except for some trapped air bubbles, are now filled with liquid. In the intermediate funicular state, the liquid is distributed partly as discrete bridges and partly in some of the filled up capillaries. In the case of the droplet bond the granule is totally enclosed in the convex surface of the enveloping liquid.

1. *Pendular Bonds*

Consider two equal spheres held together by a liquid bridge, as shown in Fig. 4. Two forces contribute to the tensile strength of the bond in an additive fashion; the pull due to surface tension at solid–liquid–gas contact line directed along the liquid surface and the negative capillary pressure or the

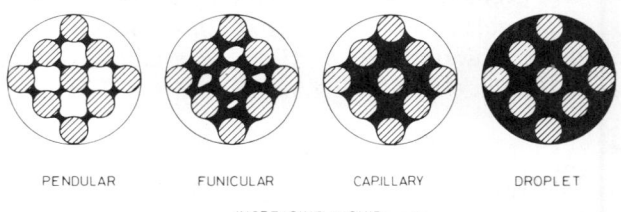

FIG. 3. Three-phase air–liquid–solid and two-phase liquid–solid regimes of a particulate ensemble.

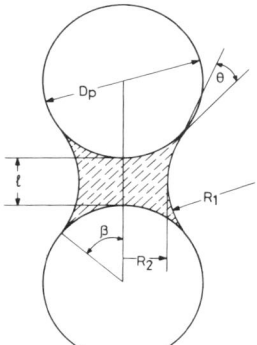

FIG. 4. Pendular bond between two spheres.

hydrostatic suction pressure existing in the liquid bridge. The shape of the projection of the liquid–gas surface can be approximated to a high degree of accuracy by a circular arc, and the bonding force is given by (P6)

$$H_p = \gamma D_p \pi \sin \beta [\sin(\beta + \theta) + (D_p/4)(1/R_1 - 1/R_2) \sin \beta] \quad (9)$$

where γ is the surface tension of the liquid, β the half-sector angle, θ the contact angle, and R_1 and R_2 the radii of curvature of the lens. The first term in the square bracket pertains to the pull exerted by surface tension and the second term is derived from the well-known Laplace equation for capillary pressure. From geometrical considerations it can be shown that

$$R_1 = [D_p(1 - \cos \beta) + l]/[2 \cos(\beta + \theta)] \quad (10)$$

and

$$R_2 = (D_p/2) \sin \beta + R_1[\sin(\beta + \theta) - 1] \quad (11)$$

It then follows that Eq. (9) can be written in an abridged form as

$$H_p/\gamma D_p = F_p^*(\theta, \beta, l/D_p) \quad (12)$$

The function F_p^*, which is a dimensionless or reduced bonding force, has been displayed graphically by Pietsch and Rumpf (P6) for various sets of variable β and parameters θ and l/D_p. The volume of the liquid in the pendular bridge V_p is given by

$$\begin{aligned}V_p = 2\pi \{ &[R_1^2 + (R_1 + R_2)^2] R_1 \cos(\beta + \theta) \\&- [R_1^3 \cos^3(\beta + \theta)]/3 \\&- (R_1 + R_2) \\&\times [R_1^2 \cos(\beta + \theta) \sin(\beta + \theta) + R_1^2(\pi/2 - \beta - \theta)] \\&- (D_p^3/24)(2 + \cos \beta)(1 - \cos \beta)^2 \} \quad (13)\end{aligned}$$

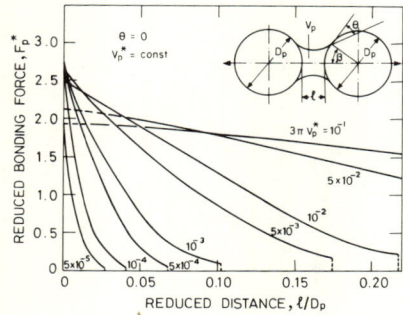

FIG. 5. Reduced pendular bonding force–reduced separation distance curves. [From Rumpf (R5).]

Again it turns out that a reduced volume V_p^* defined as

$$V_p^* = V_p/D_p^3 \tag{14}$$

is a function of β, θ, and l/D_p only. Pietsh and Rumpf (P6) have also presented a number of curves for β as a function of the reduced volume for different sets of parameters. In Fig. 5, the variation of F_p^* with l/D_p shows that the adhesion force is maximum as $l \to 0$ and drops with increase in the separation distance.

When the spheres touch each other, that is, $l/D_p = 0$, the expression for the reduced bonding force becomes simply (H6)

$$F_p^* = \pi \cos \theta - \pi R_2/D_p \tag{15}$$

For θ less than 65 degrees and R_2/D_p less than 0.25, the following expression for the reduced volume is approximately valid (H6):

$$V_p^* = \pi(R_2/D_p)^4 \tag{16}$$

Hence

$$F_p^* = \pi[\cos \theta - (V_p^*/\pi)^{1/4}] \tag{17}$$

The flatness of the computed curves in Fig. 6 shows that in the case of spheres in contact F_p^* is insensitive to rather large changes in V_p^* (or β). Mason and Clark (M3) have measured the pendular bonding force as a function of the separation distance and bridge volume. Considering the experimental uncertainty involved, their results follow the expected trend quite satisfactorily.

The tensile strength of the agglomerate in the pendular state is readily obtained by combining Eqs. (4) and (12):

$$\sigma_p = \tfrac{9}{8}[(1 - \epsilon)/\epsilon](\gamma/D_p)F_p^*(\theta, \beta, l/D_p) \tag{18}$$

Schubert (S6) has obtained good agreement between the measured strength and the model in Eq. (18) over the porosity range of 0.4–0.67.

In the simplest case both θ and l/D_p are equal to zero, and it is not unreasonable, in view of the flat curves in Fig. 6, to assign a constant value to F_p^*; hence, from Eq. (18)

$$\sigma_p = \text{const}[(1 - \epsilon)/\epsilon](\gamma/D_p) \tag{19}$$

It is convenient to express σ_p in term of the fraction saturation of void spaces S given by

$$S = V_L/\epsilon V \tag{20}$$

where V_L is the volume of the liquid in a pellet of total volume V. If the liquid is distributed equally at all interparticle coordination points then it can be shown that (P6)

$$S = 3[(1 - \epsilon)/\epsilon](k/\pi)(V_p/D_p^3) \tag{21}$$

Substitution of Eqs. (3a) and (14) gives

$$S = 3[(1 - \epsilon)/\epsilon^2]V_p^* \tag{22}$$

which is the required relationship needed to express the tensile strength as a function of liquid saturation.

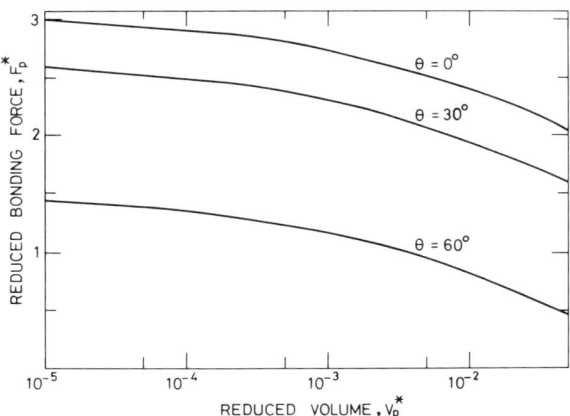

FIG. 6. Reduced pendular bonding force–reduced bridge volume curves when spheres are in contact.

2. Funicular and Capillary Pressure Bonds

It has been established (P8, R5) that when the value of S exceeds about 0.25, the liquid bridges begin to coalesce with one another and the bonding mechanism changes over from the pendular to the funicular state. When S exceeds 0.8, the existence of discrete liquid bridges is no longer possible and now the capillary pressure state alone exists. Thus, the funicular state lies in a range of saturation bounded by the lower and upper critical limits denoted by S_p and S_c, respectively.

In the capillary pressure state, where $S_c < S < 1$, the applied tensile stress has to overcome both the capillary pressure difference and surface tension of the liquid along the circumference of the pellet. It is no longer meaningful to speak of a single particle–particle bond, but only of the cohesive strength of the agglomerate as a whole. In general, the contribution of surface tension along the circumference is negligible compared to capillary suction and the tensile strength is given by (S6)

$$\sigma_c = SP_c(S) \tag{23}$$

The capillary pressure $P_c(S)$ exhibits a marked hysteresis phenomenon when the liquid is alternately withdrawn (drainage) and introduced (imbibition) into the particulate bed. Consequently, capillary pressure changes as a result of variations in saturation do not follow a unique functional relationship. In fact, the suction is always higher on the drainage side of the imbibition–drainage cycle (M8). In Fig. 7 the suction curve starts at zero when $S = 1$.

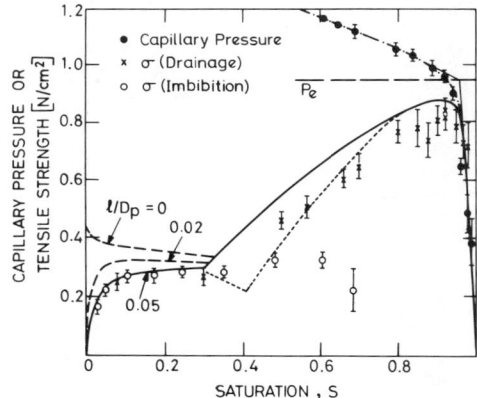

FIG. 7. Capillary pressure and tensile strength as a function of the fraction saturation. [From Rumpf (R5).]

With the progressive removal of the liquid, the curve at first rises almost vertically but subsequently tapers off to acquire a gradual slope. The intersection of the tangents extended from the steep and the flat portions of the curve occurs at about $S = 0.9$. At this saturation, the capillary pressure is known as entry suction P_e, and the pellet strength attains its maximum (R5).

In the funicular state, it is assumed that the two limiting bonding mechanisms, liquid bridges and capillary suction, are superimposed (R4). The strength may be estimated as approximately varying directly with saturation in the range $S_p < S < S_c$, as shown in Fig. 7. This figure also illustrates the tensile strength of the pellet in the pendular bond region as per the model in Eq. (18). Schubert (S6) has obtained good agreement with the theoretical strength curve over almost complete saturation range. The measurements were carried out on pellets of 71-μm limestone particles and 41.5% porosity. In the pendular state the data conformed to the theoretical curve for $l/D_p = 0.05$.

Although the capillary pressure curve cannot be predicted from the fundamental properties of the particulate bed, a crude estimate of the entry suction for wetting liquids is given by (N2)

$$P_e = \text{const}[(1 - \epsilon)/\epsilon](\gamma/D_p) \tag{24}$$

where the value of the constant ranges from 6 to 8, and D_p may be replaced by the surface-equivalent diameter in case of broad size-distribution particles. Now the tensile strength of the agglomerate in capillary regime is approximately given by (P8)

$$\sigma_c = 8[(1 - \epsilon)/\epsilon](\gamma/D_p), \qquad S \simeq 1 \tag{25}$$

and in the funicular state by

$$\sigma_f = \sigma_p(S_p) + (S - S_p)[\sigma_c - \sigma_p(S_p)]/(1 - S_p), \qquad S_p < S < 1 \tag{26}$$

From Eqs. (18) and (25), the ratio of the tensile strengths in the pendular and capillary states is

$$\sigma_p/\sigma_c = \tfrac{9}{64} F_p^* \tag{27}$$

Assuming an average value of 2.5 for F_p^*, Eq. (27) becomes

$$\sigma_p/\sigma_c \simeq 0.35 \tag{28}$$

In other words, the pellets in the capillary regime are about three times stronger than those in the pendular state.

C. Solid Bridges

In the bonding of fertilizer granules, cold-bound iron ore pellets and many other types of agglomerates, solid bridges play an important role. In fact, this mechanism tends to be much more prevalent in the dried pellets than is generally recognized because most particulate solids exhibit some solubility in water, by far the commonest bonding liquid employed; the dissolved materials crystallize out during drying of the granules and form solid bridges.

In a model (P5) of the tensile strength of an agglomerate bonded by solid bridges, it is assumed that the bridging material is uniformly distributed over the coordination points, the tensile strength of the bond is constant, and the fracture surface runs through the bridges only. Further, it is readily shown that in a multicomponent random-packed system the fractional area of a component in the cross section is on average equal to its volume fraction (D1). Starting from this assertion, it can be shown that an approximate expression for the tensile strength is

$$\sigma_b = (V_b/V_s)(1 - \epsilon)F_b$$
$$= S_b \epsilon F_b \qquad (29)$$

where V_b and V_s are, respectively, volumes of bridging material and particulate solids in a pellet, F_b is the tensile strength of the bond, and S_b is the fraction of the void space occupied by the bonding solids. For crystal-bonded pellets the utility of this relationship is somewhat limited because it has not been possible to characterize the bond strength in a satisfactory manner. The overall strength depends on the amount of the bridging solid present, its distribution in the particulate ensemble, and the intrinsic strength of the solid bridges formed. The latter two factors depend on the drying conditions and the capillary size of the porous structure (C3, C4, P5, P7). When conditions of slow drying rate prevail, large well-formed crystals are produced that contribute but little to the bond strength. Strong, solid bridges comprising fine crystallites or crystals are obtained when the drying rates are comparatively fast. Again, low drying temperatures result in the formation of salt crust with an uneven distribution of the bond. The proportion of salt deposited in the core increases as the drying rate is increased. Obviously the rate of drying at a given temperature is high when the porosity is large and the capillaries in the granule are coarse, that is, when the mean particle size is large. There is an optimal salt concentration that yields the maximum strength at a given drying temperature. With an increase in liquid saturation or salt concentration, the quantity of crystallized material in the pellet increases and, in the primary stages, so does the tensile strength.

Eventually however, a crust of increasing thickness and density degins to appear on the surface, which, along with blocking of pores due to salt deposition, hinders evaporation of the liquid. Consequently, with increasing salt content the strength attains its maximum and then drops as the crystallization rate slows down.

Capes (C4) found that the normalized compressive strength of sand granules bonded with sodium chloride is higher in the case of smaller particle size. On the other hand, Pietsch (P5) has reported that the tensile strength of pellets made from coarse powders was invariably more than that of fine powder pellets at all drying temperatures tested. These conflicting results underline the complex and still not fully understood nature of bonding by solid bridges.

D. Attractive Forces in the Absence of Material Bridges

In this category, among the molecular, electrostatic and magnetic interparticle bonds, interest is primarily centered on the van der Waals-type attractive forces that may predominate in the absence of liquid and solid bonds. The force of the van der Waals attraction between two spheres of equal size is (R4)

$$H_v = BD_p/24l^2 \qquad (30)$$

provided $l < 1000$ Å and $l \ll D_p$. The Hamakar constant B depends on particle composition and the nature of the surrounding medium. Combining this equation with the Rumpf's model in Eq. (4), we get as the theoretical tensile strength of an agglomerate

$$\sigma_v = \tfrac{9}{192}[(1 - \epsilon)/\epsilon](B/l^2 D_p) \qquad (31)$$

For example, when $\epsilon = 0.35$, $B = 10^{-12}$ ergs, $l = 10$ Å, and $D_p = 0.1$ μm, the calculated tensile strength is 0.85 kg cm^{-2}. Forces of this magnitude can lead to spontaneous pelletization in fine powders (M4), especially if the particle size is less than 1 μm. In tableting and briquetting, the mean separation distance between the particles might be reduced to less than 100 Å if the compressive pressures are high enough. Clearly, the van der Waals bonds may impart significant strength to the particulate ensemble.

The presence of an immobile adsorbed film or layer on the particle surface may lead to the formation of still stronger interparticle van der Waals bonds (R6). First, surface roughness is smoothed out, increasing the apparent particle size and contact area, and second, the separation distance is effectively reduced. Under large compaction pressures the adsorbed layers may penetrate each other, giving rise to high bonding forces. On the other hand, in nominally dry powders capillary condensation (Kelvin's effect) can result in the formation of pendular bonds at the points of contact between the particles, and this bonding mechanism may even overshadow the van der Waals bonds. Thus, the tensile strength of barium sulfate compacts was found to increase nearly fivefold from the condition of high vacuum to 90% relative humidity (R5).

Recently, Rumpf (R5) has considered the forces of attraction between a plate and a sphere and between irregular shape particles. His conclusions are that the capillary bonds are relatively insensitive to the particle shape, but the van der Waals force of attraction is extremely sensitive. Although weaker in magnitude than the two aforementioned bonds, the electrostatic bonds may persist over long separation distances.

E. Deformation of Agglomerates

The strength of the agglomerates cannot be characterized by just one parameter. In general it is necessary to have knowledge of both the tensile and shear strengths, as well as the strain behavior of the particulate ensemble. The stress–strain relationship is one of the principal factors that govern the growth mechanism in balling and granulation, as also for the survival of the agglomerates in handling and transport. A high shear-to-tensil strength ratio signifies brittleness, whereas a low ratio is associated with a tendency to plastic deformation (R5). A pellet with a small fracture strain is more prone to breakage than the one with a large fracture strain, even when the tensile strength is the same in both cases. Again, a large deformability is essential for coalescence between the agglomerates, which is an important mechanism for growth in balling and granulation.

Schubert *et al.* (S8) have carried out extensive measurements of the stress–strain behavior of agglomerates bonded by various bonding mechanisms. Some of their results are shown in Fig. 8. In the funicular and capillary regimes, the fracture is strongly dependent on whether the liquid is being imbibed or drained; markedly smaller deformation occurs under the latter condition. Dry agglomerates have a significantly smaller fracture strain than the moist pellets, and tend to undergo brittle fracture. Agglomerates bonded with solid bridges exhibit initially a linear stress–strain relationship, while in the case of other bonding mechanisms convex curvilinear characteristics are observed (see Fig. 8). Under cyclic loading and unloading conditions, the stress–strain curve invariably shows hysteresis. It is concluded that right from the beginning of the application of the stress the strain is not purely elastic, but has a plastic component also.

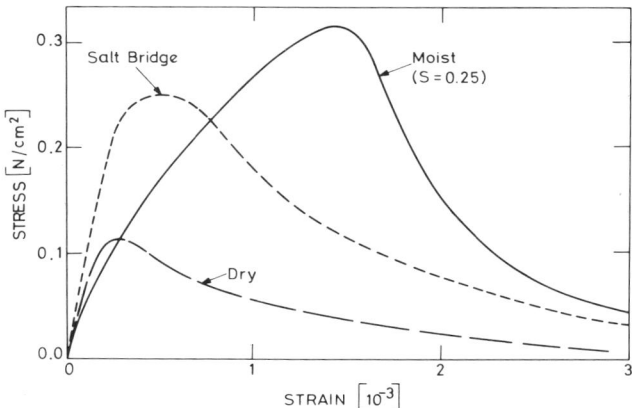

FIG. 8. Stress–strain behavior of the three kinds of agglomerates. [From Schubert et al. (S8).]

IV. Compaction and Growth Mechanisms

A. COMPACTION

It has been claimed that ordinarily pressures of the order 20 ton in^{-2} would be needed to compress a particulate mass to a density comparable with that of the iron ore pellets agglomerated in a balling drum (T3) or the uranium oxide granules compacted in a gyratory unit (G1). Strictly speaking, a direct comparison between the two-particulate ensemble size-incrementation processes, namely, pressing, briquetting, or tableting on the one hand, and balling and granulation on the other, is not very meaningful. In the former processes the pressure is rapidly degraded with the distance from the plunger and in the vicinity of the die wall; the compaction occurs under highly nonuniform conditions. Moreover, due to the constraint imposed by the die wall, the particles are severely restricted from moving and occupying optimal packing positions. Indeed, if the pressures imposed are sufficiently high, the particles may undergo extensive deformation and comminution (C12). On the other hand, the constituent particles of a growing pellet are never fully constrained but possess considerable freedom of movement, which allows them to adjust to relatively dense packing configurations. There is also no evidence of any breakage or distortion of the particles.

Tigerschoid and Ilmoni (T3) attributed the high degree of compaction attained by the granule to a cooperative action of the negative capillary pressure generated within the liquid-filled pores of the pellet and the vibratory impulses to which the particulate ensemble is subjected during the course of numerous collisions. Speaking in a somewhat more generalized manner, two kinds of forces contribute to the formation of nuclei pellets from particulate feed mass and their subsequent compaction and growth (S5). The "external" mechanical forces are associated with rolling, tumbling, agitating, chopping, and kneading actions imparted to the agglomerating charge by the balling device. These forces squeeze the particulate flocs, clusters, or crumbs and the pellet species, or the constituent particles therein, into close contact with one another, as well as shear and deform the agglomerates into a spherical shape. The "internal" interparticle bonding forces, discussed in Section III, assist in consolidating the packing configuration of the ensemble and in stabilizing the shape and structural integrity of the granules.

Although no external pressure is applied to the agglomerating mass as a whole, the pellets comprising the charge are continuously subjected to a complex but intense environment of vibration, compression, and shear. Firth (F2) had earlier drawn attention to the high static pressure exerted by the total weight of a ball on a few particles at the surface of another ball immediately underneath it. Tarján (T1) concluded from an analysis of ball dynamics in a rotating charge that the accelerating impact of a pellet on a particle under gravitational force may correspond to a pressure of several thousand atmospheres for a sufficiently vigorous collision. Actually under the impact of a compressive blow, the ensemble relaxes momentarily and permits slippage of the particles to new and tighter orientations, the agglomerate being slightly, but ordinarily irreversibly, compressed (M6). This mode of consolidation is then somewhat similar to that obtaining in the vibratory compaction (M2).

Kapur (K2), using a mercury porosimeter, has monitored the porosity of limestone granules pelletized in a small-batch balling drum. Typical results for three moisture contents of the feed charge are shown in Fig. 9. The porosity of the lightly tapped bed of loose, floclike, moist feed is about 100%. This value falls rapidly in the course of the nucleation of discrete agglomerate species. Concurrently with the increase in size, the compaction of the granules continues until a limiting value of the porosity is reached. The final porosity is closely related to the water content of the charge. In fact, the pore volume of the compacted pellet invariably turns out to be approximately equal to its liquid content (see Section VI,A).

In general, the extent of the ultimate consolidation of the granules and the rate of approach to this limiting value are functions of the size and size distribution of the particulate feed, its density and liquid content, and the

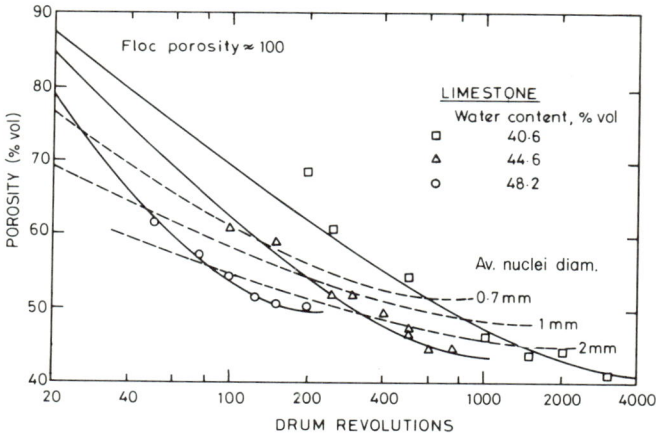

FIG. 9. Compaction of nuclei pellets in the course of agglomeration. [From Kapur and Fuerstenau (K6).]

mode and intensity of the external mechanical forces imparted by the agglomerating machines (G1, S13). Granules formed from closely sized, relatively coarse powders (e.g., sands) are rapidly compacted to the limit in a short balling time, but continue to grow in size thereafter (S3). If the final porosity attained in this case is comparatively high, it is only because, in the absence of fines filling up the void spaces, the packed bed of particles in a narrow size-distribution range has an inherently open structure. Dense, granulated structures are possible only if the feed is characterized by a widely dispersed size distribution with the requisite proportion of fines (H5). However, when fine powders are pelletized, the rate of consolidation may slow down significantly because presumably the applied mechanical forces are partially dissipated by interparticle friction, which increases with the surface area of the particles. Again, the cohesive strength of the ensemble is invariably enhanced in the presence of fines. This in turn restricts the mobility of the constituent particles for acquiring the optimal packing arrangement.

B. Growth Mechanisms

A formal description of the growth of the granules would have to incorporate the phenomena of deformation, flow, shear, and fracture of three- and two-phase agglomerates under the influence of the applied mechanical and internal bonding forces. However, at present an adequate understanding of the underlying rheological processes for embedding into this description is not available. For a practical representation it is then convenient to adopt a macroscopic or phenomenological viewpoint, commonly referred to as *growth mechanisms*. These mechanisms provide a conceptual basis for the analysis and modeling of the kinetics of balling and granulation discussed in Section V.

Excluding the nucleation phenomenon, one may classify the principal elementary growth mechanisms as: (1) snowballing; (2) crushing and layering; (3) coalescence, both random and nonrandom; and (4) abrasion transfer. At this point, for a proper elucidation of these mechanisms, it is necessary to stress the formal distinction between the two components of the agglomerating charge, namely, discrete agglomerates and particulate feed mass. The term *agglomerate* has already been defined in Section I. The liquid wet feed mass, which is considered to be an "amorphous" matrix of a random open network of particles in three dimensions, held together by the pendular bonds (K5), contributes to the formation of nuclei pellets and the growth of the agglomerate species. Further, since the transfer of material from the feed mass to a growing pellet involves either individual particles or small flocs or crumbs of particles, ideally the growth occurs in infinitesimal increments. In other words, it is convenient to assume that the feed matrix comprises particulate elements of infinitesimal size (S5).

1. *Snowballing*

In this mechanism, occasionally referred to as *onion skinning* or *layering*, the granules (also known as *seeds*) roll or otherwise move around in the loose matrix of the particulate feed. As shown in Fig. 10a, growth occurs by deposition of tightly packed layers of the feed material on the pellets, the particles adhering to each other and to the seed by means of capillary bonds (S10). If we just consider the agglomeration component of the overall pelletizing system, that is, if we exclude physical addition or removal of pellets in the input and output streams, then in a pure snowballing environment the number of pellets remains invariant but their size and mass increase continuously.

The layering feature of the snowballing mechanism can be exploited to produce cladded composite granules, for example, iron ore pellets coated with coke powder (W3). It is also possible to carry out a partial reduction in the moisture content of the pellet, along with the snowballing growth, by using a feed that is dry or at least deficient in liquid, the excess water in the seeds being sucked into the capillary network in the deposited layers (G1, S10). The snowballing mechanism invariably operates, in addition to nucleation and other growth mechanisms, in the iron ore balling drum and some

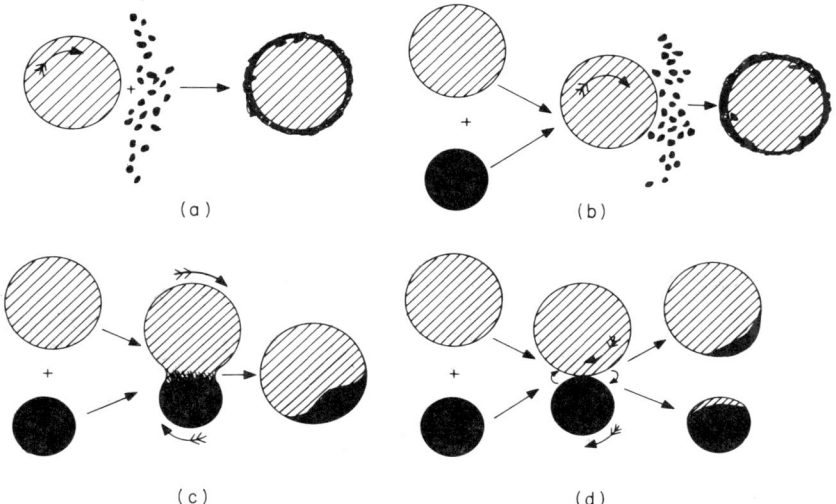

FIG. 10. Four elementary pellet growth mechanisms: (a) snowballing; (b) crushing and layering; (c) coalescence; (d) abrasion transfer. [From Sastry and Fuerstenau (S5).]

fertilizer granulation circuits (S10). The off-size agglomerates are recycled to the pelletizer as seeds. This mode of growth may predominate in a balling disk, even when operated in an open circuit, if a fine spray of water is directed over the granules at some distance away from the vicinity of the feed location (see Fig. 2). It should be noted, however, that in a steady state, closed or open balling, and granulation circuits both the pellet mass and number balance must be maintained simultaneously. Therefore, a pure snowballing mechanism, in the absence of nucleation or other seed generating mechanisms, cannot prevail in these circuits.

2. Crushing and Layering

This mechanism was first proposed by Capes and Danckwerts (C5). They agglomerated a mixture of large pellets made from white sand and small pellets made from the same sand colored blue with a dye. The resulting balls were sectioned, and it was observed that they had white interiors surrounded by layers whose colors were intermediate between that of the two sands used. This observation led to the conclusion that, as illustrated in Fig. 10b, initially the small blue granules are fragmented in collisions with the larger balls. The fragments are rapidly picked up by the remaining unbroken pellets, and kneaded or layered onto the surface of these granules by the tumbling action in the pelletizer. Subsequently, the smaller of the blue coated agglomerates are fragmented and layered, giving rise to color variations in the deposited layers. The higher rate of growth of the larger balls observed in this mode may be attributed to the fact that the destruction of a small granule is more likely when it is involved in collisions with a large pellet. The latter has a more immediate access to the broken material, being already present at the site of fragmentation (C6).

In the idealized description of an isolated crushing and layering mechanism of growth, the mass of the granules remains constant, although the population is continuously depleted. This implies that, due to its rapid pick up, the amount of the broken material in the charge is always vanishingly small (K3). Although the broken fragments may be characterized as belonging to the discrete agglomerate species (S5), both the snowballing mechanism and the layering step of the crushing and layering mechanism lead to similar results, namely, deposition of coherent and well compacted layers on the rolling pellets.

It has been pointed out that the crushing and layering mode of growth is generated when the granules are brittle and relatively weak in strength and are susceptible to breakage in the dynamic conditions prevailing in the agglomerating device (K3). Thus, it is not surprising that the granulation of coarse, closely sized sands should occur by this mechanism. This was confirmed by Linkson *et al.* (L1), who compared the balling behavior of narrow size fractions and broad size distributions using tracer techniques with colored material. The crushing and layering mechanism has also been suggested for dry pelletization of fine powders where relatively weak van der Waals forces prevail between the particles (M4). Although no direct experimental evidence has been reported, it is not unreasonable to expect that this kind of growth contributes significantly to the granulation of coarse fertilizer feed deficient in liquid content.

3. *Coalescence*

As shown in Fig. 10c, this mechanism refers to the production of large-size agglomerates by means of the clumping or fusing together of two or more granules. When pellets collide and stick or pack together in a geometric configuration favorable with respect to the rest of the moving charge, the conglomerate of balls begins to roll and eventually deforms and kneads into the next larger sphere. However, if for some reason the clump is unable to start rolling almost immediately, one or more pellets in the conglomerate may be sheared apart from the unit, and decoalescence may occur (K5). For the sake of simplicity, binary union is considered to be the elementary event in the coalescence mechanism. However, Kapur and Fuerstenau (K6) were able to produce multiple, even massive and catastrophic coalescence leading to the formation of a raspberrylike structure in appropriate experimental conditions. The coalescence events cause discrete changes in the size of the agglomerates with an overall decrease in the pellet population.

It is generally agreed that in the early stages the growth of the newly formed nuclei pellets, which are relatively porous and readily deformable, takes place primarily by the coalescence mechanism (C5, K5, L1, N2, S2). Tracer experiments have shown (L1, S5) that, during the later stages, this mechanism contributes wholly or partially to the growth of the balls made from materials of wide size distributions, as against the crushing and layering mode obtained for narrow size fractions. The realization of the coalescence mechanism depends in a complex manner on a number of factors (O1). The area of contact produced by mutual deformation of two colliding balls is a function of the deformability of the unconfined multiphase particulate ensemble under shock–stress loading conditions. The cohesive strength of the intergranule neck bond of the twin configuration formed at the instant of collision is determined by the particle size and the amount of liquid available on the granule surface. The twins comprising large balls are more likely to shear apart by the separating torque generated by the surrounding charge than those made up of small granules. In general, the coalescence mechanism is favored when the pellets exhibit plasticity, deformability, and "surface stickiness."

4. *Abrasion Transfer*

Kapur and Fuerstenau (K5) had hypothesised the possible occurrence of this agglomeration mechanism wherein small volume elements from the surface of one ball are abraded and transferred to another as two pellets roll together (see Fig. 10d). Subsequent tracer studies (L1, S5) have confirmed that this mechanism does indeed operate, at least in the later stages of balling with materials of wide size distributions. This is essentially a slow mechanism, although it may become important in the absence of other modes of growth. In the abrasion transfer mechanism the total number of pellets does not change and only changes in size occur.

C. Regions of Agglomerate Growth

Kapur and Fuerstenau (K5) have presented a unified description of the agglomeration process in which the phenomena of the compaction of the agglomerates and their passage through the various capillary regimes have

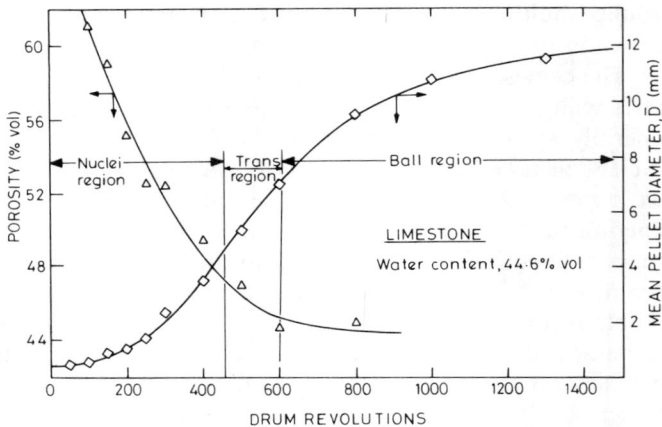

FIG. 11. Porosity and mean size of pellets, and the three regions of growth. [From Kapur (K2).]

been integrated with the prevailing growth mechanisms. Figure 11 represents the average granule size and its porosity as a function of the number of drum revolutions. The system was comminuted limestone powder pelletized in a small-batch balling drum. The distorted S-shaped curve indicates that the agglomeration proceeds through three distinct growth regions: (1) three-phase air–liquid–solid nuclei growth region; (2) intermediate transition region; and (3) two-phase liquid–solid ball growth region. The porosity of the pellets drops continuously in the nuclei region, eventually attaining a more or less stable minimum value in the vicinity of the transition region.

1. *Nucleation and Nuclei Growth Region*

During the mixing of liquid with a particulate mass, highly porous and irregularly packed networks of flocs are formed. In an agglomerator after a brief induction period during which the particles rearrange and pack together, these flocs burst or nucleate into stable discrete spherical species of nuclei. This flocs–nuclei transformation is generally quite rapid, requiring in the case of a small-batch balling drum just 2–20 drum revolutions, depending on the fineness of the material and its liquid content. In industrial operations the nuclei may be formed in the course of mechanical transport and handling of the loose moist feed. In the nuclei-growth region, comparatively porous three-phase granules are held together by bonds of the pendular–funicular type. Simultaneously with densification, they grow in size by coalescence with one another.

2. Transition Region

Near the end of the nuclei region the constricted capillaries in the agglomerate begin to fill up with liquid. Eventually, when the interstitial void volume becomes almost equal to the liquid content, the liquid is squeezed onto the surface of the pellet, whose appearance changes from semidry to wet. This movement or demixing of liquid signals the onset of the transition region. From this point onwards, apart from some pockets of trapped air, the pellet is comprised of two phases only, solid and liquid.

In the transition region the soft granule is nominally held together by the surface tension of the liquid envelope surrounding it and also by the surface and mechanical interparticle bonding forces in a well-packed particulate ensemble. By reason of its marked dilatant behavior, which is readily demonstrated because squeezing a pellet causes the liquid on the surface to recede into the pore space and form a strong capillary suction bond, the agglomerate is able to retain its integrity when subjected to large stresses. Thus, it would seem that the bonding mechanism in the pellet continuously resonates between the droplet and the capillary pressure bonds. Given the wet plastic outer shell and easy initial deformability of the pellet surface, the growth rate of the granules by coalescence is accelerated to the maximum in the transition region.

3. Ball Growth Region

Beyond the transition region, the ball growth rate is either approximately constant with time (as in the case of coarse, closely sized materials) or decreases continuously (powders with a broad size distribution). In the former case the size increment mode is predominantly by crushing and layering, whereas in the latter, the situation is much more complex. Tracer techniques have shown (L1, S5) that coalescence, abrasion transfer, and crushing and layering mechanisms contribute in varying degrees to the agglomeration process. It has also been suggested that small granules are crushed in collisions with large balls into a limited number of daughter pellets, which subsequently coalesce with the remaining balls (crushing and coalescence mechanism). The dense packing of the particles and high strength of the pellets due to capillary suction bond and relatively large ball size all contribute to the low efficiency of the coalescence mechanism in the ball growth region. For the same reasons, the character of the growth mode may alter from a random clumping, as in the nuclei and transition regions, to a nonrandom coalescence in which the uniting partners are preferably two small granules or one small pellet and one large ball.

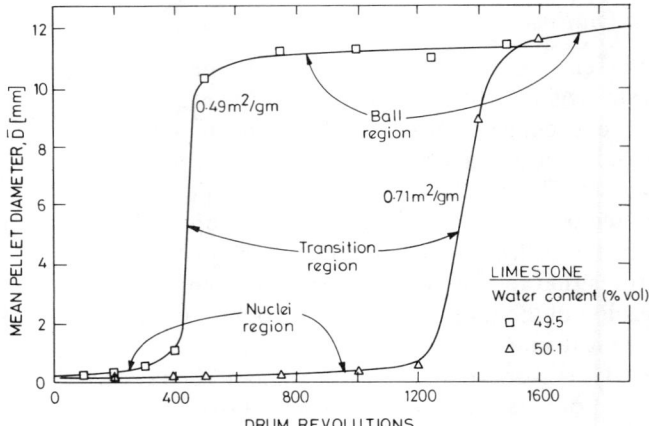

FIG. 12. Nuclei, transition and ball growth regions when fine powders are pelletized. [From Kapur (K2).]

The three regions of agglomerate growth described above are completely general phenomena; however, the demarcation between these regions may not be very sharp in some systems. When a moist coarse feed is granulated, the pellets densify readily, and as a consequence, pass rapidly through the nuclei and transition regions, which may overlap to a considerable extent. On the other hand, the growth regions can be vividly demonstrated when fine-ground limestone powders are pelletized, as shown in Fig. 12.

V. Kinetics of Balling and Granulation

The overall growth of the pellets in a given agglomeration system may occur either by a single elementary growth mechanism or by the coupling together of two or more mechanisms. Moreover, the pattern of growth may change over from one mechanism to another, and, when coupling occurs, the relative contributions of the individual elementary mechanisms may alter as the agglomerates grow in size. In many systems however we are justified in assuming that, as a first approximation, the growth mode is dominated by a single, elementary mechanism. In such cases, it is possible to formulate reasonably tractable models of the kinetics of balling and granulation, which are highly useful for analysis of the pelletizing systems including the industrial continuous circuits. The specific growth rate constant that appears in these phenomenological models provides a uniform and consistent basis for comparing the "ballability" of the particulate solids as a function of the feed characteristics and the balling device (K7, S4). In many applications of the agglomerates the pellet size, especially the dispersion in size, is rather critical. The kinetic models provide valuable insight into the evolution of the size spectrum and its statistical parameters.

A. Snowballing Kinetics

The experimental results of Capes and Danckwerts (C6) and Kanetkar (K1), using a balling drum and a disk granulator, respectively, have shown that the thickness of the snowballed layer is approximately independent of the initial size of the seed pellet. It then follows that the rate of pick up of the loose particulate material is proportional to the surface area of the granule; hence,

$$dV/dt \propto V^{2/3} \qquad (32)$$

and

$$dD(t)/dt = K_s \qquad (33)$$

Therefore

$$D(t) = D(0) + \int_0^t K_s \, dt \qquad (34)$$

or

$$D(t) = D(0) + T(t) \qquad (35)$$

where D is granule diameter, K_s linear growth velocity, which is independent of pellet size but may vary with time, and $T(t)$ the thickness of the snowballed layer. If the initial number distribution of the seed pellets is $n^0(D) \, dD$, a straightforward transformation (H1) of the size attribute in Eq. (35) gives the distribution at time t as

$$n(D, t) \, dD = n^0(D - T(t)) \, dD \qquad (36)$$

Inspection of this equation shows that in the course of the snowballing growth the size distribution curves at various times are simply shifted toward the right on the pellet size scale without any change in their shape, as demonstrated by Capes (C2) for sand pellets snowballed in a pan granulator (Fig. 13).

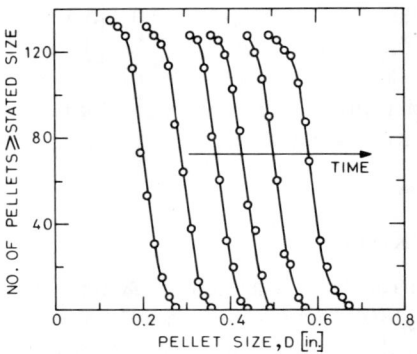

Fig. 13. Size spectra of pellets grown by snowballing. [From Capes (C2).]

For modeling a continuous pelletizer, it is advantageous to formulate the snowballing kinetics in the well-known continuity equation for the pellet species

$$\partial n(D, t)/\partial t + K_s(t)\, \partial n(D, t)/\partial D = 0 \tag{37}$$

or

$$\partial \bar{R}(D, t)/\partial t - K_s(t) n(D, t) = 0 \tag{38}$$

where the absolute cumulative distribution $\bar{R}(D, t)$ is given by

$$\bar{R}(D, t) = \int_D^\infty n(D', t)\, dD' \tag{39}$$

B. Crushing and Layering Kinetics

For a model of the crushing and layering kinetics Capes and Danckwerts (C6) have postulated that: (1) only the smallest granule in the charge is fragmented; and (2) the crushed material is redistributed among the remaining granules according to size. The increase in the diameter of any pellet is proportional to the difference between its diameter and that of the smallest remaining granule; the latter does not receive any material and hence does not grow at all. The resulting mathematical expression for the observed size distribution of the agglomerates is

$$R\left(\frac{D}{D_m}\right) = \frac{1}{2}\left[\frac{1 - (DD_m/D_m D_x)}{1 - (D_m/D_x)}\right]^{\alpha(D_m/D_x)} \tag{40}$$

where $R(D/D_m)$ is the cumulative number fraction equal to or greater than the reduced size D/D_m; D_m and D_x are, respectively, median and largest granule size in the distribution. The exponent α, a function of D_m/D_x, remains constant during the course of granulation. According to Eq. (40) the crushing and layering mechanism gives rise to a size spectrum that is self-similar or self-preserving when plotted as a function of the reduced size D/D_m (Fig. 14). Since the breakage frequency of the granules is not specified, the analysis of Capes and Danckwerts (C6) does not characterize the dynamic behavior of the agglomerating system and the evolution of the pellet spectrum in time. Moreover, as shown in Fig. 14, the similarity distribution in Eq. (40) terminates abruptly at $R = 1$, whereas their experimental data exhibit a distinct "tail" in the small-granule size range.

Kapur (K3) has formulated a set of kinetic equations for the agglomerate population and crushed material balance as follows:

$$\partial n(V, t)/\partial t = -B(V, t)n(V, t)$$
$$- (\partial/\partial V)[G(V, F)n(V, t)] \qquad (41)$$

and

$$dF(t)/dt = \int_0^\infty B(V, t)n(V, t)V \, dV$$
$$- \int_0^\infty G(V, F)n(V, t) \, dV \qquad (42)$$

where $n(V, t) \, dV$ is the number of granules in volume size range V to $V + dV$, $B(V, t)$ the fraction of granules of size V broken per unit time at

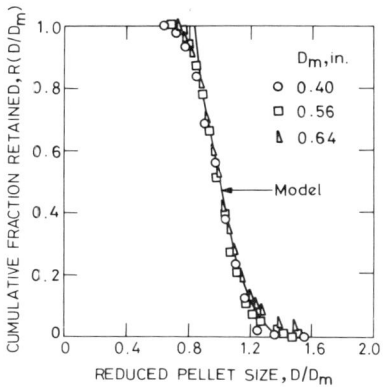

FIG. 14. Self-similar size distributions of sand pellets generated by the crushing and layering mechanism of growth. [From Capes and Danckwerts (C6).]

time t, and $G(V, t)$ is a growth function

$$dV/dt = G(V, F) \tag{43}$$

which depends on the volume of the fragmented material $F(t)$ available at any given instant. Further, it is assumed that the rate of breakage is inversely proportional to granule size,

$$B(V, t) = C/V \tag{44}$$

that the growth function is proportional to the product of granule size and the volume of the crushed material,

$$G(V, F) = GVF(t) \tag{45}$$

and that the quasi-steady-state approximation for the fragmented product is a valid assumption,

$$dF(t)/dt = 0, \qquad F(t) = 0 \tag{46}$$

On combining Eqs. (42)–(46) with Eq. (41), we get

$$[\partial n(V, t)/C\, \partial t] + n(V, t)/V + [\mu_0(t)/\mu_1](\partial/\partial V)[Vn(V, t)] = 0 \tag{47}$$

where $\mu_k(t)$, the kth moment of the distribution, is

$$\mu_k(t) = \int_0^\infty n(V, t)V^k\, dV \tag{48}$$

Clearly, $\mu_0(t)$ is the total number and μ_1 is the total volume of the agglomerates in the charge. The latter is approximately independent of time in view of the quasi-steady-state assumption in Eq. (46). Kapur (K3) has shown that an asymptotic similarity solution to Eq. (47) exists in the following form:

$$n(V, t) = [\mu_0(t)/\bar{V}(t)]Z(p) \tag{49}$$

where the mean volume of the granules $\bar{V}$ is simply

$$\bar{V}(t) = \mu_1/\mu_0(t) \tag{50}$$

and the similarity function Z is a function of a similarity variable or a reduced size p, which is defined such that

$$p = V/\bar{V}(t) \tag{51}$$

The resulting solution to Eq. (47) is

$$n(V, t) = \frac{\mu_0(t)}{\bar{V}(t)} \frac{b^{a-1}}{\Gamma(a-1)} \left[\frac{V}{\bar{V}(t)}\right]^{-a} \exp\left[-b\frac{\bar{V}(t)}{V}\right] \quad (52)$$

where

$$a = (2v_{-1} - 1)/(v_{-1} - 1) \quad (53)$$

$$b = 1/(v_{-1} - 1) \quad (54)$$

$$v_{-1} = \int_0^\infty Z(p)p^{-1} \, dp \quad (55)$$

and Γ is the gamma function. The value of the dimensionless parameter v_{-1} is restricted in the range $1 < v_{-1} < 2$. For a close agreement with the data of Capes and Danckwerts (C6) v_{-1} is approximately equal to 1.2. Further, it can be shown that the number of surviving pellets is given by the following expression (K3):

$$\mu_0^{-1}(t) = \mu_0^{-1}(t_0) + (Cv_{-1}/\mu_1)[t - t_0] \quad (56)$$

When $t \gg t_0$ and $\mu_0(t_0) \gg \mu_0(t)$, Eq. (56) may be approximated as

$$\mu_0^{-1}(t) = (Cv_{-1}/\mu_1)t \quad (57)$$

The experimental data of Capes and Danckwerts agree quite well with this expression, as is shown in Fig. 15. The slope of the plots in this figure is proportional to C, the frequency factor for breakage in Eq. (44). From the nature of the crushing and layering mechanism of growth it is evident that the rate of depletion of the granule population depends on the frequency factor alone, which is understandably higher for coarse sand granules. Substitution of Eq. (50) into Eq. (57) gives

$$\bar{V}(t) = Cv_{-1}t \quad (58)$$

Therefore, it follows that the mean volume increases linearly with granulation time. From Eq. (52) we see that the size distribution is uniquely defined by a dimensionless size $V/\bar{V}$, and in that sense it is self-similar. In order to compare with the experimental data, it is necessary to transform the size attribute from volume to diameter. From Eq. (52) it can be shown that (K3)

$$R\left(\frac{D}{D_m}\right) = \frac{3b^{a-1}r^{3-3a}}{\Gamma(a-1)} \int_{D/D_m}^\infty \left(\frac{D'}{D_m}\right)^{2-3a} \exp\left[-b\left(\frac{D'r}{D_m}\right)^{-3}\right] d\left(\frac{D'}{D_m}\right) \quad (58a)$$

where

$$r = D_m/(\bar{V})^{1/3} \quad (58b)$$

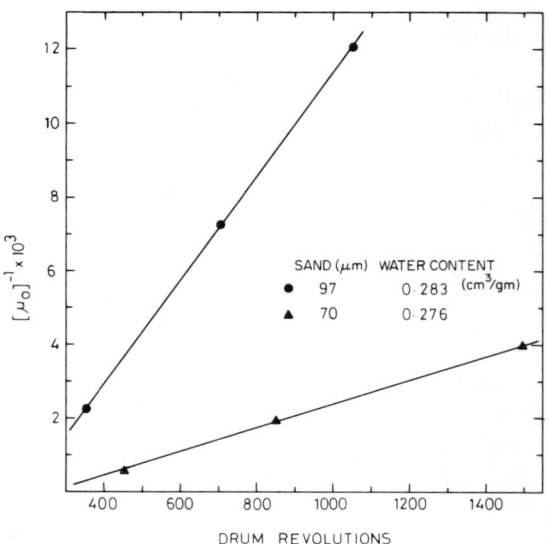

FIG. 15. Depletion of pellet population by the crushing and layering mechanism. [From Kapur (K3).]

Thus, the size distribution is again a unique function of the reduced-size D/D_m in agreement with the experimental findings (C6). The similarity distribution computed from Eq. (58a) and the observed range of 5th and 95th percentiles is illustrated in Fig. 16. In summary, Kapur's model of the crushing and layering kinetics is in conformity with the following empirical observations: (1) the ratio of final to initial diameters of the surviving agglomerates is constant; (2) the mean granule volume increases linearly with time; and (3) the pellet distribution is self-similar in a reduced size D/D_m, and 5th and 95th percentiles are 0.8 and 1.28 respectively, as compared with the average values of 0.76 and 1.26 observed for seven different sand–water combinations.

C. Random Coalescence Kinetics

Kapur and Fuerstenau (K6) have presented a discrete size model for the growth of the agglomerates by the random coalescence mechanism, which invariably predominates in the nuclei and transition growth regions. The basic postulates of their model are that the granules are well mixed and the collision frequency and the probability of coalescence are independent of size. The concentration of the pellets is more or less fixed by the packing

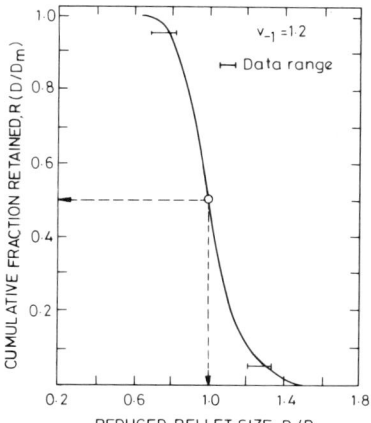

FIG. 16. Comparison between the similarity distribution in Eq. (58a) and experimental data. [From Kapur (K3).]

constraints in a loosely packed flowing charge. As a first approximation, the coordination number is not expected to change appreciably in the course of agglomeration, specially if the size distributions exhibit a self-preserving character. In this situation, the movement of a given pellet is very much restricted. The granule is likely to encounter and coalesce with its nearest neighbors, which form a cage around it. In other words, the disposition of the interacting species is such that the coalescence occurs under a so-called restricted-in-space environment (S1). In contrast, the coagulation process belongs to a free-in-space class of agglomeration, meaning that the species are present in dilute concentrations. Therefore, in thermal coagulation of the colloid phase the rate of agglomeration turns out to be proportional to the product of the number concentrations of the two reacting species (C9). In view of these observations, it is postulated that in the case of agglomeration by coalescence in the balling and granulation processes the rate is proportional to the product of the number of species of one kind with the number faction of the second kind (K6)

$$[\text{Rate}]_{i,j} \propto n_i(t)n_j(t)/\mu_0(t) \tag{59}$$

where $n_i(t)$ is the number of pellets of discrete volume size V_i. Incorporating this rate expression, the appropriate appearance–disappearance kinetic equation for the granule population is

$$\frac{dn_i(t)}{dt} = -\lambda(t)\frac{n_i(t)}{\mu_0(t)}\sum_{j=1}^{\infty} n_j(t) + \frac{\lambda(t)}{2\mu_0(t)}\sum_{j=1}^{i-1} n_j(t)n_{i-j}(t), \qquad i = 1, 2, 3, \ldots \tag{60}$$

where $\lambda(t)$ is a specific coalescence rate function, which may vary with time as the porous agglomerates are progressively compacted and the liquid is squeezed on to the surface. If at time zero all pellets are of a single size V_1, the solution to Eq. (60) is

$$n_i(t) = \mu_0(t)[1 - \exp[-\bar{\lambda}(t)]]^{i-1} \exp[-\bar{\lambda}(t)] \tag{61}$$

where

$$\bar{\lambda}(t) = \frac{1}{2} \int_0^t \lambda(t')\, dt' \tag{62}$$

The cumulative number fraction larger than size V_i is

$$R_i(t) = \{1 - \exp[-\bar{\lambda}(t)]\}^i \tag{63}$$

Moreover, the granule population is

$$\mu_0(t) = \mu_0(0) \exp[-\bar{\lambda}(t)] \tag{64}$$

Hence

$$R_i(t) = [1 - \mu_0(t)/\mu_0(0)]^i \tag{65}$$

As a first approximation the correction term for decrease in the porosity of the nuclei may be ignored and in that case (K4)

$$V_i = iV_1 \tag{66}$$

Combining Eqs. (65) and (66), and in the limit as $\mu_0(t)/\mu_0(0) \ll 1$, we have

$$R_i(t) = \exp[-V_i \mu_0(t)/V_1 \mu_0(0)] \tag{67}$$

But the volume of the granulating charge is

$$\mu_1 = V_1 \mu_0(0) \tag{68}$$

Therefore

$$R_i(t) = \exp[-V_i \mu_0(t)/\mu_1] \tag{69}$$

Substitution of Eq. (50) gives

$$R_i(t) = \exp[-V_i/\bar{V}(t)] \tag{70}$$

Clearly the granule size distribution is self-preserving in the reduced size $V_i/\bar{V}(t)$. Further, combining Eqs. (50), (64), and (68) results in the following expression for the mean volume:

$$\bar{V}(t) = V_1 \exp[\bar{\lambda}(t)] \tag{71}$$

It may turn out that the specific coalescence rate is time invariant, in that case Eq. (71) reduces to

$$\bar{V}(t) = V_1 \exp[\lambda t/2] \tag{72}$$

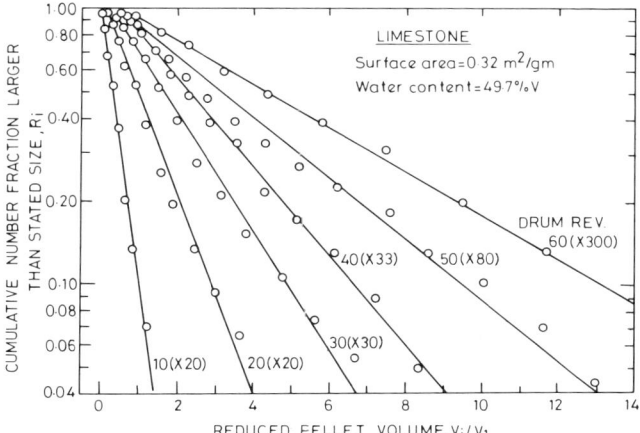

FIG. 17. Size distributions of pellets generated by the random-coalescence mechanism in the nuclei and transition regions. [From Kapur and Fuerstenau (K6).]

In Fig. 17 the cumulative distributions of limestone pellets in the nuclei and transition regions of growth are shown to be in reasonable agreement with the model. It will be noted that the dispersion in granule size increases rapidly in the random coalescence mode. In fact the variance grows approximately as the square of the mean size (K6). For coarse limestone material, the mean volume of the granules increases exponentially with balling time (Fig. 18) in conformity with Eq. (72). The time-invariant specific coalescence rate constant may be computed from the slope of the curves, and it provides a consistent and uniform basis for comparing the effect of water content on the agglomeration kinetics. On the other hand, in the case of fine limestone powders the rate function is not constant but increases rapidly with time, as illustrated in Fig. 19.

D. Nonrandom Coalescence Kinetics

Thus far it has not been possible to derive from the first principles the coalescence rate function for preferential combination of pellet species of different sizes. Kapur (K4) has proposed an ad hoc rate function in continuous sample space as follows:

$$[\text{Rate}]_{V,V'} = \lambda \frac{[V+V']^x}{[VV']^y} \frac{n(V,t)n(V',t)}{\mu_0(t)} \tag{73}$$

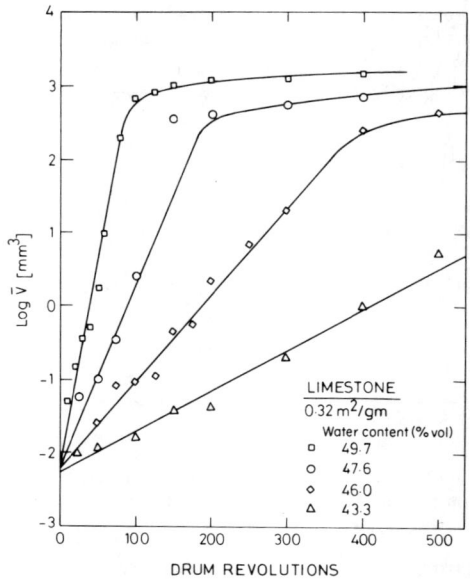

Fig. 18. Mean granule volume as a function of the agglomeration time in the random-coalescence mechanism. [From Kapur (K2).]

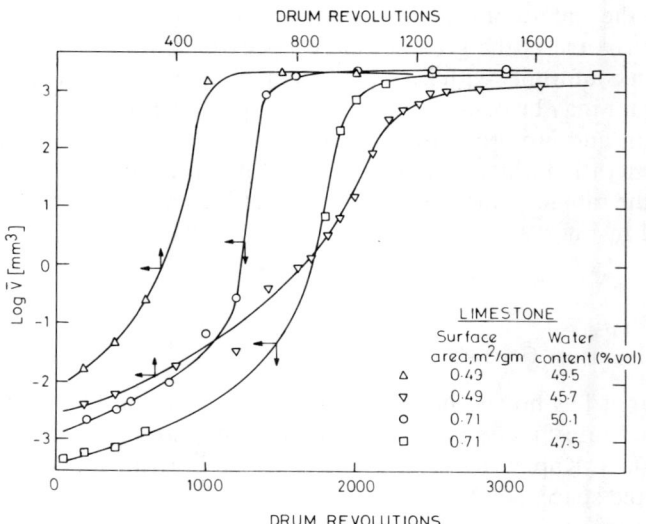

Fig. 19. Mean granule volume as a function of the agglomeration time, showing the time dependence of specific random-coalescence rate function. [From Kapur (K2).]

where λ is a function of the moist particulate feed material and the balling device, but does not depend on the granule size. The square-bracket terms reflect the nonrandom nature of the coalescence mechanism that governs the agglomeration process. The adjustable exponents x and y, as well as λ, should provide sufficient flexibility for a realistic representation of balling kinetics. The population balance equation is

$$\frac{dn(V, t)}{dt} = -\lambda \frac{n(V, t)}{\mu_0(t)} \int_0^\infty \frac{[V + V']^x}{[VV']^y} n(V', t) \, dV'$$
$$+ \frac{\lambda}{2\mu_0(t)} \int_0^V \frac{V^x}{[(V - V')(V')]^y} n(V - V', t)n(V', t) \, dV' \quad (74)$$

Kapur (K4) has shown that this integrodifferential equation admits a similarity solution given in Eqs. (49)–(51), namely,

$$n(V, t) = [\mu_0^2(t)/\mu_1]Z(p) \quad (75)$$

Substitution of this equation into Eq. (74) gives

$$-\frac{d\mu_0(t)}{\lambda \, dt}\left[\frac{\mu_1^{2y-x}}{\mu_0^{2y-x+1}(t)}\right]\left[2Z(p) + p\frac{dZ(p)}{dp}\right]$$
$$= Z(p) \int_0^\infty \frac{[p + p']^x}{[pp']^y} Z(p') \, dp' - \frac{p^x}{2} \int_0^p [pp' - p'^2]^{-y} Z(p - p')Z(p') \, dp' \quad (76)$$

Integration of Eq. (74) with respect to V from zero to infinity, followed by substitution of Eq. (75) results in an explicit expression for the rate of change of the pellet population, as follows:

$$-\frac{1}{\lambda}\frac{d\mu_0(t)}{dt}\left[\frac{\mu_1^{2y-x}}{\mu_0^{2y-x+1}(t)}\right] = Q \quad (77)$$

where the dimensionless constant Q is

$$Q = \int_0^\infty Z(p) \, dp \int_0^\infty \frac{[p + p']^x}{[pp']^y} Z(p') \, dp'$$
$$- \frac{1}{2}\int_0^\infty p^x \, dp \int_0^p [pp' - p'^2]^{-y} Z(p - p')Z(p') \, dp' \quad (78)$$

Combining Eqs. (76) and (77) results in a total integrodifferential equation in a single similarity variable p

$$\left[2Z(p) + p\frac{dZ(p)}{dp}\right]Q = Z(p)\int_0^\infty \frac{[p+p']^x}{[pp']^y} Z(p')\,dp'$$

$$- \frac{p^x}{2}\int_0^p [pp' - p'^2]^{-y} Z(p-p')Z(p')\,dp' \quad (79)$$

In principle the solution to Eqs. (77) and (79) followed by substitution in Eq. (75) provide a complete trajectory of the granule-size spectrum in time.

The growth of the mean granule volume is given by combining Eq. (50) with Eq. (77):

$$d\bar{V}(t)/dt = \lambda Q[\bar{V}(t)]^{x-2y+1} \quad (80)$$

or when $\bar{V}(t) \gg \bar{V}(0)$ and λ is time invariant, then approximately

$$\bar{V}(t) = K_n[t]^{1/(2y-x)} \quad (81)$$

where K_n is a growth constant for the nonrandom coalescence mechanism. Kapur (K4) has further shown that from Eq. (75) the cumulative number fraction distribution in the reduced size D/D_m is also self-similar:

$$R(D, t) = \bar{Z}(D/D_m) \quad (82)$$

and from Eq. (81) the median diameter varies with time in the following manner:

$$D_m(t) = K'_n[t]^{1/(6y-3x)} \quad (83)$$

The median diameter of limestone pellets in the ball growth region as a function of the granulation time on log-log scales is shown in Fig. 20. The corresponding size distributions are self-similar as illustrated in Fig. 21.

Pulvermacher and Ruckenstein (P10) have recently reexamined Kapur's similarity solution to the nonrandom coalescence equation (Eq. 74), and have established a necessary condition for the existence of a solution to Eq. (79). From the slope of the plots in Fig. 20 they have determined that the exponents x and y in Eq. (83) are related by

$$x = 2y - 1.2 \quad (84)$$

Interestingly, these authors found that the computed size spectra for various values of the exponents x and y, subject to the constraint in Eq. (84), were in all cases quite similar and in close agreement with the experimental distribu-

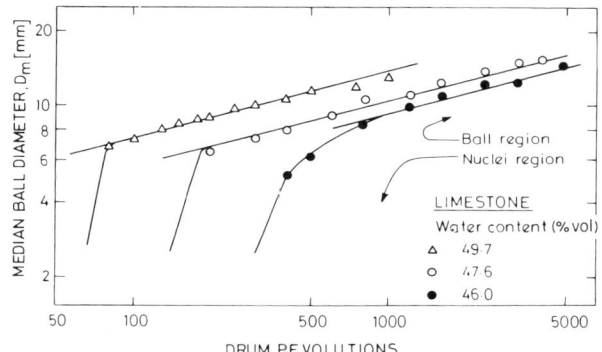

FIG. 20. Growth of the median ball diameter by the nonrandom-coalescence mechanism. [From Kapur (K4).]

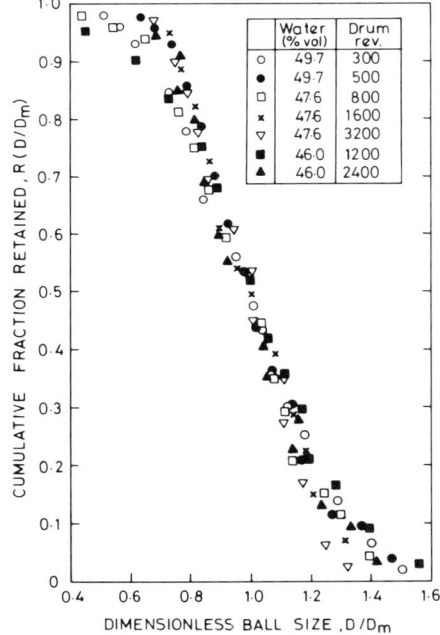

FIG. 21. Self-similar size distributions of limestone pellets generated by the nonrandom-coalescence mechanism. [From Kapur (K4).]

tions in Fig. 21. This suggests that by choosing $x = 0$ the nonrandom coalescence rate function can be modified to the following simpler expression

$$[\text{Rate}]_{V, V'} = \lambda[VV']^{-y} n(V, t) n(V', t)/\mu_0(t) \qquad (85)$$

This means that the coalescence occurs preferentially between small granules at least for the limestone system under scrutiny. Equation (83) now becomes

$$D_m(t) = K'_n[t]^{0.28} \qquad (86)$$

Simulated results of Pulvermacher and Ruckenstein (P10) have demonstrated that the similarity spectra are established in very short agglomeration times irrespective of the initial granule size distribution, and that the size dispersion decreases with an increase in the value of the exponent y.

Ouchiyama and Tanaka (O1) have presented an interesting derivation of a nonrandom coalescence rate function which is similar to the one in Eq. (73). According to these authors the frequency of loading between two pellets of diameter D and D' is:

$$\text{Frequency} \propto [D + D']^2 \qquad (87)$$

By loading is meant an application of force through the neighbors to the contact point between the two granules which are in contact with each other. Since not all couples can survive the shearing field in an agglomerating charge, there is a finite probability of coalescence, which is:

$$\text{Probability} \propto [DD']^{-y} \qquad (88)$$

Hence, the population balance equation for a restricted-in-space agglomeration process can be written as

$$\frac{dn(D, t)}{dt} = -\frac{\lambda}{\mu_0(t)} n(D, t) \int_0^\infty \frac{(D + D')^2}{(DD')^y} n(D', t) \, dD'$$

$$+ \frac{\lambda}{2\mu_0(t)} \int_0^D \frac{[D' + (D^3 - D'^3)^{1/3}]^2 D^2}{[(D^3 - D'^3)^{1/3} D']^y [D^3 - D'^3]^{2/3}}$$

$$\times n\{[D^3 - D'^3]^{1/3}, t\} n(D', t) \, dD' \qquad (89)$$

Ouchiyama and Tanaka (O1) did not solve this equation directly, but from the inspection of published data they assumed that the granule size is roughly uniformly distributed and self-preserving:

$$n(D, t) = \mu_0(t)/[D_x(t) - D_n(t)] \qquad (90)$$

where D_x and D_n are, respectively, the largest and smallest size in the distribution; the ratio D_n/D_x is constant in any one of the three growth regions which they denote as initial, middle, and later stages. Combining Eqs. (89) and (90), they derived the following relationship for the growth rate of the mean pellet size:

$$d\bar{D}(t)/dt \propto \bar{D}(t)^{3-2y} \qquad (91)$$

Ouchiyama and Tanaka (O1) further established that the ratio of D_n/D_x is uniquely related to the exponent y; hence, it is possible to compute the value of the exponent y from experimental size distributions. From the data on limestone and sand systems, these authors found that $y = 1, 2,$ and 3 respectively, in the initial, middle, and later stages. Hence, from Eq. (91), in the initial region,

$$d\bar{D}(t)/dt \propto \bar{D}(t) \qquad (92a)$$

In the middle region,

$$d\bar{D}(t)/dt \propto \bar{D}(t)^{-1} \qquad (92b)$$

In the later region,

$$d\bar{D}(t)/dt \propto \bar{D}(t)^{-3} \qquad (92c)$$

Since in the self-similar distributions

$$\bar{D}_m(t) \propto \bar{D}(t) \propto [\bar{V}(t)]^{1/3} \qquad (93)$$

therefore, the growth equation for the initial period in Eq. (92a) is the same as Kapur and Fuerstenau's relationship in Eq. (72) for the nuclei and transition regions. Equation (92c), pertaining to the latter region, also agrees quite closely with Eq. (86) for the ball-growth region.

E. EMPIRICAL KINETIC MODELS

If in the course of agglomeration the successive granule size distributions over a time interval exhibit similarity characteristics when plotted as a function of an appropriate dimensionless size, then it is reasonable to infer that there has been no change in the growth mechanism or mechanisms governing the process. Moreover, the variation of the scaling factor, for example, $\bar{V}$ or D_m with balling time, in conjunction with the similarity distribution, provides a complete trajectory of the size spectrum. Thus, it is possible to obtain an empirical quantitative description of balling and granulation kinetics, at least in a batch process, even though a mathematical model based on the growth mechanism(s) is lacking. For example, Sastry and Fuerstenau

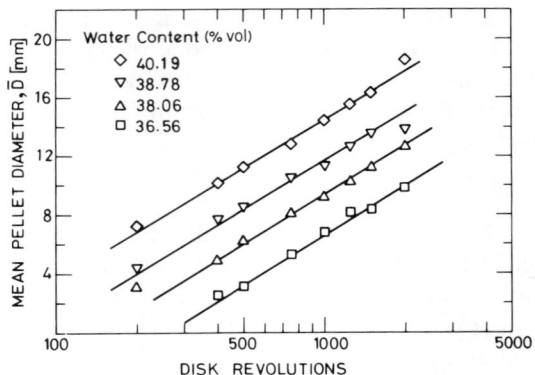

FIG. 22. Mean diameter of iron ore pellets as a function of the number of disk revolutions. [From Kanetkar (K1).]

(S3) reported that the size distributions of teconite pellets are self-preserving over a wide pelletizing interval. The scaling factor D_w, weight median diameter, is related to the agglomeration time by the following empirical equation:

$$D_w = K_e t^c \tag{94}$$

Kanetkar (K1) pelletized iron ore in a small-batch disk and found that the distributions are self-preserving in the reduced size $D/\bar{D}$ and that the mean pellet diameter increases in the following manner (Fig. 22):

$$\bar{D}(t) = z \log t + K'_e \tag{95}$$

where the intercepts K'_e is a function of the water content.

VI. Bonding Liquid and Additives

A. Water Content

Water is the most commonly used bonding liquid in the balling and granulation processes. It turns out that it has a profound influence on the agglomeration behavior. For a given particulate feed, satisfactory pelletizing is possible over a narrow range of moisture content only, and within this range the rate of growth of the granules is extremely sensitive to the amount of liquid in the charge. The percentage moisture content, defined as 100 × volume of liquid/volume of solids, depends on the size distribution of the particles and their packing characteristics. Based on the published data compiled by Linkson et al. (L1), it is in excess of 55% volume (usually 59–73% volume) for closely sized powders, and between 40% and 55% volume for materials with wide size distributions.

In theory the correct amount of water should equal or just exceed the theoretical saturated liquid content (C5, K5, N2), which is defined as equal to the void volume of the dense-packed particulate ensemble. In practice, however, the water content may range from roughly 90 to 110% of the critical liquid saturation. There are basically three aspects to this tolerance phenomenon (C5, K2). First, the granules can adjust their porosity to some extent in order to accommodate limited variations in the liquid content (see Fig. 9). Second, a fraction of the excess water that squeezes on to the surface of the pellets in and beyond the transition region may be absorbed in the deposit of material on the walls of the balling device. Third, the agglomerates invariably contain a small amount of trapped air, which may range up to 6% of the pore volume of granules made from closely sized materials or up to about 12% in the case of pellets made from powders with a wide size distribution. As a consequence, liquid and air bubbles together may just about fill or even exceed the voids of a granule, even though the amount of water is marginally less than the critical liquid saturation. The excess liquid on the pellet surface is apparently instrumental in causing the occurrence of growth at a meaningful rate.

The variation of the growth functions with added water content is shown in Fig. 23. Curve I is for granules made from narrow size sand by the crushing

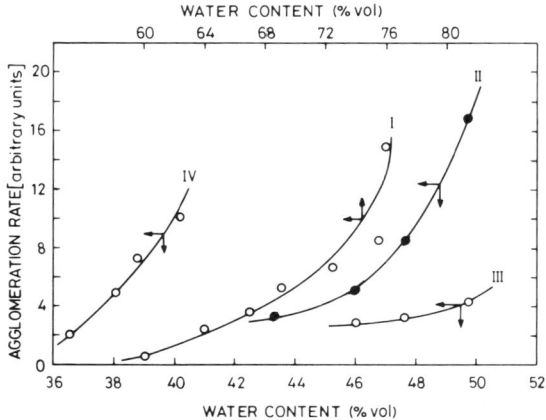

FIG. 23. (I) Effect of water content on the growth rate of agglomerates; sand granules grown by crushing and layering mechanism [from Capes and Danckwerts (C5)]. (II) Limestone nuclei by random coalescence [from Kapur (K2)]. (III) Limestone balls by nonrandom coalescene [from Kapur (K4)]. (IV) Iron ore pelletized in a disk.[From Kanetkar (K1)].

and layering mechanism; Curve II is for limestone nuclei growing by the random coalescence mode (slopes of curves in Fig. 18); Curve III pertains to limestone balls formed by the nonrandom coalescence mechanism (intercepts of the curves in Fig. 20); Curve IV pertains to iron ore pelletized in a disk (intercepts of the curves in Fig. 22). In most instances, the growth-rate constant has been found to increase exponentially with increase in the liquid content. Clearly, in large-scale balling and granulation operations a fairly close control of the moisture content is necessary if instability in the process is to be avoided.

B. BENTONITE ADDITIVE IN IRON ORE BALLING

An additive is formally defined as a component added as an aid to agglomeration or to strength development, which would not be used if satisfactory agglomerates could be made without it (M2). In terms of tonnage, the single most important additive is bentonite, which is frequently used in iron ore balling in amounts up to 10 kg ton^{-1} of iron ore and concentrate. Bentonite additive serves a multiple purpose (B2, M2, N3, R2, S14): (1) it smooths out the effect of any excess water in the feed that comes from the filter press and improves the balling characteristics, especially that of the coarse size and floated hydrophobic materials; (2) it improves the wet strength of the pellets, in particular the impact strength, so that the green balls can be screened and otherwise transported without damage; (3) it increases the resistance to decrepitation when the pellets are rapidly heated in the drying and preheating stages; (4) it provides enough dry strength to prevent the agglomerates from crumbling under their own weight before they are hardened by firing; and (5) it contributes to the fluxing action when the pellets are sintered, enhancing compressive strength and abrasion resistance of the heat indurated balls. Discussion will be restricted to the role of bentonite in the formation of green pellets, specifically to the water–bentonite interaction in the balling of iron ores and concentrates.

Model studies (K7, N3, S4) have shown that, while the rate of growth of the pellet increases markedly with increasing water content, it decreases with the addition of bentonite additive. It has been suggested that the clay mineral soaks up within its layers some water from the moist charge, and this liquid is immobilized as far as the apparent balling kinetic behavior is concerned. The growth mechanisms apparently do not undergo any change with the addition of bentonite; however, the balling time required to attain a specific size distribution of the pellets is lengthened. In other words, the trajectory of the pellet spectra remains invariant but the rate of evolution of the distributions is retarded in the presence of bentonite (K7, S4).

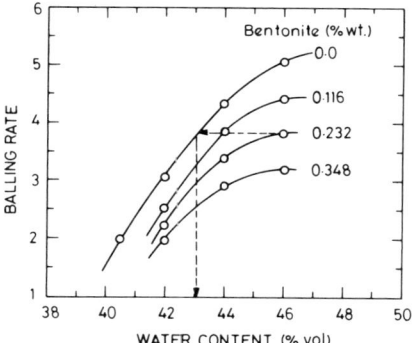

FIG. 24. Effect of bentonite additive on the balling rate of iron ore pellets. [From Kapur *et al.* (K7).]

Kapur *et al.* (K7) pelletized hematite iron ore in a small batch balling drum and found that the ball growth is in conformity with the nonrandom coalescence model in Eq. (83). In Fig. 24 the specific rate constant K'_n is plotted as a function of the moisture content of the iron ore feed, both without and with the addition of bentonite at three levels. The extent of water immobilized by bentonite can be estimated by comparing the bentonite curves with the additive-free curve at identical balling rates. The moisture retention capacity defined as the amount of water immobilized per unit weight of bentonite is shown in Fig. 25. The bentonite additive makes an excessively wet feed amenable to balling by virtue of its ability to immobilize a disproportionately larger amount of water from wetter materials. Moreover, with an increase in the amount of bentonite, the sensitivity of the balling rate to the moisture content is progressively reduced. Thus, an additional important implication of these results is that in the industrial balling

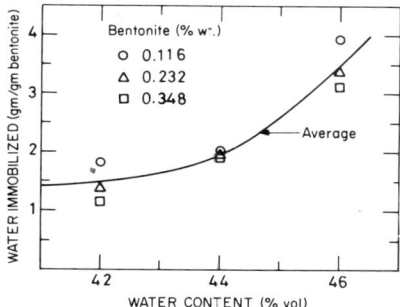

FIG. 25. Dependence of water immobilized by bentonite additive on the moisture content of feed. [From Kapur *et al.* (K7).]

circuits bentonite addition assists in a proper and controlled functioning of the pelletizing process in the presence of fluctuating moisture in the incoming feed.

In a similar investigation Sastry and Fuerstenau (S4) used up to 1.5% Wyoming bentonite in a teconite feed with 48.4, 50.3, and 52.3% volume moisture. The water retention capacity was calculated as 0.47 ± 0.11, independent of the water and bentonite contents. An evaluation of bentonites from three sources by Nicol and Adamiak (N3) indicates that the Wyoming bentonite has the highest cation exchange capacity and also the maximum retardation effect on the balling rate.

There are two principal disadvantages associated with the bentonite additive. First, a thorough and complete mixing of a small amount of finely powdered dry clay with wet feed is seldom attained in a large-scale industrial operation (J1). Indeed, Stone and Cahn (S15) found that, as a consequence of the incomplete mixing and the inverse relationship between bentonite content and balling rate, the smaller pellets invariably have a higher concentration of bentonite in a given batch. They concluded that improved mixing techniques would improve the pellet strength and ball uniformity and would at the same time reduce bentonite consumption. Second, since bentonite contains silica and alumina, its inclusion in the iron ore pellets lead to a higher consumption of limestone and coke in the blast furnace, as well as lower output of the metal.

A large number of soluble salts has been investigated as possible substitutes for bentonite, but without any demonstrable commercial success (B2, B4, R1, T2). Wada et al. (W2) have suggested that a mixture of 0.1% gum guar and calcium oxide each is an effective additive in balling of magnetite iron ore. Kramer et al. (K13), who have examined hundreds of individual organic compounds and combinations thereof, have found an adequate substitute for bentonite in certain humic acid derivatives. More recently, Roorda et al. (R2) have reported that a series of water-soluble polymers under the generic name Peridur duplicate and even exceed the desired additive characteristics of bentonite.

Pelletizing of iron ores and concentrates in the absence of a cementitious bond requires a rather fine feed size distribution, which, in some instances, may involve additional cost in grinding. In addition, sintering of these pellets entails a costly operation on a large scale. Attempts have been made in Sweden and elsewhere to produce so-called cold-bound pellets that can be fed directly to the blast furnace without sintering. Kihlstedt (K10) has reported that best results are obtained when the particle size distribution agrees with the Fuller curve for dense packing. Depending on the bonding process, the following techniques have been investigated:

1. A hydraulic setting cement is incorporated as the binder additive. The green pellets are partially hardened in silos, where they remain for about 24–36 hr, and the hardening process is completed during storage in stockpiles.
2. The binders are silica, lime, slag, or cement. The balls are somewhat dried, if necessary, and then cured in steam autoclaves. During the hydrothermal treatment lime and silica react to form hydrosilicate gels, which act as binders.
3. Hardening is effected by carbonation instead of by steam autoclaving.

VII. Granulation of Fertilizers

A typical fertilizer granulation loop is shown in Fig. 26. The feed comprises dry raw materials, raw materials in form of slurry or solution, and recycled solids. The granules from the granulator pass through a drier, where the water is removed and the capillary bonding mechanism is replaced by crystal bridges, forming strong agglomerates. The dry product is screened into onsize granule grade, which leaves the circuit. The undersize, and a fraction of the onsize if necessary, is fed back to the granulator. The oversize is also recycled after passing through a crusher. At steady state the rate of exit of the product grade granules from the circuit is equal to the rate of formation of new agglomerate entities. Irrespective of the kind of process employed, there are in general two kinds of sources of new granule formation (H3). One of these is internal to the granulator: nucleation of discrete

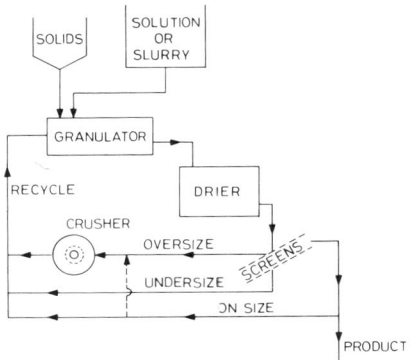

Fig. 26. Fertilizer granulation loop.

species from the particulate feed and by attrition and breakage of the tumbling charge. The other kind is external to the granulator and includes breakdown of granules in the drier and crushing in the crusher. The crusher may turn out to be the primary source of new seed species in a circuit operating in the snowballing mode.

The solution of the fertilizer salts in water, frequently referred to as the *solution phase*, acts as the bonding liquid. The solid feed to the granulator is relatively coarse: the average size often exceeds 1 mm. The amount of the liquid in the charge usually lies in the range 20–30% volume, which is definitely less than the critical saturation value. A major objective in fertilizer granulation is to maximize the product grade material in a narrow size-distribution range, typically 1–4 mm or even 1.5–3 mm (S9).

Depending on the process conditions, the fertilizer granules are formed by either random coalescence or snowballing mechanisms or both. In general, one of these growth modes is found to predominate. However, the coalescence mechanism, also known simply as agglomeration, is more common. Some fertilizer formulations that are difficult to agglomerate by coalescence may be amenable to granulation by the snowballing mechanism (also referred to as onion skinning or layering).

A. Granulation in Coalescence Mode

1. *Solution-Phase Theory*

From studies based on model batch systems by Sherrington (S9) (coarse sands bonded with saturated solutions of fertilizer salts), Butensky and Hyman (B9) (glass beads bonded with an aqueous solution of chemical grout), as well as by Newitt and Conway-Jones (N2) and Capes and Danckwerts (C5) (moist coarse sands), one may draw the following broad conclusions concerning granulation of coarse fertilizer feed in the presence of insufficient bonding liquid. In the initial stage coalescence is the primary mechanism of granule growth. In the later stage, when the specific surface area of the granules falls below a critical value, the agglomerates become "surface dry." The granules are now rigid or brittle, do not deform on collision, and do not possess sufficient surface plasticity to form a strong bond between two colliding pellets. The final product invariably contains some apparently ungranulated material in the same size range as the feed powder. Presumably, this material arises from abrasion of loosely held particles from the dry surface of the large granules, and perhaps by the destruction of the weak pellets. The abraded powder and broken fragments are continuously picked up by snowballing and layering. The granules eventually attain a steady-state size distribution, which is effectively invariant of agglomeration time, charge loading, and the technique employed for mixing liquid with the feed solids. The mean size of the viable granules in dynamic equilibrium is a function of the amount of liquid, provided the granule-to-particle size ratio is not too large. The mean size increases and the proportion of the ungranulated material decreases with an increase in the solution phase content.

Sherrington (S9) and Butensky and Hyman (B9) have presented a model that relates the equilibrium granule size to the amount of the granulating liquid present. Consider a spherical agglomerate comprising a number of single size spherical particles, whose idealized cross section is shown in Fig. 27. It is assumed that the granule surface consists of an outer spherical shell of thickness g times the particle radius, where g is an unknown factor whose value is about unity. A fraction S of the void space is filled with liquid, the remainder being entrapped air. The weight of the bonding liquid is

$$(\pi/6)S\epsilon\rho_L(D - gD_p)^3 \tag{96}$$

where ρ_L is the liquid density. The weight of the solids per granule is

$$(\pi/6)(1 - \epsilon)\rho_S D^3 \tag{97}$$

Therefore, the weight of liquid per unit weight of solids is

$$W = W_\infty[1 - g(D_p/D)]^3 \tag{98}$$

where

$$W_\infty = S\epsilon\rho_L/(1 - \epsilon)\rho_S \tag{99}$$

is the limit of W when the relative diameter or the size enlargement ratio D/D_p approaches infinity. In other words, W_∞/S is the weight of liquid per unit weight of solids required to fill the voids in an infinitely large granule. Equation (98) by Butensky and Hyman (B9) reduces to the expression given by Sherrington (S9) when D/D_p is large and $S = 1$. Obviously, the model is

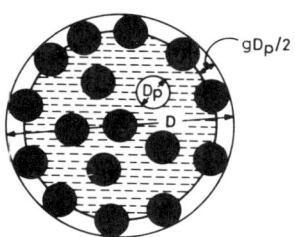

FIG. 27. Idealized cross section of a granule that is deficient in liquid. [From Butensky and Hyman (B9).]

valid only when the size enlargement ratio exceeds a minimum value of 3 or 4 (B9).

Actual measurements of the liquid content have shown that the larger granules have more liquid associated with them than the average and the smaller granules have less, in agreement with Eq. (98). Assuming that S and g are independent of pellet size, the liquid content may be related to the pellet size distribution in the following manner (B9):

$$W = W_\infty \sum_i [1 - g(D_p/D_i)]^3 w_i \tag{100}$$

where w_i is the weight faction of solids in granules of size D_i. Next, we define an appropriate mean granule size D^* so that

$$W = W_\infty [1 - g(D_p/D^*)]^3 \tag{101}$$

Combining Eqs. (100) and (101) and solving for the mean relative size D^*/D_p, we get

$$D^*/D_p = \frac{g}{1 - \left\{\sum_i [1 - g(D_p/D_i)]^3 w_i\right\}^{1/3}} \tag{102}$$

A comparison of the measured mean relative size with the model in Eq. (101) is shown in Fig. 28. The material was monosize glass beads. The data fit the model quite well, with the exception of fine 0.038-mm powder. It is evident that the steady-state size distribution is a function of the liquid content, and consequently, as shown by Sherrington (S9), there is an optimal granulating liquid for maximum granulation efficiency, that is, percentage of the product-grade material.

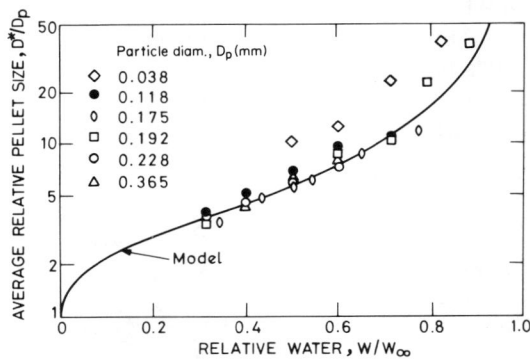

FIG. 28. Effect of the relative liquid content on the average relative pellet size. [From Butensky and Hyman (B9).]

2. *Granulation Loop*

The presence of the recycle loop in fertilizer granulation process makes it prone to surging and drifting, which coupled with long lag times give rise to difficulties in maintaining a stable operation of the plant. A rational strategy for countering the surging and drifting effects can be established from the solution phase theory (S9). For example, if there is a momentary increase in the liquid feed to the granulator, this will cause an excess production of the oversize granules for a short period (Eq. 101). This, in turn, may result in a higher average feed size when this impulse of excess oversize recycles and completes the loop. As a consequence, a follow-up surge, leading again to increased production of oversize, will travel through the loop even though the liquid feed has settled down to its normal level. It may take several hours before this kind of surging smoothens out. Again, if for some reason the crusher product becomes even slightly coarser, a steadily accelerating drift in the process will occur, leading to increasing production of the oversize granules. The drifting can be quite serious unless corrective measures, including turn-down of the liquid feed to the granulator, are taken promptly.

Control of the liquid phase to the granulator is the primary control variable available to the plant operator. The minimum amount of water that enters the feed along with the new raw material is determined first by the fertilizer formulation. It may be less or more than that required to form sufficient liquid phase for optimal granulation efficiency. In the former case an adjustable extra water supply is all that is needed; the recycle is now determined by the granulation efficiency attained in the system and the plant is "granulation limited" (S9). In the latter case it may be necessary to recycle some of the onsize granules, preferably after crushing, in order to reduce the average feed size to the granulator. The plant is now "water balance limited" and much more difficult to control (S9).

B. Granulation in Snowballing Mode

1. *Model Studies*

In circuits operating in the snowballing mode, there is an increased tendency for large-size feed particles and large recycled aggregates of particles to act as seeds rather than constituents of the deposited layer surrounding another granule. Moreover, individual particles and their clusters of size greater than about 0.7–1.2 mm always act as seeds rather than agglomerate with each other (S10).

Sherrington (S10) has carried out interesting simulation studies on laboratory scale using 1.2–1.5-mm glass beads as the seeds and finer sand as the layering material. In the absence of glass beads the fraction of oversize material in the granule size distribution (> 3.5 mm) increases steadily with increasing liquid content. When the beads are present and liquid content is low, the oversize production is quite small and a large amount of sand remains ungranulated. Initially, the principal effect of adding increasing amounts of the liquid phase is to increase the fraction of on-size granules at the expense of ungranulated sand. When the liquid content required to layer all the sand is exceeded, growth occurs by coalescence with a rapid increase in the formation of oversize granules. It is estimated that 30% seeds in the charge is about the lower limit below which snowballing cannot be achieved. As shown in Fig. 29, addition of seeds to the sand increases the maximum proportion of onsize agglomerates at the optimal liquid content which ranges from about 40 to 80% volume. The maximum yield of the onsize granules is not very sensitive to the proportion of seeds present, but the liquid content for the maximum yield decreases with an increase in the number of seeds. These results suggest that the realization of coalescence or snowballing as the principal growth mechanism, as well as the dynamic behavior of the system in the latter case, depends in a complex manner on the interactions between recycle ratio, size of recycled material, and liquid-to-solid feed ratio.

The liquid-phase requirement for the snowballing mechanism may be computed in the following manner (S10). It is assumed that the liquid withdrawal is negligible (i.e., $g = 0$) and the fractional saturation in the deposited

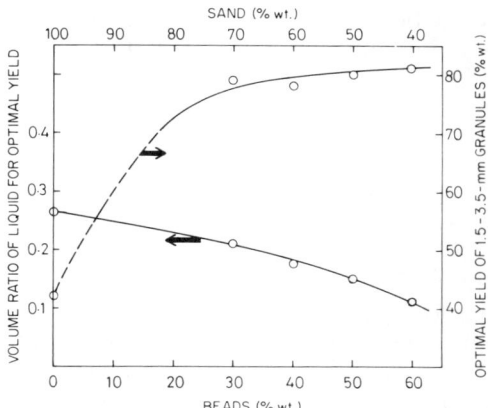

FIG. 29. Optimal liquid content and optimal yield of on-size granules for different fractions of seeds in the charge. [From Sherrington (S10).]

layer is unity. Let subscript 1 pertain to the snowballed granule or the deposited layer portion of the granule. The weight of the layered solids in a granule of size D_1 is

$$(\pi/6)(D_1{}^3 - D^3)\rho_{S,1}(1 - \epsilon_1) \tag{103}$$

and the weight of the liquid in the granule is

$$(\pi/6)(D_1{}^3 - D^3)\rho_L\epsilon_1 + (\pi/6)D^3\rho_L\epsilon S \tag{104}$$

where S, the fraction of pores in the seed filled by liquid, is the degree of ingress of liquid into the seed. Let L be the weight ratio of seed to layered solids:

$$L = D^3\rho_S(1 - \epsilon)/(D_1{}^3 - D^3)\rho_{S,1}(1 - \epsilon_1) \tag{105}$$

From these relationships the weight of the liquid per unit weight of solids in a granule is

$$W = \rho_L(\rho_S J_1 + L\rho_{S,1}JS)/(\rho_{S,1}\rho_S + L\rho_{S,1}\rho_S) \tag{106}$$

where

$$J = \epsilon/(1 - \epsilon) \tag{107}$$

and

$$J_1 = \epsilon_1/(1 - \epsilon_1) \tag{108}$$

The snowballed granule diameter is

$$D_1 = D[1 + \rho_S(1 - \epsilon)/L\rho_{S,1}(1 - \epsilon_1)]^{1/3} \tag{109}$$

When $\rho_{S,1} = \rho_S$, which is generally valid, Eq. (106) becomes

$$W = (\rho_L/\rho_S)(J_1 + LSJ)/(1 + L) \tag{110}$$

and when $S = 0$

$$W = (\rho_L/\rho_{S,1})[J_1/(1 + L)] \tag{111}$$

Equations (106) and (109) give expressions for the liquid content and granule size in a snowballing granulation system.

2. Granulation Loop

Sherrington (S10) has presented a static model of the granulation loop operating in the snowballing mode. The model combines the solution phase theory in Eq. (106) with the material balance across the granulator. The recycle ratio defined as recycle/raw solid feed, is given by

$$U = (2\rho_S/SJ_1\rho_L)(1 + X) + J_1YX/SJ - J_1/SJ \tag{112}$$

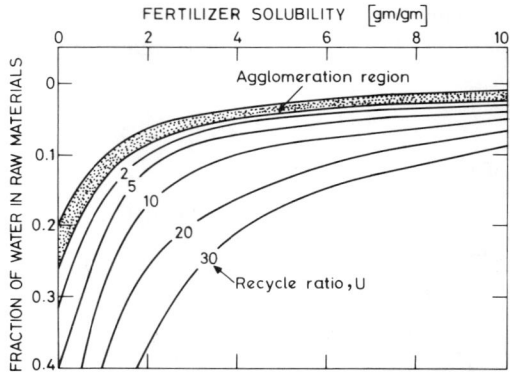

FIG. 30. Simulated behavior of fertilizer loop operating in the snowballing mode. [From Sherrington (S10).]

where ρ_S is now density of the fertilizer, X the weight ratio of fertilizer to water in the fertilizer solution at the granulation temperature, and Y the weight ratio of water to fertilizer in the incoming raw materials. Results of numerical simulation using typical values for the various parameters are presented in Fig. 30 in terms of the variation of the recycle ratio with fertilizer solubility and fraction of water with the raw materials. In the snowballing region the recycle ratio is always larger than in the coalescence region, and it increases with increasing solubility and increasing water fraction. The recycle contours are bunched together when the fertilizer solubility is high. The hatched area pertains to the coalescence region in which snowballing is not possible, although agglomeration can occur outside this region.

Han (H3) and Han and Wilenitz (H4) have also presented steady-state models of fertilizer granulators based on population balance on the granules in the process loop operating in the snowballing mode. From the viewpoint of process control some interesting interrelationships between various recycle ratios, crusher speed, crusher product size, and the granule growth rate have been established.

VIII. Miscellaneous Topics

In this section three miscellaneous topics in the area of agglomeration shall be discussed, namely, dry pelletization, spherical agglomeration in liquid suspension, and spontaneous or inadvertent agglomeration of fine particles.

A. Dry Pelletization

Beds of very finely divided solids are difficult to handle in processing and usage. Because of the presence of the van der Waals and electrostatic interparticle bonding forces, these beds do not flow readily under gravity, exhibit a tendency to bank and form arches in bins, and normally possess large void volume. For any significant compaction very high compressive forces are needed in order to overcome the friction between the particles. For these reasons, light dusty and fluffy fine powders, such as carbon black, dyes, pigments, and plastics, are pelletized in dry state into relatively large and dense agglomerates of considerable crushing strength, which exhibit the normal flow and packing characteristics of granular materials.

A number of studies (F3, I2, M4, V2) have shown that, provided the constituent particles are small enough, any particulate material will pelletize by systematic agitation without the use of a binder. The rate of agglomeration is greatly accelerated in the presence of relatively large-size solid particles or agglomerates. These seeds, nominally plus 200 mesh in size, are usually the recycled pellets of the powder itself.

Typical results of the dry pelletizing process when a charge of carbon black and its plus 52-mesh seeds is tumbled in a drum (I2) are shown in Figs. 31 and 32. Initially, the seeds grow in size rapidly by snowballing. The dip in the bulk density of the seeds (Fig. 32) suggests that the deposited layers are quite porous in the beginning but get compacted in due course to

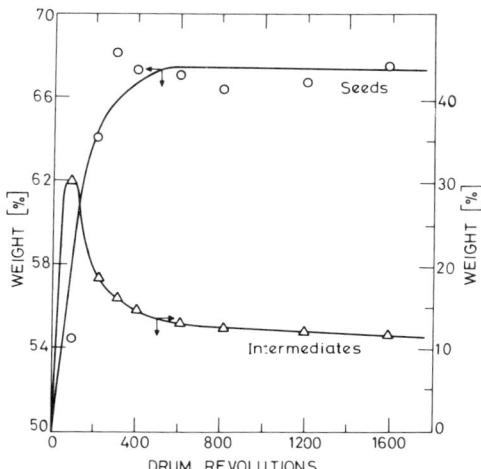

Fig. 31. Variations in the weights of carbon black seeds and intermediate granules with pelletizing time. [From Israel and Ventakeswarlu (I2).]

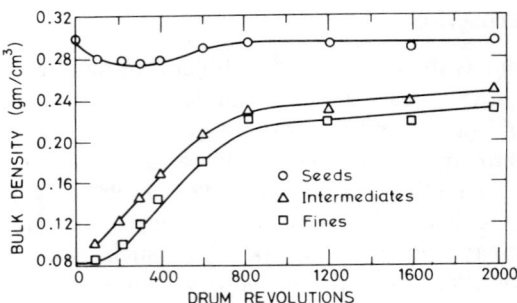

FIG. 32. Bulk densities of seed and intermediate granules and fines as a function of the pelletizing time. [From Israel and Ventakeswarlu (I2).]

almost the same extent as the original core agglomerates. The increase in the density occurs by compressive blows to the pellets in collisions with other agglomerates and the walls of the pelletizer, and by the shear forces to which they are subjected during the tumbling of the charge. The intermediate-size pellets, which are formed by nucleation from the matrix of fine powder and also by the breakup of larger balls, are rapidly depleted by crushing brought on by the impact with larger and denser pellets. The resulting fragments are picked up by the surviving granules. Some intermediates survive the impacts and eventually acquire the status of seeds, and as a consequence a change in the total number of balls occurs. The bulk densities of the intermediate nuclei and fines increase rapidly in the initial stage and more slowly thereafter. In the case of the fines, densification is assisted by the seed pellets.

Meissner *et al.* (M6) have pelletized dry zinc oxide powder in the presence of seeds. According to them the process can be divided into two stages. In Stage 1, the rate of disappearance of the fines is greater than in Stage 2, but the ball population is almost constant. The bulk densities of both balls and fines undergo maximum increase in the initial stage. The duration of this stage is shortened and the amount of fines at the end of this stage is diminished with an increase in the ratio of seeds to fines in the feed charge. Intermediate-size pellets first appear in Stage 2. Although the depletion rate of the remaining fines is now slowed down markedly, the whole charge may be converted into dense pellets given a sufficiently long period of pelletizing. The transition between the two stages may not be sharp, specially if the proportion of fines is high.

In the initial stage the amount of fines consumed per unit charge weight depends only on the total number of drum revolutions, independent of the rotational speed (I2, M6). The disappearance rate per unit charge weight and per drum revolution of the fines varies directly with seed density, the square of the seed diameter, the cube of the volume fraction balls, and inversely with a function of the fines density (M6).

In the absence of seeds, the rate of change of fines density is extremely slow, and the appearance of discrete granule species occurs only after a long tumbling time (M6). It would then seem that the nucleation of new pellets is greatly facilitated by the compaction of the particulate bed in the presence of seeds. In this connection it is interesting to note that Ford and Shennan (F3) ball-milled uranium dioxide powder and uranium dioxide–carbon mixtures for varying lengths of time prior to dry pelletization. Undoubtedly, significant compaction of the fines took place in the ball mill, which, in turn, assisted in the nucleation of the granules.

B. Spherical Agglomeration in Liquid Suspension

In this process finely divided solids in liquid suspension are agglomerated by suitable agitation with addition of a small amount of a second liquid (bridging or bonding liquid), which is immiscible with the first liquid and preferentially wets the particles (F1, S11). Apparently the second liquid partially or wholly displaces the suspension medium from the surface of the solids. When two or more suspended particles collide, adhesion occurs, owing to the formation of pendular bonds between the particles. Agitation of the resulting flocs produces compact spheres from which much of the original liquid is ejected and replaced by the wetting liquid. The kind of agglomerated species obtained depends on the wetting properties of the solids, its size distribution, and the amount of the bridging liquid. The scope of the process is claimed to be very wide because, in general, the wetting properties of the solid phase can be suitably tailored through selective adsorption of surfactants. Small clusters or microagglomerates are formed when a small amount of bridging liquid is used with vigorous agitation of the suspension. Large, compact spheres similar to those produced in the conventional balling and granulation processes, are formed if more bonding liquid is employed in conjunction with a gentle tumbling or shaking action (C7). Possible applications of the spherical agglomeration technique include liquid–solid separation, fractionation of mixtures of solids by preferential agglomeration, and formation of microspheroids and large agglomerates (F1, S11).

Kawashima and Capes (K8) have studied the kinetics of microagglomeration, using as a model system a suspension of narrow size-distribution sands in carbon tetrachloride and 20% aqueous calcium chloride solution as the bonding liquid. The mean diameter of the pellet produced ranges from about 0.5 mm to almost 1.5 mm. The size distribution are log normal in a reduced pellet size $D/\bar{D}$ under all conditions tested, independent of particle size, charge loading, agitator torque, etc. As shown in Fig. 33, the population density of the pellets in suspension decreases in the manner of a first-order kinetic process as follows:

$$\log[(\mu_0(t) - \mu_{0,e}(t))/(\mu_{0,f} - \mu_{0,e})] = -K_1 t \tag{113}$$

where $\mu_{0,e}$ is the number of granules per unit volume of the suspension at steady state and $\mu_{0,f}$ is a constant denoting the initial flocculated state of the solid phase. The specific rate constant K_1 increases steeply with increasing agitator torque, bonding liquid content and particle size in the following empirical manner:

$$K_1 \propto [T_q]^{3.37}[\bar{D}_p]^{2.84}[S]^{2.45} \tag{114}$$

where T_q is the agitator torque.

Using sands of narrow and wide size distributions suspended in various organic liquids, Capes and Sutherland (C7) have shown that large, compact agglomerates are formed if the amount of the bridging liquid is sufficient to occupy about 44–88% of the pore space in a densely compacted bed of sand particles. The final size distribution attained represents a balance between

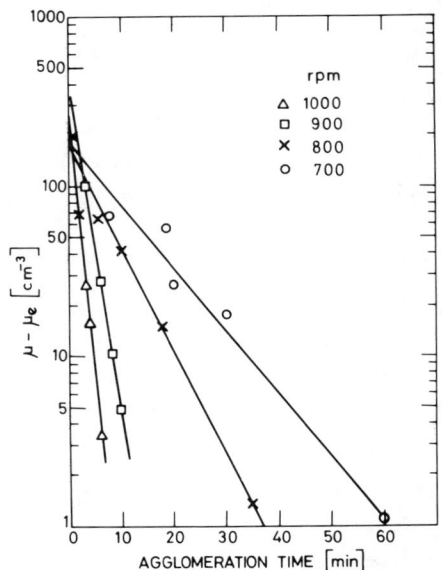

FIG. 33. First-order kinetics of microagglomeration in liquid suspension. [From Kawashima and Capes (K8).]

the destructive and cohesive forces acting on the pellets. The equilibrium sphere size varies more or less directly with the ratio of the interfacial tension to the sand particle, inversely with 0.8 power of the agitator speed, and increases linearly with the weight of sand being agglomerated.

An interesting method for solid agglomeration is aqueous suspension, in which a suitable flocculant replaces the bridging liquid, has been described by Yusa and Gaudin (Y1). In a 3.69% volume suspension of kaolinite, 118 mg of partially hydrolyzed polyacrylamide long-chained polymer (HPAM) in solution was added for each gram of clay. On gentle mixing in a drum, the flocs compacted into dense, strong pellets that could be readily wet screened.

C. Inadvertent Agglomeration

Given the right conditions, the natural tendency of fine particles to form agglomerates may turn out to be a great disadvantage in many processes dealing with particulate matter. In general, spontaneous and inadvertent agglomeration is not desired in comminution, separation, mixing, handling, or storage of finely divided solids. The following discussion is taken mostly from Pietsch's (P4) review of this subject, which includes an extensive literature survey.

Spontaneous agglomeration occurs as a rule in fine grinding of dry solids in tumbling mills. This phenomenon results in what is known as *grinding equilibrium*, where a major proportion of the grinding energy is spent in crushing the strong granules, which are continuously being formed in the rotating charge. Thus, the grinding time is increased, and a limit is imposed on the degree of fineness that can be attained. Many auxiliary grinding agents (antiagglomerants) have been suggested, especially for the grinding of cement clinkers. Uncontrolled agglomeration is highly detrimental to the efficiency of many separation processes, such as screening and sieving, size analysis by sedimentation, classifying and sorting in liquid suspension by size and density, and flotation.

It may not be possible to realize a thoroughly homogenized, random mixture if agglomerates are formed during the mixing of dry or moist powders. Preventive measures include the use of antiagglomerants and mixing in an environment of pronounced shear, impact, and chopping to break up the granules. On the other hand, demixing can occur in the handling and conveying of dry, homogenized beds of particles whose constituents differ greatly in size or density. This may be prevented by controlled agglomeration of the mixture with a suitable binder.

Many auxiliary chemicals that perform the functions of antiagglomeration, grinding aid, and flow conditioner have been employed. The list includes, among others, acetone, sodium stearate, naphthenic acid, triethanolamine, talc, starch, calcium carbonate, carbon black, kaolin, and fine silica. Nash et al. (N1) have studied the effect of the last-mentioned agent (Cab-O-Sil) on the physical properties of finely divided high molecular weight polyethylene glycol (Carbowax 6000). In 1% concentration Cab-O-Sil reduces the shear and tensile strengths of the powder bed by about 30 and 60%, respectively, increases the bulk density by approximately 30%, and neutralizes the electrostatic charge on the powder. Rumpf (R5) has shown that the van der Waals force of attraction between a large particle and a small, irregularly shaped particle exhibits a pronounced dip when the separation distance approaches 10^{-1}–10^{-2} μm. It would seem that the role of the antiagglomerant is to impair the interparticle bonding mechanism and thus discourages the formation of the agglomerates.

Acknowledgment

The Education Development Centre, I. I. T. Kanpur, provided financial aid for preparing this chapter. The author thanks Mohini Mullick for correcting the language of the text.

Nomenclature

a	Dimensionless constant (Eq. 53)	D_l	Diameter of layered pellet
a'	Ratio of number of particle pairs per unit area to the same per unit volume	D_m	Median pellet diameter
		D_n	Smallest pellet size in a distribution
B	Hamakar constant	D_p	Particle diameter
$B(V, t)$	Fraction of pellets of size V broken per unit time	$\bar{D}_p$	Mean effective particle diameter or volume–surface mean particle diameter
b	Dimensionless constant (Eq. 54)		
b'	Ratio of overall area of contact per particle pair to the surface area of smaller particle of the pair	D_w	Weight median diameter
		D_x	Largest pellet size in a distribution
C	Constant of proportionality (Eq. 44)	D^*	Mean pellet size defined in Eq. (101)
c	Exponent (Eq. 94)	E	Energy parameter
D	Pellet diameter	F	Amount of fragmented material
$\bar{D}$	Mean pellet diameter	F_b	Tensile strength of solid bridge bond
D_d	Disk diameter (m)	F_c	Attractive force per unit overall contact area
D_i	Discrete pellet diameter		

F_p^*	Dimensionless reduced bonding force	$\bar{R}(D)$	Cumulative number of pellets equal to or greater than pellet size D
G	Constant of proportionality (Eq. 45)	$R(D/D_m)$	Cumulative number fraction equal to or greater than D/D_m
$G(V, F)$	Growth function in crushing and layering model	r	Ratio of D_m to the cube root of V
g	Factor for thickness of outer dry shell	S	Fraction saturation of void space
$H(D_p)$	Interparticle bonding force	S_b	Fraction of void space filled by solid bond
H_p	Bonding force of pendular bond	S_c, S_p	Upper and lower limits of fraction saturation for funicular bond
H_v	Force of van der Waals attraction between two spheres		
J	Ratio of voids to solid fractions	$\bar{s}$	Mean effective surface area per particle
J_1	Ratio of voids to solid fractions in deposited layer		
K_e, K_e'	Empirical pellet growth rate constants	T	Thickness of snowballed layer
		T_q	Agitator torque
K_1	Empirical pellet growth rate constant (Eq. 113)	t	Agglomeration time
		t_0	Initial time
K_n, K_n'	Growth constants in nonrandom coalescence kinetics	U	Recycle ratio
		V	Pellet volume
		$\bar{V}$	Mean pellet volume
K_s	Linear pellet growth velocity in snowballing	V_b	Volume of bridging material
		V_i	Discrete volume size of pellet
k	Coordination number	V_L	Volume of liquid in pellet
L	Weight ratio of seed to layered solids	V_p	Volume of liquid in pendular bridge
l	Surface separation or mean surface separation of particle pair	V_p^*	Reduced volume of liquid in pendular bridge
l_0	Mean surface separation of particle pair when tensile strength is zero	V_S	Volume of particulate solids in pellet
		$\bar{v}$	Mean volume per particle
m	Exponent (Eq. 2)	v_{-1}	-1th moment of similarity function
n_c	Critical speed of disk (rpm)		
$n(D)$	Number distribution of pellets in diameter	W	Weight ratio of liquid to solids in pellet
$n^0(D)$	Initial number distribution of pellets	W_x	Limiting weight ratio of liquid to solids in pellet
n_i	Number of pellets of size V_i	w_i	Weight fraction of pellets of size D_i
$n(V)$	Number distribution of pellets in volume		
		X	Weight ratio of fertilizer to water in solution
P_c	Capillary pressure		
P_e	Entry suction	x	Exponent in nonrandom coalescence rate function
p	Similarity variable		
Q	Dimensionless constant (Eq. 77)	Y	Weight ratio of water to fertilizer in feed raw materials
R_1, R_2	Radii of curvature of liquid bridge		
		y	Exponent in nonrandom coalescence rate function
R_i	Cumulative number fraction larger than size V_i		
		Z	Similarity function

$\bar{Z}$	Cumulative similarity distribution	μ_k	kth moment of $n(V)$
		ρ	Bulk density of pellet
z	Empirical constant (Eq. 95)	ρ_0	Bulk density of pellet when tensile strength is zero

GREEK LETTERS

α	Exponent (Eq. 40)	ρ_L	Density of liquid
β	Half sector angle in liquid bridge	ρ_S	Density of particles
β_d	Inclination angle of disk	$\rho_{S,1}$	Density of layered solids
γ	Surface tension of liquid	σ	Tensile strength of pellet
Γ	Gamma function	σ_b	Tensile strength of pellet bonded by solid bridges
ϵ	Volume fraction voids in pellet		
ϵ_1	Volume fraction voids in snowballed layer	σ_c	Tensile strength of pellet in the capillary state
θ	Contact angle	σ_f	Tensile strength of pellet in the funicular state
λ	Coalescence rate function		
$\bar{\lambda}$	Integrated coalescence rate function	σ_p	Tensile strength of pellet in the pendular state
μ_0	Total number of pellets	ψ	Reduced interparticle force per unit overall area of contact
μ_1	Total volume of pellets		

References

A1. Anonymous, *Chem. Eng. News* **42**, 78 (1964).
A2. Ashton, M. D., Cheng, D. C. H., Farley, R., and Valentin, F. H. H., *Rheol. Acta* **4**, 206 (1965).
B1. Ball, D. F., *JISI* **192**, 40 (1959).
B2. Ball, D. F., Dawson, P. R., and Fitton, J. T., *Inst. Min. Met. Trans.* **79**, C189 (1970).
B3. Bazilevich, S. V., *Stal Engl. (USSR)* **8**, 551 (1960).
B4. Beale, C. V., Appleby, J. E., Butterfield, P., and Young, P. A., *Iron Steel Inst. (London)*, Spec. Rep. No. 78 (1964).
B5. Berquin, Y. F., *Genie Chim.* **86**, 45 (1961).
B6. Bhrany, U. N., Johnson, R. T., Myron, T. L., and Pelezarski, E. A., *in* "Agglomeration" (W. A. Knepper, ed.), p. 229. Wiley (Interscience), New York, 1962.
B7. Brociner, R. E., *Chem. Eng. (London)* **46**, CE227 (1968).
B8. Browning, J. E., *Chem. Eng.* **74**, 147 (1967).
B9. Butensky, M., and Hyman, D., *Ind. Eng. Chem. Fundam.* **10**, 212 (1971).
C1. Cahn, D. S., *Trans. AIME* **250**, 173 (1971).
C2. Capes, C. E., *Chem. Eng. (London)* **45**, CE78 (1967).
C3. Capes, C. E., *Powder Technol.* **4**, 77 (1970-71).
C4. Capes, C. E., *Powder Technol.* **5**, 119 (1971-72).
C5. Capes, C. E., and Danckwerts, P. V., *Trans. Inst. Chem. Eng.* **43**, T116 (1965).
C6. Capes, C. E., and Danckwerts, P. V., *Trans. Inst. Chem. Eng.* **43**, T125 (1965).
C7. Capes, C. E., and Sutherland, J. P., *Ind. Eng. Chem. Process. Des. Develop.* **6**, 146 (1967).
C8. Cavanagh, P. E., *Can. Min. Metall. Bull.* **50**, 692 (1957).
C9. Chandrasekhar, S., *Rev. Mod. Phys.* **15**, 1 (1943).
C10. Chari, K. S., and Sivashankaran, V. S., Semin. Particle Technol., I.I.T. Madras, Preprint C4 (1971).
C11. Cheng, D. C. H., *Chem. Eng. Sci.* **23**, 1405 (1968).
C12. Cooper, A. R., and Eaton, L. E., *J. Am. Ceram. Soc.* **45**, 97 (1962).

C13. Corney, J. D., *Brit. Chem. Eng.* **8**, 405 (1963).
D1. Debbas, S., and Rumpf, H., *Chem. Eng. Sci.* **21**, 583 (1966).
E1. Engelleitner, W. H., AIME, SME Preprint. 65-B-309 (1965).
E2. Engelleitner, W. H., *Am. Ceram. Soc. Bull.* **54**, 206 (1975).
E3. English, A., and Greaves, M. J., *Trans. AIME* **226**, 307 (1963).
F1. Farnand, J. R., Smith, H. M., and Puddington, I. E., *Can. J. Chem. Eng.* **39**, 94 (1961).
F2. Firth, C. V., *Proc. AIME Blast Furnace Coke Oven Raw Mater.* **4**, 46 (1944).
F3. Ford, L. H., and Shennan, J. V., *J. Nucl. Mater.* **43**, 143 (1972).
G1. Garrett, K. H., Records, F. A., and Stevenson, D. G., *Chem. Eng. (London)* **46**, CE216 (1968).
H1. Hahn, G. T., and Shapiro, S. S., "Statistical Methods in Engineering." Wiley, New York, 1967.
H2. Hakala, R. W., *J. Phys. Chem.* **71**, 1880 (1967).
H3. Han, C. D., *Chem. Eng. Sci.* **25**, 875 (1970).
H4. Han, C. D., and Wilenitz, I., *Ind. Eng. Chem. Fundam.* **9**, 401 (1970).
H5. Haughey, D. P., and Beveridge, G. S. G., *Can. J. Chem. Eng.* **47**, 130 (1969).
H6. Heady, R. B., and Cahn, J. W., *Met. Trans.* **1**, 185 (1970).
H7. Hignett, T. P., in "Chemistry and Technology of Fertilizers" (V. Sauchelli, ed.), p. 269. Van Nostrand-Reinhold, Princeton, N.J., 1960.
I1. Illig, H. J., *Silikattechnik* **22**, 7 (1971).
I2. Israel, R., and Venkateswarlu, D., *J. Inst. Eng. (India)* **51**, 67 (1971).
J1. Jones, H. A., *Inst. Min. Met. Trans.* **80**, C52 (1971).
K1. Kanetkar, V. V., Ph.D. thesis, Indian Institute of Technology, Bombay, 1974.
K2. Kapur, P. C., Ph.D. thesis, University of California, Berkeley, 1967.
K3. Kapur, P. C., *Chem. Eng. Sci.* **26**, 1093 (1971).
K4. Kapur, P. C., *Chem. Eng. Sci.* **27**, 1863 (1972).
K5. Kapur, P. C., and Fuerstenau, D. W., *Trans. AIME* **229**, 348 (1964).
K6. Kapur, P. C., and Fuerstenau, D. W., *Ind. Eng. Chem. Process. Des. Develop.* **8**, 56 (1969).
K7. Kapur, P. C., Arora, S. C. D., and Subbarao, S. V. B., *Chem. Eng. Sci.* **28**, 1535 (1973).
K8. Kawashima, Y., and Capes, C. E., *Powder Technol.* **10**, 85 (1974).
K9. Kayatz, K., *Zem. Kalk Gips* **17**, 183 (1964).
K10. Kihlstedt, P. G., *Proc. Int. Min. Congress, 9th, Prague, 1970*, 307 (1970).
K11. Klatt, H., *Zem. Kalk Gips* **11**, 144 (1958).
K12. Kočova, S., and Pilpel, N., *Powder Technol.* **7**, 51 (1973).
K13. Kramer, W. E., Ward, W. J., and Young, W. E., Pelletizing iron ore with organic additives, Undated Rep. Nelco Chem. Res. Center, Chicago.
K14. Krupp, H., *Adv. Colloid Interface Sci.* **1**, 111 (1967).
L1. Linkson, P. B., Glastonbury, J. R., and Duffy, G. J., *Trans. Inst. Chem. Eng.* **51**, 251 (1973).
M1. Macavei, G., *Brit. Chem. Eng.* **10**, 610 (1965).
M2. Madigan, D. C., Aust. Min. Develop. Lab. Bull. No. 9 (1970).
M3. Mason, G., and Clark, W. C., *Chem. Eng. Sci.* **20**, 859 (1965).
M4. Meissner, H. P., Michaels, A. S., and Kaiser, R., *Ind. Eng. Chem. Process. Des. Develop.* **3**, 197 (1964).
M5. Meissner, H. P., Michaels, A. S., and Kaiser, R., *Ind. Eng. Chem. Process. Des. Develop.* **3**, 202 (1964).
M6. Meissner, H. P., Michaels, A. S., and Kaiser, R., *Ind. Eng. Chem. Process. Des. Develop.* **5**, 10 (1966).
M7. Misra, V. N., Sinvhal, R. C., and Khangaonkar, P. R., *Trans. Indian Inst. Metals*, 24, Aug. (1973).

M8. Morrow, N. R., *Ind. Eng. Chem.* **62**, 32 (1970).
N1. Nash, J. H., Leiter, G. G., and Johnson, A. P., *Ind. Eng. Chem. Process. Des. Develop.* **4**, 140 (1965).
N2. Newitt, D. M., and Conway-Jones, J. M., *Trans. Inst. Chem. Eng.* **36**, 422 (1958).
N3. Nicol, S. K., and Adamiak, Z. P., *Inst. Min. Met. Trans.* **82**, C26 (1973).
O1. Ouchiyama, N., and Tanaka, T., *Ind. Eng. Chem. Process. Des. Develop.* **13**, 383 (1974).
P1. Papadakis, M., and Bombled, J. P., *Rev. Mater. Construct.* **549**, 289 (1961).
P2. Peck, W. C., *Chem. Ind. (London)* 1674, Dec. (1958).
P3. Pietsch, W., *Aufbereit. Tech.* **7**, 177 (1966).
P4. Pietsch, W., *Staub Reinhalt. Luft* **27**, 24 (1967).
P5. Pietsch, W., *Aufbereit. Tech.* **8**, 297 (1967).
P6. Pietsch, W., and Rumpf, H., *Chem. Ing. Tech.* **39**, 885 (1967).
P7. Pietsch, W., and Rumpf, H., *Can. J. Chem. Eng.* **46**, 287 (1968).
P8. Pietsch, W., Hoffman, E., and Rumpf, H., *Ind. Eng. Chem. Prod. Res. Develop.* **8**, 58 (1969).
P9. Pilpel, N., *Chem. Process. Eng.* **50**, 67 (1969).
P10. Pulvermacher, B., and Ruckenstein, E., *Chem. Eng. J.* **9**, 21 (1975).
R1. Ridgion, J. M., Cohen, E., and Lang, C., *JISI* **117**, 43 (1954).
R2. Roorda, H. J., Burghardt, O., Kortmann, H. A., Jipping, M. J., and Kater, T., *Proc. Int. Min. Proc. Congress, 11th, Cagliari*, Preprint 6 (1975).
R3. Rumpf, H., *Chem. Ing. Tech.* **30**, 144 (1958).
R4. Rumpf, H., *in* "Agglomeration," (W. A. Knepper, ed.), p. 379. Wiley (Interscience), New York, 1962.
R5. Rumpf, H., *Chem. Ing. Tech.* **46**, 1 (1974).
R6. Rumpf, H., and Turba, E., *Ber. Dtsch. Keram. Ges.* **41**, 78 (1964).
S1. Sastry, K. V. S., and Fuerstenau, D. W., *Ind. Eng. Chem. Fundam.* **9**, 145 (1970).
S2. Sastry, K. V. S., and Fuerstenau, D. W., *Trans. AIME* **250**, 64 (1971).
S3. Sastry, K. V. S., and Fuerstenau, D. W., *Proc. Inst. Briquetting Agglomeration* **12**, 113 (1971).
S4. Sastry, K. V. S., and Fuerstenau, D. W., *Trans. AIME* **252**, 254 (1972).
S5. Sastry, K. V. S., and Fuerstenau, D. W., *Powder Technol.* **7**, 97 (1973).
S6. Schubert, H., *Chem. Ing. Tech.* **45**, 396 (1973).
S7. Schubert, H., *Powder Technol.* **11**, 107 (1975).
S8. Schubert, H., Herrmann, W., and Rumpf, H., *Powder Technol.* **11**, 121 (1975).
S9. Sherrington, P. J., *Chem. Eng. (London)* **46**, CE201 (1968).
S10. Sherrington, P. J., *Can. J. Chem. Eng.* **47**, 308 (1969).
S11. Sirianni, A. F., Capes, C. E., and Puddington, I. E., *Can. J. Chem. Eng.* **47**, 166 (1969).
S12. Stirling, H. T., *in* "Agglomeration" (W. A. Knepper, ed.), p. 177. Wiley (Interscience), New York, 1962.
S13. Stoev, S. M., and Watson, D., *Inst. Min. Met. Trans.* **77**, C14 (1968).
S14. Stone, R. L., *Trans. AIME* **238**, 284 (1967).
S15. Stone, R. L., and Cahn, D. S., *Trans. AIME* **241**, 533 (1968).
T1. Tarján, G., *Aufbereit. Tech.* **7**, 28 (1966).
T2. Tigerschoid, M., *JISI* **117**, 13 (1954).
T3. Tigerschoid, M., and Ilmoni, P. A., *Proc. AIME Blast Furnace Coke Oven Raw Mater.* **9**, 18 (1950).
T4. Tonry, J. R., *in* "Agglomeration" (W. A. Knepper, ed.), p. 1. Wiley (Interscience), New York, 1962.
T5. Turner, G. A., and Balasubramanian, M., *Powder Technol.* **10**, 121 (1974).
V1. Violetta, D. C., and Nelson, J. C., AIMM & PE Preprint 66-B-71 (1966).

V2. Voyutski, S. S., Zaionchkovskii, A. D., and Rubina, S. I., *Kolloid Zh.* **14**, 28 (1952).
W1. Wada, M., and Tsuchiya, O., *Proc. Int. Min. Congr., 9th, Prague, 1970*, 23 (1970).
W2. Wada, M., Tsuchiya, O., and Okada, S., *Bull. Res. Inst. Min. Dressing Met. Tohoku Univ.* **22**, 109 (1966).
W3. Westen Starratt, F., *J. Metals* **8**, 1546 (1956).
Y1. Yusa, M., and Gaudin, A. M., *Am. Ceram. Soc. Bull.* **43**, 402 (1964).

Supplementary References

Section III. Bonding Mechanisms in Agglomerates

Adorjan, L. A., Theoretical prediction of strength of moist particulate materials, *in* "Agglomeration 77" (K. V. S. Sastry, ed.), p. 130. Am. Inst. Min. Metall. Pet. Eng. Trans., New York, 1977.

Ayers, P., Development of dry strength in pellets made with soluble salt binders, *Inst. Min. Metall., Trans.* **85**, C177 (1976).

Rumpf, H., Particle adhesion, *in* "Agglomeration 77" (K. V. S. Sastry, ed.), p. 97. Am. Inst. Min. Metall. Pet. Eng. Trans., New York, 1977.

Schubert, H., Tensile strength and capillary pressure of moist agglomerates, *in* "Agglomeration 77" (K. V. S. Sastry, ed.), p. 144. Am. Inst. Min. Metall. Pet. Eng. Trans., New York, 1977.

Section V. Kinetics of Balling and Granulation

Ouchiyama, N., and Tanaka, T., The probability of coalescence in granulation kinetics, *Ind. Eng. Chem., Process Des. Develop.* **14**, 286 (1975).

Ramabhadran, T. E., On the general theory of solid granulation, *Chem. Eng. Sci.* **30**, 1027 (1975).

Sastry, K. V. S., Similarity size distribution of agglomerates during their growth by coalescence in granulation or green pelletization, *Int. J. Miner. Process.* **2**, 187 (1975).

Sastry, K. V. S., and Fuerstenau, D. W., Kinetics of green pellet growth by the layering mechanism, *Trans. Am. Inst. Min. Eng.* **262**, 43 (1977).

Section VI. Bonding Liquid and Additives

Goksel, M. A., Fundamentals of cold bond agglomeration processes, *in* "Agglomeration 77" (K. V. S. Sastry, ed.), p. 877. Am. Inst. Min. Metall. Pet. Eng. Trans., New York, 1977.

Section VII. Granulation of Fertilizers

Hicks, G. C., McCamy, I. W., and Norton, M. M., Studies of fertilizer granulation at TVA, *in* "Agglomeration 77" (K. V. S. Sastry, ed.), p. 847. Am. Inst. Min. Metall. Pet. Eng. Trans., New York, 1977.

Kapur, P. C., Role of similarity size spectra in balling and granulation of coarse, liquid deficient powders, *in* "Agglomeration 77" (K. V. S. Sastry, ed.), p. 156. Am. Inst. Min. Metall. Pet. Eng. Trans., New York, 1977.

Section VIII. Miscellaneous Topics

Capes, C. E., McIlhinney, A. E., and Sirianni, A. F., Agglomeration from liquid suspension—research and application, *in* "Agglomeration 77" (K. V. S. Sastry, ed.), p. 910. Am. Inst. Min. Metall. Pet. TEng. Trans., New York, 1977.

Kawashima, Y., and Capes, C. E., Further studies of the kinetics of spherical agglomeration in a stirred vessel, *Powder Technol.* **13**, 279 (1976).

Puddington, I. E., and Sparks, B. D., Spherical agglomeration processes, *Miner. Sci. Eng.* **7**, 282 (1975).

PIPELINE NETWORK DESIGN AND SYNTHESIS

Richard S. H. Mah

Department of Chemical Engineering
Northwestern University
Evanston, Illinois

and

Mordechai Shacham

Ben Gurion University of the Negev
Beersheva, Israel

I. Introduction	126
II. Steady-State Pipeline Network Problems: Formulation	127
A. Description and Characterization of Flow Networks	127
B. Modeling of Network Elements	136
C. Alternative Problem Formulations	140
D. Problem Specifications	144
E. Comparison with Electrical Circuits	146
III. Steady-State Pipeline Network Problems: Methods of Solution	148
A. Numerical Methods	148
B. Techniques for Large Networks	160
C. Networks with Regulators and Other Nonlinear Elements	168
IV. Design Optimization and Synthesis	170
A. Sensitivity Analysis	173
B. Design Optimization	175
C. Synthesis	185
V. Transient and Compressible Flows in Pipeline Networks	190
A. Governing Equations	190
B. Methods of Solution	192
VI. Concluding Remarks	198
Appendix: Description of Test Problems	200
Nomenclature	203
References	205

I. Introduction

Pipeline networks constitute major bulk carriers for crude oil, natural gas, water, and petroleum products. Each day approximately 18 million barrels of oil and 70 billion cubic feet of natural gas travel through one form or another of pipeline networks from the source to the user. Within refineries and chemical complexes, process fluids are conveyed through piping networks from one unit to another as they undergo physical and chemical transformations. Networks of pipes and valves form an integral part of pressure-relieving and fire-water systems which are designed to handle contingencies in the operation of process units. More recently solids in the form of slurry are also being transported in pipelines. A coal-carrying pipeline has been operating in Arizona since 1970; another from the Rocky Mountain states to the Pacific Northwest has been proposed.

Although chemical engineers have long been acquainted with such research areas as two-phase flow, until recently relatively little has appeared in chemical engineering literature on the systems and computational aspects of pipeline network design and analysis. The purpose of this review is to bring to the attention of chemical engineers the similarity between this field and the design and synthesis of traditional processes and the opportunities for research and innovation in this area.

Some typical questions raised in pipeline network design and analysis are
  i. What is an optimal design for a network linking a given set of sources to a given set of consumers with certain design specifications?
  ii. How can an optimal network configuration be synthesized?
  iii. How can an existing pipeline network be modified to meet certain new specifications on pressures and flow rates?
  iv. How would the performance of a given pipeline system be affected by an unusually large and sudden demand in one section?
  v. How long would it take for internal pressures to go below certain critical levels in a distribution network in the event of supply failures and what is the best operating strategy in such an event?

Not all these questions can be completely and satisfactorily answered at present.

In this review the status of the relevant technology will be assessed with particular reference to formulation of problems and methods of solution. We shall, for the most part, be concerned with the technical literature of the last ten years. Since that period corresponds to the total eclipse of analog simulation which had been previously used, to some extent, in modeling pipeline networks (R3), we shall focus exclusively on digital computation methods. However, we shall not be content with a mere catalog of the different methods investigated in the pipeline network literature. Computational evaluations, some based on our own investigations, will be used to assess both the strengths and the limitations of the methods whenever possible. Some indications of economic incentives will also be given in the last section of this review.

Inasmuch as the nature of pipeline elements sets these networks apart from electrical networks (more commonly referred to as electrical circuits) we shall review briefly the modeling of these elements. We shall, however, limit ourselves to the correlations developed for single-phase fluid flow; the modeling of two-phase flow is a subject of sufficient diversity and complexity to merit a separate review.

The introduction of graph theory imparts to the analysis of pipeline network problems a unified viewpoint which is aesthetically very pleasing. However, our chief justification for its inclusion is its power to elucidate the analysis of complex problems, which often leads to highly efficient computational schemes. Although isolated applications of graph theory have appeared in pipeline network literature, the appreciation of the full potential of this useful branch of mathematics is not widespread. For this reason we shall begin with a brief review of those aspects of graph theory which, in our experience and judgment, have proved or are likely to be of greatest utility in pipeline network design and synthesis.

II. Steady-State Pipeline Network Problems: Formulation

A pipeline network is a collection of elements such as pipes, compressors, pumps, valves, regulators, heaters, tanks, and reservoirs interconnected in a specific way. The behavior of the network is governed by two factors: (i) the specific characteristics of the elements and (ii) how the elements are connected together. The first factor is determined by the physical laws and the second by the topology of the network.

A. DESCRIPTION AND CHARACTERIZATION OF FLOW NETWORKS

1. *Graphs and Digraphs*

The mathematical abstraction of the topology of a pipeline network is called a graph which consists of a set of *vertices* (sometimes also referred to as nodes, junctions, or points)

$$V = \{v_1, v_2, \ldots, v_N\} \quad (1)$$

together with a set of *edges* (sometimes also referred to as arcs, branches, or lines)

$$E = \{e_1, e_2, \ldots, e_P\} \qquad (2)$$

In this abstraction each edge corresponds to a pipeline network element and each vertex corresponds to a junction connecting two or more elements. It is often convenient to refer to the formal definition of a graph G as the sets

$$G = (V, E) \qquad (3)$$

Just as the junctions of a real pipeline network are mutually distinguishable by reason of differences in kind, in position in the network, and in location in the terrain spanned by the network, each vertex in the graphs that we shall consider in this discussion is distinguishable from the others. The vertices will be individually labeled, but the exact label assignment is immaterial as long as each vertex bears the same label throughout the discussion, and, in engineering applications, corresponds throughout to the same junction in the physical network. Such a graph is sometimes referred to as a *labeled graph*. Thus, in Fig. 1 the vertices of the graph are labeled with letters of the alphabet, and the edges with arabic numerals. Strictly speaking, the edge label is redundant, since it can always be uniquely identified by the associated vertices. For instance, an alternative label of edge **1** is $\{\mathbf{a}, \mathbf{b}\}$. More generally for edge e_i we may write

$$e_i = \{v_j, v_k\} \qquad (4)$$

The vertices v_j and v_k are said to be *incident with* the edge $\{v_j, v_k\}$ or *adjacent* to each other. It should be noted that the pictorial rendition of the set of vertices and edges of an actual network is not always unique. However, if there is a one-to-one correspondence between the vertices and between the edges of two graphs, their graph-theoretic properties will be identical and the two graphs are then said to be *isomorphic* to each other.

The formalism of graph theory lends itself to a number of very useful definitions. One useful concept is the *degree* $d(v)$ of a vertex v, which is defined as the number of edges with which the vertex is incident. Another is a

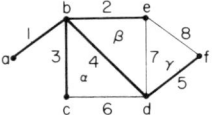

FIG. 1. A graph.

path of G from $\mathbf{v}_i$ to $\mathbf{v}_j$, which is a sequence of edges $\{\mathbf{v}_{i1}, \mathbf{v}_{i2}\}, \{\mathbf{v}_{i2}, \mathbf{v}_{i3}\}, \ldots,$ $\{\mathbf{v}_{i(k-1)}, \mathbf{v}_{ik}\}$, such that any two consecutive edges have a common vertex, $\mathbf{v}_i \equiv \mathbf{v}_{i1}$ and $\mathbf{v}_j \equiv \mathbf{v}_{ik}$. A path from $\mathbf{v}_i$ to $\mathbf{v}_j$ is a *cycle* if $\mathbf{v}_i \equiv \mathbf{v}_j$. A path (cycle) is *simple* (or elementary), if all the edges it contains are distinct[1] and no vertex is passed through more than once. Thus, in Fig. 1 the degrees of vertices, **a**, **b**, **c**, **d**, **e** and **f** are 1, 4, 2, 4, 3 and 2 respectively. One path from **b** to **e** is $\{\mathbf{b}, \mathbf{d}\}$, $\{\mathbf{d}, \mathbf{f}\}$ and $\{\mathbf{f}, \mathbf{e}\}$, another is $\{\mathbf{b}, \mathbf{e}\}$. The sequence $\{\mathbf{c}, \mathbf{b}\}, \{\mathbf{b}, \mathbf{d}\}, \{\mathbf{d}, \mathbf{e}\}, \{\mathbf{e}, \mathbf{b}\}, \{\mathbf{b}, \mathbf{c}\}$ is a cycle, but it is not a simple cycle, because the edge $\{\mathbf{c}, \mathbf{b}\}$ (alias $\{\mathbf{b}, \mathbf{c}\}$) is repeated.

A graph is *connected* if, for any two of its vertices $\mathbf{v}_i$ and $\mathbf{v}_j$, there is a path from $\mathbf{v}_i$ to $\mathbf{v}_j$. A connected graph which has no simple cycles is called a *tree*. A graph $G' = (V', E')$ is a *subgraph* of G if $V' \subseteq V$ and $E' \subseteq E$. Finally, a *spanning tree* of a graph G is a subgraph of G, which is a tree and has the same vertex set as G. For instance, the edges, $\{\mathbf{1}, \mathbf{2}, \mathbf{3}, \mathbf{4}, \mathbf{5}\}$, form a spanning tree in Fig. 1. As we shall see later, many of the interesting graph-theoretic properties of pipeline networks may be expressed in terms of the spanning trees of the graph.

In an *undirected* graph such as shown in Fig. 1, the order of the vertex pair is immaterial, i.e., $\{\mathbf{a}, \mathbf{b}\} \equiv \{\mathbf{b}, \mathbf{a}\}$. This type of representation is appropriate for a physical network of pipes. However, for most design applications we will be concerned with fluid flows whose directions are either known or convenient to assume. Such a structure may be represented by a directed graph (or *digraph*) whose edges are ordered pairs of vertices. We should like to point out that the direction selected for an edge is merely a convenience and need not correspond to the direction that the fluid flows in the corresponding pipe segment. Should the actual flow be in an opposite direction, the computed flow will carry a negative sign. We denote a directed edge from $\mathbf{v}_j$ to $\mathbf{v}_k$ by $\langle \mathbf{v}_j, \mathbf{v}_k \rangle$. In contrast to the undirected edge, $\langle \mathbf{v}_j, \mathbf{v}_k \rangle = -\langle \mathbf{v}_k, \mathbf{v}_j \rangle$. Fig. 2a and b shows a cyclic graph and a cyclic digraph.

Although most properties of a graph (e.g., degree of vertex, path, cycle) have counterparts in a digraph and vice versa, there are some notable exceptions (e.g., a tree). In this discussion we will not attempt to enumerate all the properties of graphs and digraphs, but merely highlight those properties that seem to us to be most useful and relevant. We shall be concerned mainly with digraphs, but the analysis sometimes requires us to examine certain properties of the underlying graphs. To avoid any confusion we shall carefully qualify the discussion of each property and hereafter use the term "graph" to mean "undirected graph."

---

[1] Self-loops (around the same vertex) and parallel edges (joining the same two vertices) will be excluded in our discussions.

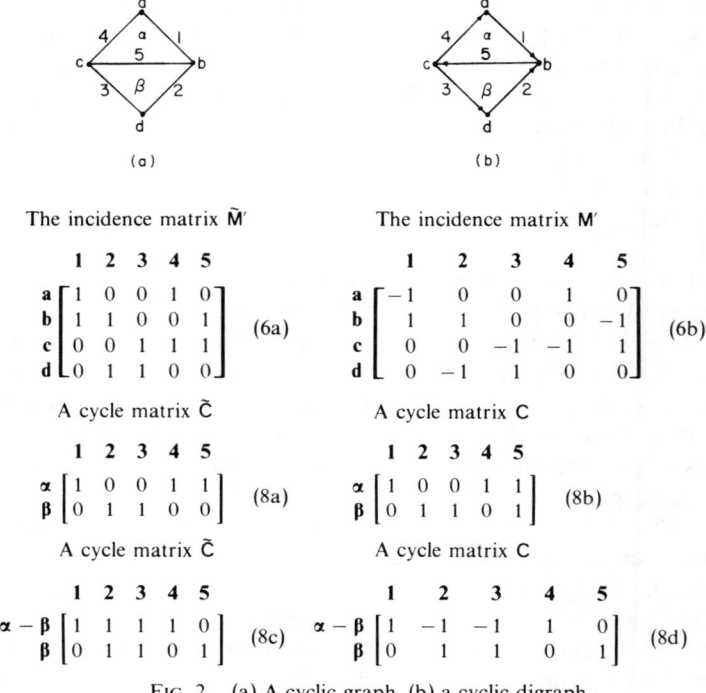

FIG. 2. (a) A cyclic graph, (b) a cyclic digraph.

2. Matrix Representation of Graphs and Digraphs

Although pictorial representation of graphs and digraphs is very convenient for visual analysis, for any but the simplest applications it is often necessary to enlist the aid of a digital computer. The algebraic representations of graphs and digraphs in terms of matrices containing only elements of 0, 1 and −1 lend themselves to direct computer processing. But in large applications a further mapping into a suitable data structure often precedes the actual computer implementation to take full advantage of the structure and sparsity of these matrices.

To have a better appreciation of the utility of these representations let us first consider the laws that govern flow rates and pressure drops in a pipeline network. These are the counterparts to Kirchoff's laws for electrical circuits, namely, (i) the algebraic sum of flows at each vertex must be zero; (ii) the algebraic sum of pressure drops around any cyclic path must be zero. For a connected network with N vertices and P edges there will be $(N - 1)$ independent equations corresponding to the first law (Kirchoff's current law) and C independent equations corresponding to the second law (Kirchoff's voltage law), where C, the *cyclomatic number* is given by

$$C = P - (N - 1) \tag{5}$$

The important point to note is that these laws govern the collective behavior of the network over and above the physical laws which govern the behavior of each individual network element.

The two types of governing equations can be readily expressed in terms of incidence and cycle matrices of a digraph, respectively. The incidence matrix **M′** is an $N \times P$ matrix in which each edge is represented by a column and each vertex by a row. The element m_{ij} is assigned a value of -1, if vertex **i** is the emitting vertex of edge **j**, a value of $+1$, if vertex **i** is the receiving vertex of edge **j**, and a value of 0, if edge **j** is not an incident edge of vertex **i**. Thus, with reference to the cyclic digraph, depicted in Fig. 2b, the incidence matrix **M′** is given by Eq. (6b). It should be noted that each column of **M′** contains exactly two nonzero entries, a $+1$ and a -1 and because of this special structure, the rank of an N-row incidence matrix **M′** is $(N - 1)$. Physically, this property of **M′** corresponds to the fact that there are only $(N - 1)$ independent balances around the vertices in addition to the overall balance in an N-vertex cyclic network. If we omit one of the rows of **M′** to form a reduced incidence matrix **M** of the same rank, the conservation equations for an isolated cyclic network may be restated as

$$\mathbf{Mq} = 0 \tag{7}$$

where **q** is the column vector of stream flow rates. We should like to emphasize that the reduced incidence matrix **M** refers to a digraph and that a flow rate will have a positive numerical value if the flow is in the direction assigned to an edge, and a negative numerical value if the flow is in the opposite direction. Strictly speaking, the conservation equations should be applied only to mass flow rates. However, for a liquid or a compressible fluid undergoing modest changes of pressure, the density is either constant or nearly so. In such cases for all practical purposes we may apply conservation equations to volumetric flow rates as well. Since these constitute the majority of cases of interest to us, we shall conduct our discussion in terms of volumetric flow rates for the sake of brevity, with the understanding that they should be replaced by the corresponding mass flow rates when large pressure changes in a compressible flow are being considered.

Similarly, in a cycle (or circuit) matrix **C** the columns and rows represent the edges and cycles respectively. A link in cycle **i** in the same direction as edge **j** is denoted by $+1$, a -1 would signify that the link and the edge are in opposite directions and a 0 would be an indication that edge **j** is not in the path of cycle **i**. It should be noted that the cycle matrix of a digraph is clearly not unique, even though we restrict the representation to simple cycles only. However, for any complete set of independent cycles, the cycle matrix **C** is of rank C. We shall have more to say about finding independent cycles after the concept of a spanning tree is introduced. A cycle matrix **C** corresponding to the digraph in Fig. 2b is given by Eq. (8b). Another, formed by subtracting the second row from the first, is given by Eq. (8d). Other cycles may be similarly formed from linear combinations of rows, but they may not be simple cycles. In terms of a cycle matrix the equations corresponding to Kirchoff's voltage law may now be restated as

$$\mathbf{C\sigma} = 0 \tag{9}$$

where σ is the vector of pressure drops along each of the edges. It can easily be shown that the incidence matrix **M** and the cycle matrix **C** span two orthogonal and complementary subspaces, that is

$$\mathbf{MC}^T = 0 \tag{10}$$

and

$$\mathbf{CM}^T = 0 \tag{11}$$

where the transpose of a matrix is indicated by the superscript T. Although these two equations may be derived from Kirchoff's laws, they are in fact direct consequences of the properties of the graph. Their validity may be established irrespective of any physical laws.

In the foregoing discussion the properties of the incidence matrix and the cycle matrix were illustrated in terms of a cyclic digraph, but the results on the ranks of these matrices actually hold true for any connected digraph with N vertices. For an undirected graph, $\tilde{\mathbf{M}}$ and $\tilde{\mathbf{C}}$ contain only 0 and 1 (sometimes referred to as binary matrices), mathematical relations of identical form are obtained except that modulo 2 arithmetic[2] is used instead of ordinary arithmetic. The ranks of $\tilde{\mathbf{M}}$ and $\tilde{\mathbf{C}}$ defined in terms of modulo 2 arithmetic are $N - 1$ and C, as before, and Eqs. (10) and (11) are modified to read

$$\tilde{\mathbf{M}}\tilde{\mathbf{C}}^T \cong 0 \;(\text{mod } 2) \quad \text{and} \quad \tilde{\mathbf{C}}\tilde{\mathbf{M}}^T \cong 0 \;(\text{mod } 2) \tag{12}$$

---

[2] Two integers are said to be congruent "modulo 2" if they differ only by an integral multiple of 2. In modular arithmetic we disregard this integral multiple. Hence, $0 \cong 2$ (mod 2) $\cong 4$ (mod 2), for example. Such an algebraic system is referred to in mathematical literature as Galois field modulo 2. The notation $\mathbf{AB} = \mathbf{C}$ (mod 2) indicates that the matrix multiplication is to be carried out using modulo 2 arithmetic i.e. each element $c_{ij} = \sum_k a_{ik} b_{kj}$ (mod 2). A more extended discussion of this subject may be found in Deo (D4, pp. 118–120).

Equation (6a) shows the incidence matrix $\tilde{M}'$ corresponding to the graph in Fig. 2a, and Eqs. (8a) and (8c) show two cycle matrices corresponding to the same graph.

Now the equations derived from Kirchoff's first law are essentially material balances around each of $(N - 1)$ vertices. As an alternative, balances could also be drawn up around groups of such vertices. Is there a special way of grouping the vertices, which will yield a particularly advantageous formulation? Also, as we have noted, the selection of cycles is not unique, but the cycles must be independent. How can we generate an independent set of cycles? Are some of these independent sets more fundamental than others? If so, how many fundamental sets are there? To answer these questions we must explore further the properties of a graph.

For a graph with n vertices any two of the following three conditions will define a tree: (i) it is connected, (ii) it is acyclic, and (iii) the number of edges is $(n - 1)$. For an N-vertex graph G a spanning tree is a tree for which $n = N$. Thus for the graph in Fig. 1, $T_1 = \{1, 2, 3, 4, 5\}$ is a spanning tree, so is $T_2 = \{1, 3, 6, 7, 8\}$. The total number of spanning trees, according to Binet-Cauchy theorem (D4, pp. 218–219) is given by $\det(MM^T)$, which increases rapidly with the size of the graph. For the graph in Fig. 1, it works out to be 21.

The edges in a spanning tree are called *tree branches* or *branches*. All other edges of G are called chords. Thus, with reference to T_1, the chords are $\{6, 7, 8\}$. Because there is one and only one path between any two vertices of T_1, the addition of any chord to T_1 will create exactly one cycle. Such a cycle is called a *fundamental cycle*. It follows that there are as many fundamental cycles as there are chords $(P - N + 1 = C)$. Thus for the graph in Fig. 1 the fundamental cycles are $\{3, 4, 6\}$, $\{2, 4, 7\}$, and $\{2, 4, 5, 8\}$. Notice that the fundamental cycles are defined only with respect to a given spanning tree. If more than one chord is added to T_1 at the same time, cycles which are not fundamental cycles will also be created. For instance, simultaneous addition of chords 7 and 8, will create not only the last two fundamental cycles but also $\{5, 7, 8\}$ which is not a fundamental cycle. Since each chord occurs only once in a set of fundamental cycles, it should be evident that the rows of a cycle matrix corresponding to the fundamental cycles will be linearly independent and the rank of the cycle matrix will be $(P - N + 1)$. Such a matrix will be referred to as a fundamental cycle matrix.

Since the permutation of rows and columns is immaterial, we can always arrange the columns so that the first $(N - 1)$ correspond to the tree branches and the last C correspond to the chords. Hence a fundamental cycle matrix $\tilde{\Gamma}$ can always be written as

$$\tilde{\Gamma} = [\tilde{T} \vdots I_C] \tag{13}$$

		Branches					Chords		
(a)	Cycles	1	2	3	4	5	6	7	8
	α	0	0	1	1	0	1	0	0
	β	0	1	0	1	0	0	1	0
	β + γ	0	1	0	1	1	0	0	1

		Branches					Chords		
(b)	V_A or V_B	1	2	3	4	5	6	7	8
	a or {b, c, d, e, f}	1	0	0	0	0	0	0	0
	e or {a, b, c, d, f}	0	1	0	0	0	0	1	1
	c or {a, b, d, e, f}	0	0	1	0	0	1	0	0
	{d, f} or {a, b, c, e}	0	0	0	1	0	1	1	1
	f or {a, b, c, d, e}	0	0	0	0	1	0	0	1

FIG. 3. (a) Fundamental circuit matrix, (b) fundamental cut-set matrix.

where I_C is the identity matrix of order C and $\tilde{T}$ is a $C \times (N-1)$ binary matrix corresponding to the tree branches. Fig. 3a shows such a matrix with respect to the graph and the tree (indicated by heavy lines) in Fig. 1.

We shall now turn to the other question which we have posed earlier on the grouping of vertices. To facilitate the discussion let us introduce three additional concepts and definitions of graph theory. A connected subgraph of a graph G, which is maximal (i.e., it is not a subgraph of any other subgraph of G) is called a *component* of G. Thus, the component of a connected graph is the graph itself. An edge of a graph G is a *bridge*, if the graph resulting from the deletion of that edge contains more components than G. Finally, in a connected graph G, a *cut-set* is a minimal set of edges whose deletion from G separates some vertices from others (resulting in an increase in the number of components). Thus, in Fig. 1, the edge **1** is both a bridge and a cut-set, but none of the other edges is a bridge. The set of edges, {**2, 3, 4**} which separate vertices {**a, b**} from vertices {**c, d, e, f**} is a cut-set, and so is {**2, 4, 6**}. But {**2, 3, 4, 6**} is not a cut-set because it is not minimal.

It follows that every branch in a tree is a bridge. Its removal creates two components containing subsets of vertices, V_A and V_B. For instance, with respect to the tree, $T_1 = \{1, 2, 3, 4, 5\}$ in Fig. 1 the removal of branch **1** creates two components containing vertex subsets, $V_A = \{a\}$ and $V_B = \{b, c, d, e, f\}$. Similarly, the removal of branch **4** creates two components containing vertex subsets, $V_A = \{a, b, c, e\}$ and $V_B = \{d, f\}$. Now consider the original graph G. The two components containing V_A and V_B are linked by a unique cut which contains one and only one tree branch together with possibly some chords. There are $(N-1)$ such cuts corresponding to the $(N-1)$ tree branches. If we now construct an incidence matrix based on edges incident with each of the $(N-1)$ vertex subsets, V_A (or V_B) then we obtain the so-called cut-set matrix $\tilde{K}$,

$$\tilde{K} = [I_{N-1} \vdots \tilde{B}] \qquad (14)$$

where I_{N-1} is the identity matrix of order $(N-1)$ and $\tilde{B}$ is an $(N-1) \times C$ binary matrix corresponding to the chords. As before, we have so arranged the columns that the first $(N-1)$ correspond to the tree branches and the last C correspond to the chords. Figure 3b shows the cut-set matrix corresponding to the graph G and the tree T_1 in Fig. 1.

Just as the incidence matrix $\tilde{M}$ is related to material balances around each of $(N-1)$ vertices, the cut-set matrix $\tilde{K}$ is clearly related to material balances around the $(N-1)$ vertex subsets V_A (or V_B, since material conservation holds over the whole network). In this case the $(N-1)$ vertex subsets happen to be $\{a\}$, $\{e\}$, $\{c\}$, $\{d, f\}$, and $\{f\}$.

Now clearly the rows of $\tilde{K}$ are linear combinations of rows of $\tilde{M}$, and since they are obviously linearly independent, the row spaces of $\tilde{K}$ and $\tilde{M}$ must be the same. Hence, from Eq. (12) we have

$$\tilde{K}\tilde{\Gamma}^T \cong 0 \text{ (mod 2)} \quad \text{and} \quad \tilde{\Gamma}\tilde{K}^T \cong 0 \text{ (mod 2)} \qquad (15)$$

and

$$\tilde{B} \cong \tilde{T}^T \qquad (16)$$

The reader can easily verify this important result by noting that the submatrix containing the last three columns of Fig. 3b is the same as the transpose of the submatrix containing the first five columns of Fig. 3a. For a digraph the fundamental cycle matrix Γ and the cut-set matrix K may be similarly constructed based on a spanning tree of the underlying graph. Only in this case the defining equations, (13) and (14), may contain -1 entries in submatrices T and B. The corresponding relation to Eq. (16) now becomes

$$B = -T^T \qquad (17)$$

Equations (16) and (17) show the important link between fundamental cycles and cut-sets of a graph. Thus, a spanning tree provides a convenient starting point for formulating a consistent set of governing equations for steady-state pipeline network problems.

More extended treatment of graph theory is to be found in the books by Berge (B6), Deo (D4), Harary (H2), and Ore (O2).

A number of papers (C4, D2, E2, M1) concerning pipeline networks have reported the use of spanning trees in generating independent cycles or in ordering equations and variables. Mah (M1) presented an algorithm for generating spanning trees. Spanning trees also occur naturally in an important class of pipeline network problems (see Sections IV,B and C). Lam and Wolla (L1) implemented an input processor which made use of some properties of the graph. Enger and Feng (E1) discussed flowgraph analysis of pipeline networks. Epp and Fowler (E2) and Mah (M1) investigated the selection of a best set of fundamental cycles and devised special algorithms for this purpose. But the relationship between fundamental cycles and cutsets, although well-known to graph theorists, does not appear to have been exploited in pipeline literature.

In this brief review we attempted to bring together the key elements of graph theory which have a direct bearing on the analysis of pipeline network problems. Although much progress on pipeline network problems has undoubtedly been made without the benefit of graph theory, its value in bringing together many isolated results in diverse applications can hardly be overrated. The introduction of graph theory into the domain of pipeline network problems is a significant recent development in this field.

B. Modeling of Network Elements

We shall now turn to the second factor in a pipeline network: the network element. In particular we would like to focus on the mathematical modeling of different types of network elements. As we have pointed out earlier, there are many types of pipeline network elements. But viewing each element as an edge $\{i, j\}$ in a graph, the two edge variables of interest to us are (i) the flow rate through the element, q_{ij}, and (ii) the pressure drop across the element, $\sigma_{ij} = (p_i - p_j)$. They are sometimes referred to as the through variable and the cross variable associated with the edge, $\{i, j\}$. In addition, we also have the variables associated with the physical characteristics of the element, such as the diameter d, the length l, and the roughness factor ϵ of a pipe, and the variables associated with the properties of the fluid, such as density ρ and viscosity μ. All these variables are interrelated through the model of the element.

1. Pipe Sections

For liquids flowing in pipes the pressure drop is commonly taken proportional to a power of the flow rate, usually around 2. One of the simplest correlations used in water distribution network calculation is the Hazen–Williams formula,[3]

$$p_i - p_j = \alpha_1 (l/d^{4.87})(q_{ij}/\epsilon_1)^{1.852} \qquad (18)$$

[3] Boldface parentheses will be used to demarcate a factor in multiplication, e.g. $g_c(p_i - p_j)$, as opposed to functional dependency, e.g., $\sigma_k(I_k)$.

where ϵ_1 is a dimensionless pipe roughness coefficient, and $\alpha_1 = 0.937$, if the pressures, p_i and p_j are expressed in lbs/in.^{2}, the flow rate q_{ij} in gallons per minute, the pipe diameter d in inches, and the pipe length l in feet. Typical values of ϵ_1 are given by Rosenhan (R4). The properties of water are implicitly incorporated in this correlation, which is widely used in water distribution (C4, C10, D3, E2, L1, L2, L5, L8, N1, N2, R4, S2, T2, W2).

Another correlation used for water flowing in pipes is Manning's formula (B3, C4, E2, F1),

$$p_i - p_j = \alpha_2 \epsilon_2{}^2 l q_{ij}^2 / d^{16/3} \tag{19}$$

where α_2 is a constant dependent upon the units used and ϵ_2 is Manning's roughness coefficient. Numerical values of these coefficients are given by Bauer et al. (B3) for various units and different types of flow channels.

A correlation which takes explicit account of fluid properties is the equation (F9),

$$p_i - p_j = 8 f l \rho q_{ij}^2 / g_c \pi^2 d^5 \tag{20}$$

in which Fanning summarized the experimental and analytical results of Chezy, Darcy, Weisbach, Prony, and other earlier investigators after subtracting pressure losses caused by entrance and exit effects, expansions in the passage, and the like. In this equation, f is a dimensionless friction factor, g_c is the gravitational conversion factor and d, l, p, q, ρ are expressed in consistent units. For laminar flow (Reynolds number, Re < 2000)

$$f = 64/\text{Re} \tag{21}$$

and the substitution of Eq. (21) in Eq. (20) yields Poiseuille's equation. For turbulent flow (Re > 3000), the friction factor may be calculated from the Moody correlation

$$1/f = \{0.86859 \ln[0.27\epsilon_3/d + 2.51/(\text{Re}\sqrt{f})]\}^2 \tag{22}$$

where ϵ_3 is a pipe roughness factor. Between these limits the values of f are sometimes interpolated (B5, I1). Various revisions of this correlation under the names of Darcy–Weisbach, Colebrook–White, and Moody have been used by many investigators (C3, C4, C10, J2, N1, S7, Y2).

Because Eq. (22) is implicit in f, an iterative method of solution must be employed. However, the convergence of this iteration usually presents no difficulty (B5, D2). Bending and Hutchison (B5) noted that in pipeline network calculations it was not necessary to calculate exact friction factors for each overall network iteration. In their experience the problem could be solved satisfactorily with single updating of factors for each overall iteration. This truncation results in significant reduction of computing time. In all three aforementioned formulas [Eqs. (18)–(20)], it is assumed that $q_{ij} \geq 0$. If this is not true, then the direction of pressure drop will be reversed. In our subsequent discussions it will be tacitly assumed that $p_i \geq p_j$ and $q_{ij} \geq 0$, unless it is explicitly stated otherwise.

Equation (20) is sometimes used for calculations involving short runs of gas piping or when pressure drop is not a significant fraction (< 0.1) of the total pressure. A more accurate correlation for gases is given by the Weymouth formula (A2, K3),

$$q_{ij} = 433.45 d^{2.667} \left(\frac{\theta^\circ}{p^\circ}\right)\left(\frac{p_i^2 - p_j^2}{l\rho\theta}\right)^{1/2} \quad (23)$$

where θ is the temperature of the gas in °R, ρ is its specific gravity with respect to air, and the flow rate q_{ij} in ft.^{3}/day is measured at the reference temperature θ° (°R) and absolute pressure p° (lbs/in.^{2}). According to the GPSA "Engineering Data Handbook" (A2), the result calculated by the "Weymouth formula agrees more closely with metered rate than those calculated by any other formula." It is recommended for short pipelines and gathering systems and tends to give a conservative design.

For long distance gas transmission and large diameter pipelines, various forms of the "Panhandle formula" are widely used. One variation, the modified Panhandle formula, is given by

$$q_{ij} = 737 \eta d^{2.530} (\theta^\circ/p^\circ)^{1.02} \left[\frac{p_i^2(1 + \tfrac{2}{3}Yp_i) - p_j^2(1 + \tfrac{2}{3}Yp_j)}{\rho^{0.961} l\theta}\right]^{0.510} \quad (24)$$

where the pipeline efficiency η varies from 0.94 for clean "pipeline quality" gas in new pipe to 0.88 for "rich gas in old pipe," and Y is related to the compressibility factor by the following equation:

$$Yp = 1/Z - 1 \quad (25)$$

Various Panhandle equations have been used in natural gas transmission applications (F4, R5, S5, Z1).

Of the five correlations which we have reviewed, the first three are explicit in pressure drop, $(p_i - p_j)$. Of these Eqs. (18) and (19) can easily be rewritten explicitly for q_{ij}. On the other hand, the last two correlations, Eqs. (23) and (24) are explicit only in q_{ij}. This distinction becomes important, when we come to selection of alternative problem formulation.

2. Valves and Regulators

Valves and regulators are used in pipeline networks to perform a variety of functions. Isolating valves are used to interrupt flows and to shut off sections of a network. Gate, plug, ball, and butterfly valves are most commonly used for this purpose. Check valves may be used to prevent back flows in certain parts of a network. These valves are kept open by the flowing fluid. They close either by gravity or flow reversal. "Lift checks" and "swing checks" are the two major types of check valves in common usage. Pressure relief valves are used to prevent equipment damage or failure caused by excessive pressure. A common device of this type is the rupture disk which ruptures when a predetermined pressure limit is exceeded.

All three types of valves described above are "on–off" type devices (binary devices). A basic requirement in the design of these valves is that they offer minimum resistance to flow when open. For many types of calculations it is often justifiable to neglect pressure losses through such devices. But such a simplification cannot be applied to flow control valves. With these devices the control of flow is accomplished either by a constriction or by a diversion. In either case an additional resistance to the flow is introduced. Globe, angle, cross, and needle valves are typical devices of this type. Specific pressure drop correlations should be developed for such devices and used whenever possible. One such correlation used by Stoner (S5) takes the form

$$q_{ij} = \alpha_{ij}[(p_i - p_j)p_j]^{1/2} \qquad (26)$$

where α_{ij} is a resistance factor. This equation is apparently a simplification of a correlation given in Perry and Chilton (P1)

$$q_{ij}\rho_i = \alpha\zeta a[2g_c(p_i - p_j)\rho_i]^{1/2} \qquad (27)$$

The physical devices used to regulate pressure are called regulators. There are two types: the downstream regulators (or pressure reducing valves) and the upstream regulators (or pressure retaining valves). The idealized downstream regulator may be modeled by the following equation (D8)

$$p_j = \min\{p_i, p_s\} \qquad (28)$$

where p_s is the regulator set-point pressure. The valve is closed, when $p_j > p_i$. For the idealized upstream regulator, we have

$$p_i = \max\{p_j, p_s\} \qquad (29)$$

and the valve is closed, when $p_i < p_s$.

Regulators have interesting implications for pipeline network calculations, which will be explored in Section III,C.

3. *Other Network Elements*

Pumps and compressors are also commonly encountered in pipeline network calculations. For simulation purposes, the common practice (D8, E2, J1, J2) is to fit the performance curve with a polynomial equation,

$$p_i - p_j = \alpha_0 + \alpha_1 q_{ij} + \alpha_2 q_{ij}^2 + \cdots \qquad (30)$$

where the coefficients α_0, α_1, α_2, ... are regression parameters. The parameter α_0 represents the maximum pressure difference (head) developed by the pump. Another form of the pump correlation given by Alexander *et al.* (A1) is

$$p_i - p_j = \alpha_0 - \alpha_1 q_{ij}^{\alpha_2} \qquad (31)$$

For compressors the more commonly used form (B7, F4, S5, W12) is

$$q_{ij} = \frac{hp}{\alpha_0 (p_j/p_i)^{\alpha_1} - \alpha_2} \qquad (32)$$

where hp is the compressor horsepower. Pumps and compressors are commonly equipped with check valves to permit flows in one direction only.

Although reservoirs and tanks are not network elements in the sense discussed above, they do enter into pipeline network calculations. For many types of calculations, the impact on the network behavior may be modeled by treating a reservoir or a tank as a constant pressure vertex. On the other hand, the storage field deliverability curve (S5) is sometimes represented by

$$q_{ij} = \alpha_0 (p_i^2 - p_j^2)^{\alpha_1}, \qquad p_i > p_j \qquad (33)$$

In the foregoing discussion on network elements, we have not accounted for the elevation difference between the two vertices adjoining the network element. This difference, although negligible in many applications, is important in the transportation of liquids as, for example, in water distribution networks. In these applications a term $\rho g(z_j - z_i)$ should be added to the pressure drop equation.

C. Alternative Problem Formulations

We are now in a position to formulate the steady-state pipeline network problem based on the laws governing the behavior of the network and its elements. As it turns out, there is more than one way of formulating the problem, and since the computational efforts required for the solution are unequal, it behooves us to examine the ramifications of these formulations.

It should be clear from Eq. (7) that since mass conservation applies to the flow network as a whole, there can be no net inflow or outflow of material. This requirement is obviously met for an isolated network. However, most applications of practical interest will include external inputs and outputs, representing feed sources and delivery sinks. To conform to the requirement of Eq. (7) the digraph for such an open network must include an "environment vertex" (M10) from which all inputs are derived and to which all outputs are returned. The underlying graph of a nontrivial flow network is therefore always cyclic.

For computational purposes it is useful to distinguish the directed edges associated with the external inputs and outputs, which are usually specified, from those associated with the flows internal to the network. Denoting the net output from vertex **i** by w_i and including henceforth only the internal edges in the incidence matrix $\hat{M}$, we may restate Eq. (7) as

$$\hat{M}\hat{q} = w \qquad (34)$$

or

$$\sum_{j \in V_{Ai}} q_{ji} - \sum_{k \in V_{Bi}} q_{ik} = w_i, \qquad i = 1, 2, \ldots, N-1 \qquad (35)$$

where V_{Ai} is the subset of vertices associated with the incident edges directed toward vertex **i** and V_{Bi} is the subset of vertices associated with incident edges directed away from vertex **i**.

For each of the P network elements we also have an equation of the form,

$$p_i - p_j = \sigma_{ij}(q_{ij}), \qquad \langle \mathbf{i}, \mathbf{j} \rangle \in E \qquad (36)$$

or more concisely, if edge **k** denotes $\langle \mathbf{i}, \mathbf{j} \rangle$,

$$p_i - p_j = \sigma_k(q_k), \qquad \mathbf{k} \in E \qquad (37)$$

If all external flows, w_i and the pressure at one vertex are specified, then Eqs. (35) and (37) constitute a set of $(P + N - 1)$ equations for the $(P + N - 1)$ variables, p_i and q_k. Notice that by including p_i explicitly, Kirchoff's second law is automatically satisfied. We shall refer to this as formulation A.

Now let C_i denote a fundamental cycle. Then substituting Eq. (37) in Eq. (9), we may restate Kirchoff's second law as

$$\sum_{k \in C_i} \sigma_k(q_k) = 0, \qquad i = 1, 2, \ldots, C \qquad (38)$$

By eliminating p_i, we now have the P flow rates, q_k, in $N - 1 + C(=P)$ equations [Eqs. (35) and (38)]. We shall refer to this as formulation B.

If the network element equations [Eq. (36)] are explicit in flow rates, we may similarly substitute Eq. (36) in Eq. (35) and obtain a set of $(N-1)$ equations in the $(N-1)$ unknown pressures, p_i,

$$\sum_{j \in V_{Ai}} q_{ji}(p_j, p_i) - \sum_{k \in V_{Bi}} q_{ik}(p_i, p_k) = w_i, \qquad i = 1, 2, \ldots, N-1 \qquad (39)$$

We shall call this formulation C. Notice that in this formulation the conservation equation around each vertex is expressed in terms of the pressures at the adjacent vertices, V_A and V_B. The structure of these equations is related to that of the underlying graph in an interesting manner.

Let us construct a binary matrix A whose rows correspond to the equations and whose columns correspond to the variables. Let the element a_{ij} be 1 if variable j occurs in equation i, and let it be zero otherwise. Such a matrix is called an occurrence matrix. For the special case of Eq. (39) the occurrence matrix is symmetric. It reflects the structure of the underlying graph, since $a_{ij} = 1$, if and only if the graph contains an edge $\{i, j\}$. If we now introduce the notation # to denote the operation which assigns a value of one to a variable if its numerical value is nonzero, and a value of zero if otherwise, that is to say, for any variable x

$$x^\# = \begin{cases} 0, & \text{if } x = 0 \\ 1, & \text{if } x \neq 0 \end{cases} \tag{40}$$

then the occurrence matrix A is related to the incidence matrix by the following equation:

$$\mathsf{A} = (\mathsf{M}'(\mathsf{M}')^T)^\# \tag{41}$$

For the example shown in Fig. 2b and Eq. (6b),

$$\mathsf{A} = \left(\begin{bmatrix} -1 & 0 & 0 & 1 & 0 \\ 1 & 1 & 0 & 0 & -1 \\ 0 & 0 & -1 & -1 & 1 \\ 0 & -1 & 1 & 0 & 0 \end{bmatrix} \begin{bmatrix} -1 & 1 & 0 & 0 \\ 0 & 1 & 0 & -1 \\ 0 & 0 & -1 & 1 \\ 1 & 0 & -1 & 0 \\ 0 & -1 & 1 & 0 \end{bmatrix} \right)^\#$$

$$= \begin{bmatrix} 2 & -1 & -1 & 0 \\ -1 & 3 & -1 & -1 \\ -1 & -1 & 3 & -1 \\ 0 & -1 & -1 & 2 \end{bmatrix}^\# = \begin{bmatrix} 1 & 1 & 1 & 0 \\ 1 & 1 & 1 & 1 \\ 1 & 1 & 1 & 1 \\ 0 & 1 & 1 & 1 \end{bmatrix}$$

Now the matrix $\hat{\mathsf{M}}$ in Eq. (34) possesses all the attributes of an incidence matrix except that the underlying graph need not be cyclic. By a process analogous to the development of Eqs. (14) and (16), we can construct an analogous cut-set matrix $\hat{\mathsf{K}}$ and $\hat{\mathsf{T}}$ with respect to a spanning tree of this underlying graph. Now let the flow rates associated with the tree arcs and with the chords be denoted by q_T and q_C respectively. Then corresponding to Eq. (34) we have

$$\hat{\mathsf{K}}\hat{\mathsf{q}} = \mathsf{I}_{N-1}q_T - \hat{\mathsf{T}}^T q_C = \mathsf{w} \tag{42}$$

where w_i now represents the net output from the vertex subset V_{Ai}. Hence,

$$q_T = w + \hat{T}^T q_C \tag{43}$$

or

$$\hat{q} = \begin{pmatrix} q_T \\ \hline q_C \end{pmatrix} = \begin{pmatrix} w \\ \hline 0 \end{pmatrix} + \begin{pmatrix} \hat{T}^T \\ \hline I_C \end{pmatrix} q_C = w' + \hat{\Gamma}^T q_C \tag{44}$$

Equation (44) shows that the flows in a network can always be expressed in terms of the flows in the chords. Since each chord corresponds to a fundamental cycle, q_C is the vector of "mesh flows."

If we now substitute Eq. (44) in Eq. (38) we obtain a set of C equations in C variables, q_C. For nonzero input/output vector, w' the resultant equation set is not homogeneous in q_C and the flows in the chords and the network can be determined uniquely. We shall term this new set of equations formulation D. The formulations used in different literature references are summarized in Table I along with their characteristics.

TABLE I

FORMULATION OF STEADY-STATE PIPELINE NETWORK PROBLEMS

Formulation	Equations	Variables	Dimension	Reference
A	(35), (37)	Pressures and flow rates	$P + N - 1$	Carnahan and Christensen (C3). Bending and Hutchison (B5)
B	(35), (38)	Flow rates	P	Cross (C11), Wood and Charles (W11), Mah (M1), Williams (W7), Jeppson and Tavallaee (J2)
C	(39)	Pressures or heads	$N - 1$	Warga (W1), Shamir and Howard (S2), Liu (L8), Gay and Middleton (G1), Lam and Wolla (L2), Lemieux (L5), Carnahan and Wilkes (C2), Donachie (D8), Collins and Johnson (C10)
D	(38) with substitution of (44)	Mesh flows	C	Epp and Fowler (E2), Gay and Middleton (G1), Gay and Preece (G2)

The four alternative formulations which we have just discussed involve equation sets of different dimensions. For a connected graph, $P \geq N - 1$. The equality applies when it is a tree, but for most pipeline networks, $P > N - 1$. On the other hand the number of independent cycles $C (= P - N + 1)$ is usually much less than N. Hence the four formulations are ranked roughly in the order of decreasing dimensions. For large networks involving 1000 vertices or more (B10), the difference in the computing efforts required using different formulations can be quite significant. It is well known that in matrix inversion, the storage requirement increases in proportion to the square of the matrix dimension and the computing time increases in proportion to the cube of the matrix dimension. Why then would one not use formulation D for every problem?

The answer to this question involves several factors. The most obvious observation is that not all the equations are linear. And all nonlinear equations are not equally difficult to solve. Equations (35) in formulations A and B are linear. On the other hand, the cycle equations in formulations C and D are almost always nonlinear, although the symmetry in these formulations is clearly an advantage. As a rule, formulations A and B require more computation per iteration but fewer iterations to converge than formulations C and D.

The second factor is the form of Eq. (36). If it is not explicit in flow rate or pressure, then either formulation C or formulations B and D are infeasible. In this connection we note that since Eqs. (18) and (19) are explicit in both variables, the simulation of water distribution networks using these equations presents no difficulties in this respect.

The characteristics of the network also enter into the considerations in selecting the problem formulation. If the network is acyclic, are the formulations involving cycle equations still appropriate? Epp and Fowler (E2) suggested that if the pressures at two connected vertices are given, a cycle equation can always be written for a "pseudo-loop" which contains a fictitious edge linking these two vertices. In this way formulation D may be modified to accommodate networks that are not completely cyclic. However, for other types of specifications, formulations A and C are clearly more appropriate. The choice of a formulation is thus closely intertwined with the nature of problem specification which is the next topic of our discussion.

D. Problem Specifications

1. *Admissible Specification Sets*

In Section II,C we have deliberately chosen a simple set of problem specifications for our steady-state pipeline network formulation. The specification of the pressure at one vertex and a consistent set of inputs and outputs (satisfying the overall material balance) to the network seems intuitively reasonable. However, such a choice may not correspond to the engineering requirements in many applications. For instance, in analyzing an existing network we may wish to determine certain input and output flow rates from a knowledge of pressure distribution in the network, or to compute the parameters in the network element models on the basis of flow and pressure measurements. Clearly, the specified and the unknown variables will be different in these cases. For any pipeline network how many variables must be specified? And what constitutes an admissible set of specifications in

the sense that all flows and pressures in the network are uniquely defined?
 These questions were addressed by Shamir and Howard (S2) who gave a useful empirical rule: For any vertex at least one of the following should be left unspecified: (i) the input/output flow rate, (ii) the pressures at the vertex and at adjacent vertices, or (iii) the parameter of a network element incident to the vertex. A more comprehensive treatment of those topics is given by Cheng (C7).

Let us consider the problem specification discussed in Section II,C. For a network with M external flows (inputs and outputs), the specification introduces the following additional equations:

$$w_{i_j} = w_{i_j}^\circ, \qquad j = 1, 2, \ldots, M-1 \tag{45}$$

$$\sum_{j=1}^{M} w_{i_j} = 0 \tag{46}$$

$$p_k = p_k^\circ \tag{47}$$

$$\beta_i = \beta_i^\circ, \qquad i = 1, 2, \ldots, b \tag{48}$$

the β's being the parameters in the network element models. The reader can easily verify that Eqs. (35), (37), (45)–(48) form a consistent and independent set of equations for the variables, $q_i(i = 1, 2, \ldots, P)$, $w_{i_j}(j = 1, 2, \ldots, M)$, $p_k(k = 1, 2, \ldots, N)$ and $\beta_i(i = 1, 2, \ldots, b)$.

In general for a set of nonlinear equations the necessary condition for determinancy is that there exists at least one admissible set of output variables for the equations (C7, S4). We can think of an output variable as that variable for which a given equation is solved either by an iteration process or by an elimination process. The set of all such assigned pairs of variables and equations is called the output set. Clearly an admissible output set must satisfy the conditions: (i) each equation has exactly one output variable, (ii) each variable appears as the output variable of exactly one equation, (iii) each output variable must occur in the assigned equations in such a manner that it can be solved for uniquely. Such an output set (circled entities) is illustrated in terms of an occurrence matrix in Fig. 4. Algorithms for finding output sets have been published by Steward (S4) and Gupta et al. (G8).

The introduction of a new specification (e.g., pressure at vertex $\mathbf{k}'$) brings in a new equation (e.g., $p_{k'} = p_{k'}^\circ$) and necessitates the elimination of one of the specification equations [Eqs. (45)–(48)]. The elimination of an existing specification equation avoids over-specification of the set of equations as a whole. But the new set of governing equations must again satisfy the additional requirement of possessing an admissible output set in order for the specifications to be admissible. Similarly, if a parameter, β_i is to be

$$f_1(x_2, x_4) = 0$$
$$f_2(x_1, x_2, x_3) = 0$$
$$f_3(x_1, x_3, x_5) = 0$$
$$f_4(x_1, x_5) = 0$$
$$f_5(x_3, x_4, x_5) = 0$$

(a)

	x_1	x_2	x_3	x_4	x_5
1	0	①	0	1	0
2	①	1	1	0	0
3	1	0	①	0	1
4	1	0	0	0	①
5	0	0	1	①	1

(b)

FIG. 4. Structural representation of equations: (a) equations, (b) an occurrence matrix.

computed, then the corresponding specification equation (48) must be replaced by a new specification equation, and the modified set of governing equations must possess an admissible output set.

2. Effect on Problem Structure

One of the interesting consequences of changing specifications is the effect on the equation structure. With formulations C and D, the occurrence matrix is symmetric. But if external flows, w_i and model parameters, β_i are introduced as the unknown variables, the symmetry may be destroyed. One way of preserving the local symmetry is to augment the system of equations and to bifurcate the variables in terms of state and design variables (M2).

By the same token certain types of specifications seem to render the solution much easier computationally. For instance, with Eqs. (45)–(48), the computation can be carried out sequentially for the acyclic portions of the network. Beginning with the specified inflows and outflows at vertices of degree one, the flow rates may be computed vertex by vertex using material balances [Eq. (35)]. Vertices and edges in any cyclic subnetwork may be aggregated and treated as an aggregated vertex (or "pseudo-node"), but all flows external to these vertices can be so determined. Similarly, starting with the computed flow rates and the reference pressure at one vertex, the pressures at other vertices may be computed sequentially. Only in the computation of conditions within the cyclic subnetworks will solution of simultaneous nonlinear equations be required. Clearly an efficient computational scheme can be devised for this particular set of specifications [Eqs. (45)–(48)]. But can we relate this solution to a problem with a different type of specifications? This question is addressed by Cheng (C7) and will be treated in Section III.

E. COMPARISON WITH ELECTRICAL CIRCUITS

In our treatment so far we have dwelt on the description, formulation, and specification of steady-state pipeline network problems. As we stated at the beginning, the behavior of such a network is influenced by both the topology of the network and the nature of the network elements. An important characteristic of a pipeline network is that its elements are nonlinear. The types of nonlinear relations encountered range from "quasi-linear" for pipes and valves, in the sense that the flow through the element and the pressure drop across the element are nondecreasing functions of each other, to discontinuities for pumps, compressors, and regulators. The influence of these nonlinear elements on the development of pipeline network analysis is best understood by contrast to the treatment of electrical circuits.

The direct analogy to a steady-state pipeline network would be a direct current electrical circuit, in which edge current, edge voltage, current source, and node voltage correspond to the variables q_k, σ_k, w_i and p_k in our pipeline network discussion. Indeed the earlier analogs for pipeline network simulation were resistor circuits of this type (S7). But the more instructive comparison is the transient analysis of an electrical network containing resistors, inductors, and capacitors, the so-called RLC network. For such a network, because the elements are lumped, linear, and time-invariant, the analysis may be carried out in the frequency domain after Laplace transformation of all functions and variables. Thus, corresponding to Eq. (34) we have

$$\hat{M}\hat{q}(s) = w(s) \tag{49}$$

and corresponding to the model of a network element we have

$$\hat{q}(s) = Y(s)\hat{\sigma}(s) \tag{50}$$

where $\hat{q}(s)$, $w(s)$, and $\hat{\sigma}(s)$ are the Laplace transforms of the corresponding vectors, $\hat{q}(t)$, $w(t)$, and $\hat{\sigma}(t)$, and $Y(s)$ is a diagonal edge admittance matrix. Since the edge-voltage vector $\hat{\sigma}(s)$ can always be expressed in terms of the node-voltage vector $p(s)$,

$$\hat{\sigma}(s) = \hat{M}^T p(s) \tag{51}$$

it follows that

$$\hat{M}Y(s)\hat{M}^T p(s) = w(s) \tag{52}$$

where $[\hat{M}Y(s)\hat{M}^T]$ is the so-called admittance matrix. Thus, the problem of determining node voltage amounts to solving the set of linear equations [Eq. (52)] symbolically, followed by inverse transformation.

Symbolic solution of Eq. (52) using determinant and cofactors is extremely inefficient because large numbers of terms generated are later cancelled out. It turns out to be useful to invoke the Binet–Cauchy theorem once again, because each nonvanishing term is in fact the product of the edge admittances corresponding to a spanning tree. The sum of all "tree admittance products" is the determinant. This result, known as Maxwell's formula, permits the node admittance determinant to be evaluated without cancellations from a knowledge of spanning trees. Similarly, the cofactor may be evaluated using 2-trees of the graph (D4). The graph-theoretic procedure involving the "loop impedance matrix" follows from an analogous development (D4). Thus for RLC networks because the elements are lumped, linear, and time-invariant, it is possible to take the analysis further before resorting to numerical computation.

III. Steady-State Pipeline Network Problems: Methods of Solution

A. Numerical Methods

Under all but laminar flow conditions, the steady-state pipeline network problems are described by mixed sets of linear and nonlinear equations regardless of the choice of formulations. Since these equations cannot be solved directly, an iterative procedure is usually employed. For ease of reference let us represent the steady-state equations as

$$f(x) = 0 \tag{53}$$

where x is the vector of state (or unknown) variables. The successive iterations generate a sequence of approximations, $x_0, x_1, x_2, \ldots$ with x_0 as the vector of initial guesses. The procedure converges, if and only if

$$\lim_{k \to \infty} x_k = x^* \tag{54}$$

where x^* is a solution to Eq. (53).

As a matter of expediency, the iterative procedure cannot be allowed to continue indefinitely. The criterion by which the procedure is terminated is usually based on

(i) the magnitude of the residual $f(x)$. For instance, terminate the iterations, if

$$\|f(x_k)\| < \epsilon \tag{55}$$

where ϵ is a specified error tolerance.

(ii) the magnitude of changes in x. For instance, terminate the iteration, if for each state variable $x_k > 0$,

$$(x_k - x_{k-1})/x_k < \epsilon \tag{56}$$

and for each state variable $x_k \approx 0$,

$$x_k - x_{k-1} < \epsilon \tag{57}$$

(iii) a predetermined upper bound on the number of iterations, k.

In many computer programs more than one of the conditions listed above are implemented as the termination criteria.

In this section we shall review the numerical methods used for solving Eq. (53). These are chiefly iterative methods. They differ in the procedures according to which the iterates are generated.

By itself a steady-state pipeline network problem rarely poses a formidable computing problem. Even for a large and complex network, the central processor time scarcely ever exceeds a few minutes per case. However, in design and operational studies it is often necessary to compute numerous parametric cases. Citing from experience in water distribution network analysis King (K7) noted that up to 250 parametric cases might be solved in a given network study. It is for this reason that the effectiveness and efficiency of the method used are important considerations from a practical engineering viewpoint.

The numerical methods used by different investigators are summarized in Table II. In addition to the performance data reported in the literature, we also evaluated a number of iterative methods on two benchmark problems using the same initial estimates and the termination criterion Eq. (55) with $\epsilon = 10^{-5}$. The results of our numerical experiments are summarized in Tables III and IV and a description of these two problems is given in Appendix A. In the following paragraphs we shall comment briefly on the key features of each method and its numerical performance as indicated by the available data, giving greater coverage to the more promising or less well-known methods.

1. *The Newton–Raphson Method*

This method is based on the Taylor expansion of $f(x)$ about the kth iterate, x_k,

$$0 = f(x^*) = f(x_k) + \left\{\frac{\partial f_i}{\partial x_j}\right\}(x^* - x_k) + \cdots \tag{58}$$

By neglecting the higher order (nonlinear) terms in $x^* - x_k$ and replacing x^* by the $(k + 1)$th approximation, x_{k+1}, we have

$$x_{k+1} = x_k - \{\partial f_i/\partial x_j\}^{-1} f(x_k) \tag{59}$$

The matrix of partial derivatives, $\{\partial f_i/\partial x_j\}$, is commonly referred to as the Jacobian and sometimes denoted by $\{J_{ij}\}$ or J. In Ej. (59) it is to be evaluated at $x = x_k$.

The chief merit of this method is its rapid rate of convergence starting with a set of good initial guesses. Since the discussion of its quadratic convergence

TABLE II

ITERATIVE METHODS FOR STEADY-STATE PIPELINE NETWORK PROBLEMS

Method	Formulation[a]	Largest network solved		References
		N	P	
Newton–Raphson	B	22	38	Stuckey (S7), Mah (M1)
	C	70	100	Warga (W1), Martin and Peters (M4), Lemieux (L5), Shamir and Howard (S2), Stuckey (S7), Brameller et al. (B10), Stoner (S5), Carnahan and Wilkes (C2), Donachie (D8)
Broyden	D	170	307	Epp and Fowler (E2), Brameller et al. (B10)
Wolfe	C	30	50	Lam and Wolla (L2)
Hardy Cross	A	28	35	Carnahan and Christensen (C3)[b]
Balancing heads	D	289	544	Cross (C11), Kniebes and Wilson (K8), Ingels and Powers (I1), Daniel (D2), Stuckey (S7), Williams and Pestkowski (W8), Williams (W7), Gerlt and Haddix (G3)
Balancing flows	C	20	33	Knights and Allen (K9), Liu (L8),[c] Stuckey (S7), Rosenhan (R4)
Linearization	A	22	38	Bending and Hutchison (B5)
	B	46	57	Wood and Charles (W11), Jeppson and Tavallaee (J2)
	C	57	75	Collins and Johnson (C10)

[a] See Table I.
[b] Tearing was used.
[c] A modified version (the Newton–Cross method) was used.

TABLE III

Computational Evaluation Based on Sample Problem A ($N = 10$, $P = 12$)

Iterative method	Formulation[a]	Number of iterations	CP time[b] (seconds)	Comments
Newton–Raphson	B	6	1.07	
	C	17		Reported by Carnahan and Wilkes (C2)
Broyden	B	70	1.52	Terminated after 70 iterations. Convergence not attained.
Hardy Cross				
Balancing heads	D	57	0.73	
Linearization	B	9	0.27	
	D	15	0.23	

[a] See Table I.
[b] The central processor (CP) time is based on a CDC Cyber 73 computer.

characteristics in the neighborhood of a solution is given in standard textbooks on numerical analysis (C1, I2) we shall not belabor the point. A second advantage which is less well known is that the method lends itself very readily to sensitivity analysis, a subject which will be treated in Section IV.

The Newton–Raphson method has been applied to pipeline network problems since 1954 (W1). Its performance has been generally very good, although convergence difficulties have been reported (S2), when starting from inappropriate initial guesses. In some cases large oscillations around

TABLE IV

Computational Evaluation Based on Sample Problem B ($N = 22$, $P = 38$)

Iterative methods	Formulation[a]	Number of iterations	CP time[b] (seconds)
Newton–Raphson	B	4	8.2
Hardy Cross			
Balancing heads	D	59	2.4
Linearization	B	7	3.43
	D	17	1.05

[a] See Table I.
[b] The central processor (CP) time is based on a CDC Cyber 73 computer.

the solution occurred (D8, L2) when it was used in conjunction with formulation C. In practice Eq. (59) is almost always modified to include a damping (or step scaling) factor, τ:

$$x_{k+1} = x_k - \tau\{\partial f_i/\partial x_j\}^{-1} f(x_k) \tag{60}$$

where $0 < \tau \leq 1$. The value of τ is chosen to ensure a reduction in the residual $f(x_k)$. One strategy is to carry out a one-dimensional search in τ to minimize the norm of the residual, $\|f(x_{k+1})\|$. This strategy has been used by Lemieux (L5) and investigated by Broyden (B13), Lam and Wolla (L2), and Donachie (D8). Broyden (B13) concluded that it is less efficient than the alternative of simply picking a τ to ensure that

$$\|f(x_{k+1})\| < \|f(x_k)\| \tag{61}$$

while Donachie (D8) adopted the simple strategy of halving the step size ($\tau = \frac{1}{2}$) every time an oscillation occurred.

Unlike the other alternative methods, analytical expressions of partial derivatives are required and the Jacobian must be evaluated in the Newton–Raphson method. These requirements sometimes prove to be the undoing when the method is applied to complicated equations. Brown (B12) has developed a modification to the Newton–Raphson method, which requires only some of the partial derivatives to be calculated. We have tested Brown's method on our sample problems but have found that it actually required more computing time than the unmodified Newton–Raphson method.

2. Generalized Secant Methods

Although the evaluation of partial derivatives is not usually an insurmountable obstacle in networks involving one-phase flow in pipes, several investigators (C3, L2) have explored alternative iterative methods which do not require direct evaluation of partial derivatives. These methods are generally based on linearized approximations using "secants" rather than "tangents."

a. *Broyden's Method.* Broyden (B13) proposed a class of modified Newton–Raphson methods which are sometimes referred to as the quasi-Newton methods. One of these methods has been applied to water distribution network problems by Lam and Wolla (L2).

In this method the inverse of the Jacobian is approximated by $-H$ and the successive approximations are generated according to the following formula:

$$H_{k+1} = H_k - (\tau_k z_k + H_k y_k) z_k^T H_k / z_k^T H_k y_k \tag{62}$$

where

$$z_k = H_k f(x_k) \tag{63}$$

$$y_k = f(x_{k+1}) - f(x_k) \tag{64}$$

In place of Eq. (60) we now have

$$x_{k+1} = x_k + \tau H f(x_k) \tag{65}$$

The advantage of this method is that it avoids both the evaluation of partial derivatives and the inversion of the Jacobian. To start the iterations, an initial estimate H_0 is also required; an identity matrix is frequently used for this purpose.

In comparison with the Newton–Raphson method, Lam and Wolla (L2) have found the quasi-Newton method to be less sensitive to the initial guesses and to require less computing time for the larger problems ($N = 30$, $P = 50$) tested. It should be pointed out that for simple network element models and large networks, the advantages offered by this method may only be marginal, since H is usually a fairly dense matrix and many multiplications are involved in generating an update using Eqs. (62)–(65).

b. *Wolfe's Method.* For a set of n equations, Eq. (53), an $(n + 1)$ point generalized secant method was proposed by Wolfe (W9). A version of this method has been evaluated by Carnahan and Christensen (C3) using formulation A with tearing. The basis of Wolfe's method is best understood by reference to the development of the one-dimensional secant method. The one dimensional secant approximation to a truncated Taylor expansion of $f(x^*)$ about x_2 is

$$0 = f(x^*) = f(x_2) + \frac{f(x_2) - f(x_1)}{x_2 - x_1}(x^* - x_2) \tag{66}$$

nonlinear terms of $(x^* - x_2)$ being neglected. It follows that if

$$\pi_1 = \frac{x_2 - x^*}{x_2 - x_1} \quad \text{and} \quad \pi_2 = \frac{x^* - x_1}{x_2 - x_1} \tag{67}$$

then

$$\pi_1 + \pi_2 = 1 \tag{68}$$

$$\pi_1 f(x_1) + \pi_2 f(x_2) = 0 \tag{69}$$

$$\pi_1 x_1 + \pi_2 x_2 = x \tag{70}$$

In Wolfe's method $n + 1$ initial "trial solutions," $x_1, x_2, \ldots, x_{n+1}$ are required. The weights, π_j, are found by solving the n-dimensional equivalent of Eqs. (68) and (69):

$$\sum_{j=1}^{n+1} \pi_j = 1 \tag{71}$$

$$\sum_{j=1}^{n+1} \pi_j f_i(x_j) = 0, \qquad i = 1, 2, \ldots, n \tag{72}$$

and a new approximation $\bar{x}$ is obtained according to

$$\bar{x} = \sum_{j=1}^{n+1} \pi_j x_j \tag{73}$$

To complete the iteration a new set of trial solutions is formed by replacing one of the trial solution, x_k say, by $\bar{x}$. A usual replacement criterion is that $\|x_k\| = \max \|x_j\|$.

Unlike the previous method Wolfe's method requires the inversion of an $(n + 1) \times (n + 1)$ matrix at each iteration, but a short-cut procedure may be used to take advantage of the fact that at each iteration only one column of this matrix is modified. Another disadvantage of this method is that it cannot use the information from a previous case (e.g., the Jacobian) to obtain a better starting point. To the best of our knowledge there is no demonstrable advantage to recommend the use of this method in pipeline network problems.

3. The Hardy Cross Method

In contrast to the methods described in Sections III,2 and 3, which are generally applicable to all nonlinear equations, the Hardy–Cross methods have been developed specifically for pipeline network calculations. These methods, devised by Cross (C11) long before the advent of modern computers, are still widely used (D2, D6, G3, G7, H5, I1, K8, V2). Although the original publication (C11) is readily accessible to most readers, more concise accounts of these two methods may be found in publications by Fair and Geyer (F1) and Stuckey (S7).

a. *Method of Balancing Heads.* In this method a set of initial flows, which satisfies the material balance Eq. (35), is assumed. Flow corrections are then applied one cycle at a time until Kirchoff's second law is satisfied for all cycles. For network elements modeled by the equation,

$$\sigma_k = \alpha_k q_k |q_k|^{n-1} \tag{74}$$

the flow correction for cycle C_i is given by

$$\delta q_{c_i} = -\sum_{k \in C_i} \alpha_k q_k |q_k|^{n-1} \bigg/ \sum_{k \in C_i} n\alpha_k |q_k|^{n-1} \tag{75}$$

where the summations are taken in a consistent direction around the cycle C_i. This method is equivalent to the application of Newton's method to formulation D one equation at a time.

Because the interactions between cycles are neglected, the convergence rate of this method may be significantly altered by the selection of cycles (D2, G1). It works best when the overlaps between cycles are minimal (D2). Gay and Middleton (G1) investigating the effect of cycle selection on convergence found that for three cases of increasing cycle overlaps starting from the minimum overlap, the number of iterations were 15, 35, and 50, respectively. Convergence was never attained in the last of these cases. Modifications of this method to incorporate the use of convergence acceleration factors were introduced by Barlow and Markland (B2), and by Williams (W7) who claimed up to 80% reduction in the number of iterations.

Although many iterations are usually required in the method of balancing heads, the computation per iteration is simple and fast and the storage requirement is minimal, since no matrix operation or storage is involved. These factors favor the selection of the method of balancing heads, particularly when the program is to be run on a small computer.

b. *Method of Balancing Flows.* The method of balancing heads is applicable to a network with specified external flows. If these specifications are replaced by pressures instead, the flow distribution may be determined by the method of balancing flows. In this method the pressures at all vertices are either known or assumed, and pressure corrections are applied one vertex at a time to satisfy all the nodal balances [Eq. (39)]. This method is equivalent to the application of Newton's method to formulation C one equation at a time, the interactions between vertices being neglected. The general consensus (C11, G1, K9) seems to be that this method is slower than the method of balancing heads. Given the range of alternative methods available, it would seem that this method is only of historical interest.

A modified version of this method, the so-called Newton–Cross method, was proposed by Liu (L8).

4. *The Linearization Method*

A significant development in pipeline network computation in the last few years was the introduction of the so-called linearization method first by Wood and Charles (W11) and later, independently, by Bending and Hutchison (B5). This method has been applied to both formulations A and B and appears to be quite competitive in the applications so far reported.

In the development given by Wood and Charles (W11), formulation B was used and the network elements were modeled by Eq. (74). Clearly the only nonlinear terms are those in Eq. (38) introduced by the network element models. By replacing Eq. (74) by the following equation,

$$\sigma_k = \alpha_k q'_{k,(r)} | q_{k,(r-1)}|^{n-1} \tag{76}$$

where the subscripts in parentheses indicate the iteration number, the model is linearized with estimates of q_k from the $(r-1)$th iteration. In this manner, Eqs. (35) and (38) become a set of linear equations which can now be solved to obtain the new estimates, $\mathbf{q}'_{(r)}$. It was observed, however, that this simple procedure tended to promote oscillations and that the best results were obtained using the average of the two estimates,

$$q_{k,(r)} = [q'_{k,(r)} + q_{k,(r-1)}]/2 \tag{77}$$

For the special case for which $n = 2$, it can be shown that the linearization method defined above becomes identical to the Newton–Raphson method. The result may be generalized to apply to any homogeneous function of degree n.

While it is technically erroneous to claim that the linearization method does not require any initialization (J2), it is true that the initialization procedure used appear to be quite effective. A more comprehensive discussion of initialization procedure will be given in Section III,A,5. With this initialization procedure, the linearization method appears to converge very rapidly, usually in less than 10 iterations for formulations A and B. Since the evaluation of $\mathbf{f}(\mathbf{x})$ and its partial derivatives is not required, the method is also simpler and easier to implement than the Newton–Raphson method.

On the debit side, the linearization method is quite sensitive to the form of the network element model. Jeppson and Tavallaee (J2) reported that convergence rate was slow when the usual pump and reservoir models were incorporated, but they obtained significant improvements after the models had been suitably transformed. Although the number of iterations required is small using formulations A and B, the dimension of the matrix equation is substantial. Hence, it becomes essential to use sparse computation techniques if the method is to retain its competitive edge in larger problems.

5. Initialization Procedures

In all the aforementioned methods it is crucial to start with good initial guesses. How does one obtain good initial guesses? And how sensitive are the methods to the initial guesses? We have already briefly alluded to the second question in conjunction with Lam and Wolla's (L2) findings. Very little else has been reported on a quantitative basis on this subject. We shall now summarize the investigations to date on the first question. We should point out that most of the reported initialization procedures involve networks with only pipes and valves. Since the variables to be initialized depend on the problem formulation, it is convenient to group our discussions accordingly.

For formulations A and B, one general procedure is to solve the laminar flow equations which are linear and use the solution as the initial guesses for the nonlinear equations. Variations of this procedure have been used by Bending and Hutchison (B5), Wood and Charles (W11), and Jeppson and Tavallaee (J2) in conjunction with the linearization method.

The initialization procedure for formulation C was investigated by Lam and Wolla (L2) who reported successful convergence with the "point of inflection" method. The points of inflection of Eq. (53) are defined as those values of x for which the second derivatives of $f(x)$ vanish. They are found at

$$p_1 = p_2 = \cdots = p_N \qquad (78)$$

for Eq. (39) and correspond to pressures in a lossless network. According to Lam and Wolla's procedure the initial guess of pressure p_i is calculated by solving the nodal balance equation (39) assuming Eq. (78) to hold for all other pressures. Lam and Wolla (L2) found that the average of these nodal pressures provided a still better set of initial guesses and claimed that in their experience convergence was always obtained using this procedure in conjunction with Broyden's method.

Finally, for formulation D the flows in the tree branches can be computed sequentially assuming zero chord flows. This initialization procedure was used by Epp and Fowler (E2) who claimed that it led to fast convergence using the Newton–Raphson Method.

6. *Comparison of Iterative Methods*

On the subject of comparing iterative methods a word of caution is in order. Clearly in any quantitative comparison, the termination criteria should be comparable and the benchmark problems should be run on the same computer. Yet even for simple problems and methods, these two requirements prove to be difficult to enforce and insufficient to ensure meaningful comparisons. To allow for the fact that different methods do not terminate at exactly the same point even when the same termination criterion is used, Broyden (B13) introduced a mean convergence rate, R', which is expressed in terms of the number of function evaluations, r, and the initial and final norms of f:

$$R' = \frac{1}{r} \ln(\|f(x_0)\|/\|f(x_n)\|) \tag{79}$$

A similar measure was defined by Varga (V1).

Even when comparison is made on the same computer, significant discrepancies could arise from the differences in programming skills and the selection of programming languages. It is also well known that there is usually some trade-off between computing time and computing storage in implementing a given iterative method. Except when the performance disparity between two methods is large, either of these two factors alone can give a misleading picture. When evaluating different programs for use in the same installation, computing cost is sometimes a more realistic basis of comparison, since it is a composite of many factors which should be considered in the final decision.

The comparison of different methods is sometimes reported in terms of number of iterations (G1). The danger of such a comparison lies in the fact that the computation per iteration may be vastly different for different iterative methods. Results presented in this manner should always be carefully qualified.

Finally, a special point to look for in comparing iterative methods for pipeline network problems is to use the same problem formulation for both methods; otherwise the results may reflect differences in formulations as well as iterative methods.

With these cautionary notes and the understanding that the expectation of simple answers may be incompatible with the complexity of the subject and the paucity of data, we shall now offer a few observations.

With reference to the iterative methods reviewed, it is perhaps easier to pick the losers than the winners. With the methods now available, we see little reason to prefer Brown's method, Wolfe's method, the method of balancing flows, and the Newton–Cross method. Of the remaining alternatives the choice is not clear cut. The Newton–Raphson method with step scaling is very effective when analytical derivatives are readily available, but it tends to require more computing time than some of the alternatives. Broyden's method, as it stands, is probably only competitive on small and medium size problems, because it does not lend itself readily to sparse computation techniques. When computing storage is a serious limitation, the method of balancing heads offers an attractive alternative but with some obvious disadvantages which have already been mentioned. Finally, the experience so far is quite positive on the linearization method. Like the Newton–Raphson method it is well suited to sparse computation techniques and with this modification it could be quite effective for large networks. In Tables III and IV we present some of our own computational results on these four methods. Included among them is the linearization method using formulation D, which turns out to be the best method for the two problems tested.

7. *Other Methods*

In principle, the steady-state pipeline network problems can always be solved by the transient solution methods after allowing sufficient time steps for the solution to reach steady state. This possibility was discussed by Nahavandi and Catanzaro (N1) who made a comparison of a transient solution method with the Cross method of balancing flows (R4). For the particular 35-node and 45-branch hydraulic network problem tested, the transient solution method took 108 seconds as compared with the 134 seconds required by the Hardy–Cross method. (See also Section V,A,2.)

A radically different approach to the steady-state problem was investigated by Hsing (H6). In this approach the steady-state flow problem was formulated as the following constrained minimization problem:

$$\min_{\mathbf{q}} \left\{ \phi = \sum_{k \in E} \int_0^{q_k} \sigma_k(q_k)\, dq_k \right\} \tag{80}$$

subject to the nodal balances

$$\sum_{k \in E_{Bi}} q_k - \sum_{k \in E_{Ai}} q_k = w_i, \qquad i = 1, 2, \ldots, N-1 \tag{81}$$

where E_{Ai} is the subset of incident edges directed toward vertex **i** and E_{Bi} is the subset of incident edges directed away from vertex **i**.

To show that this constrained minimization is indeed equivalent to the steady-state formulation, let us adjoin the equality constraints to the objective function to form the Lagrangian function,

$$\psi(\mathbf{q}, \boldsymbol{\lambda}) = \sum_{k \in E} \int_0^{q_k} \sigma_k(q_k)\, dq_k + \sum_{j=1}^{N-1} \lambda_j \left(\sum_{k \in E_{Bj}} q_k - \sum_{k \in E_{Aj}} q_k - w_j \right) \tag{82}$$

The necessary conditions for minimization are

$$\partial \psi / \partial \lambda_j = 0, \qquad j = 1, 2, \ldots, N-1 \tag{83}$$

$$\partial \psi / \partial q_k = 0, \qquad k = 1, 2, \ldots, P \tag{84}$$

The first set of conditions yields Eq. (81) and the second set yields

$$\sigma_{ij} = \lambda_i - \lambda_j \tag{85}$$

which gives rise to the interpretation of the Lagrange multiplier λ_i as the pressure at vertex **i**. For $\sigma_k(q_k)$ which are monotonic nondecreasing functions of q_k, it can be shown that $\{\partial^2\phi/\partial q_j\,\partial q_k\}$, the Hessian matrix of ϕ is everywhere positive semi-definite and that ϕ is a convex function. Hence, Eqs. (83) and (84), which are equivalent to Eqs. (35) and (37), are also the sufficiency conditions for minimization.

A particularly efficient method of solving the minimization problem [Eqs. (80)–(81)], is the so-called out-of-kilter algorithm (F8) which was evaluated by Hsing (H6). Compared with the Hardy–Cross method of balancing heads, the out-of-kilter algorithm showed better consistency in convergence but required significantly longer computing time in all four problems tested by Hsing (H6).

B. TECHNIQUES FOR LARGE NETWORKS

In our discussion of numerical methods so far we have deliberately focused our attention on the numerical characteristics of these methods. These considerations predominate in the selection of numerical methods for small network applications. However, for medium and large networks, the computing time and computing storage can often be greatly reduced by taking advantage of the problem structure, and the different methods are not equally amenable to such adaptations. In this section we shall discuss the techniques for handling large network applications. The techniques we are about to discuss all make use of the fact that the system of equations is sparse. By a sparse system we mean that the number of nonzero elements in the occurrence matrix is relatively small compared to the total number of elements in the matrix. For large networks, the fraction of variables occurring in any one equation is usually small. Consequently, equations for large problems are nearly always sparse.

1. *Tearing and Diakoptics*

A well-known class of techniques for reducing the number of iterates is the use of tearing (L4). We shall illustrate this procedure by way of an example taken from Carnahan and Christensen (C3). Let us consider the two-loop network shown in Fig. 5 and assume that formulation A is used. To abbreviate the notation let us denote the material balance around vertex **i** [Eq. (35)] by $f_i = 0$ and the model of the element $\langle \mathbf{i}, \mathbf{j} \rangle$ [Eq. (36)] by $f_{ij} = 0$. Then assuming all external flows and one vertex pressure, p_i say, are specified, we have a set of 12 equations that must be solved simultaneously. But if we now assume a value for q_{12}, the remaining equations may be solved sequentially one at a time to yield the variables in the following

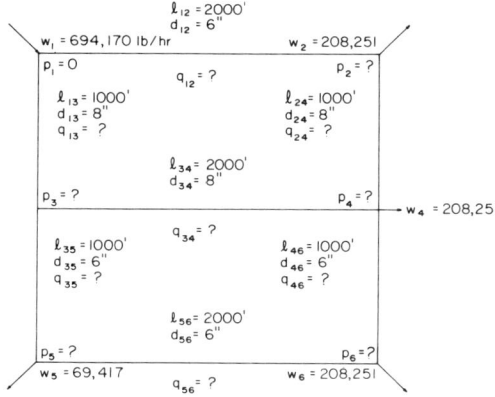

FIG. 5. A simple network. Reproduced from Carnahan and Christensen (C3).

order: $f_{12} \to p_2, f_2 \to q_{24}, f_{24} \to p_4, f_1 \to q_{13}, f_{13} \to p_3, f_{34} \to q_{34}, f_3 \to q_{35}, f_{35} \to p_5, f_4 \to q_{46}, f_{46} \to p_6$ and $f_6 \to q_{56}$. Notice that in this procedure we have left out the model equation $f_{56} = 0$, which would be satisfied, if the assumed value of q_{12} is correct. If it is not, then a process of iteration will be required. The variable q_{12} in the above example is the tear variable and the equation $f_{56} = 0$ is sometimes referred to as the tear equation.

The effect of tearing is to delete the tear variables and tear equations from the original set and to solve them iteratively external to the remaining set of equations and variables. In order for tearing to be a viable strategy, the number of tear variables required must be small and the tear equations must not be too difficult to solve. In this example, after tearing the iteration will involve only one equation, assuming the model equations are pressure explicit.

The feasibility of applying tearing to pipeline network problems was demonstrated by Carnahan and Christensen (C3) on examples up to 35 edges and 28 vertices using formulation A. No data on computing time or storage requirements were given by these authors. However, it is known from previous investigations (C9, M1) that tearing can give rise to a smaller set of equations with a solution time actually greater than that of the original set of equations.

More recently, Cheng (C7) showed that to apply tearing effectively one must take into consideration the topology of the network and the nature of problem specifications. For instance, if the network consists of a number of cyclic subnetworks imbedded in an acyclic framework and if all the external flows and one reference pressure are specified, the flows and the pressures external to the cyclic subnetworks may be computed sequentially and only the equations pertaining to each of the cyclic subnetworks need be solved simultaneously, using formulation B (see Section II,D,2). If instead we now specify 3 pressures and $(M - 3)$ external flows, then an effective strategy is to tear all the flows connecting the cyclic subnetworks. On the other hand, if the overall network is cyclic to start with, a more effective strategy is to use formulation C: the 2 additional pressure specifications reduce the number of simultaneous equations to be solved from $(N - 1)$ to $(N - 3)$, provided that the specifications are consistent and that the correlations used are flow-explicit.

A special tearing procedure known as diakoptics (K10) was investigated by Brameller et al. (B10) and by Gay and Middleton (G1). According to Brameller et al. (B10), this procedure lends itself very readily to the exercise of engineering judgment. However, computational performance of this method reported by Gay and co-workers (G1, G2) has not been too impressive. The more recent work of Gay and Preece (G2) indicates that the "nodal method of diakoptics" is outperformed by an alternative method based on formulation D for problems up to 100 edges and 50 cycles.

2. *Column Row Reordering and Cycle Selection*

Unlike tearing, the techniques we are about to describe have no deleterious effects on the convergence rate of the iterative procedure. In fact the application of cycle selection techniques may even enhance the convergence rate. Although we have grouped them together for convenience, the two types of techniques are in fact quite different. Column and row reordering algorithms are basically data processing techniques which seek to enhance storage and computing efficiency by reordering the rows and columns of the coefficient matrix, whereas the cycle selection algorithms indirectly modify the problem formulation.

a. *Column and Row Reordering.* As we pointed out earlier (see Section II,C), the coefficient matrices in formulations C and D are symmetric, if the unknowns are vertex pressures and chord (loop) flows, respectively. Hence approximately only half of the matrix elements need to be stored and the Cholesky decomposition may be applied in solving these equations (D6, E2). Furthermore, by judiciously reordering the rows and columns, the matrix becomes banded, i.e., all nonzero elements lie within a band about the diagonal of the matrix. Let us consider the following band matrix

$$A = \begin{bmatrix} A_{11} & A_{12} & 0 \\ A_{12}^T & A_{22} & A_{23} \\ 0 & A_{23}^T & A_{33} \end{bmatrix} \tag{86}$$

where the submatrices are of dimensions: $A_{11}(1 \times 1)$, $A_{12}(1 \times B)$, $A_{22}(B \times B)$, $A_{23}(n \times \overline{n-1-B})$ and $A_{33}(\overline{n-1-B} \times \overline{n-1-B})$. During the pivotal condensation (elimination) with row 1, only the submatrix A_{22} is modified. Hence if we know the structure of the matrix beforehand, only an $(\overline{B+1} \times \overline{B+1})$ submatrix is needed in fast storage and the number of arithmetic operations will be proportional to B^2 instead of n^2. The aim of the reordering algorithm is therefore to minimize the bandwidth, $B(a_{ij} = 0$ for all $|i - j| > B)$. King (K7) reported one spectacular case of computing time reduction from 72 seconds to 6.5 seconds (including 2 seconds on reordering) after applying bandwidth reduction to a 676-vertex network.

Bandwidth reduction algorithms for formulations C and D have been proposed by King (K7) and Epp and Fowler (E2), respectively. Neither of these two algorithms guarantees minimum bandwidth, since they are both based on heuristic considerations. In fact, King (K7) even disclaims any guarantee of improvement. However, this modest disclaimer should not detract from the fact that these one-pass algorithms are simple to apply and that they can produce very worthwhile improvements in computing efficiency.

b. *Cycle Selection Algorithms.* An important feature of formulations B and D is that they require that a set of independent cycles be found either by hand or by computer. As we pointed out in Section II,A the independent cycle set is usually not unique. Many investigators (C11, D2, F3, G1, H7) have observed the significant influence of cycle selection on the convergence behavior of the method of balancing heads. Similar effects on the Newton–Raphson method using formulation B and D were investigated by Mah (M1) and Epp and Fowler (E2), respectively.

The impact of cycle selection on the structure of the equations is most readily seen with reference to formulation B. The nonzero terms in Eq. (38) correspond to the links (edges) in the cycle C_i. Consider the network in Fig. 6. Any two of the following sets of edges may be chosen as the two independent cycles:

(i) $(1, 2, \ldots, n, n + 1, n + 2, n + 3)$
(ii) $(1, 2, \ldots, n, n + 4)$
(iii) $(n + 1, n + 2, n + 3, n + 4)$

If cycles (i) and (ii) are chosen $(2n + 4)$ edges or terms will be involved. But if cycles (ii) and (iii) are chosen instead, the total will be $(n + 5)$. The disparity is quite significant even for such a simple network, if n is large, say 200.

If the network is known to be dominated by an acyclic branch such as the n-link branch in Fig. 6, a close approach to a minimal formulation is ob-

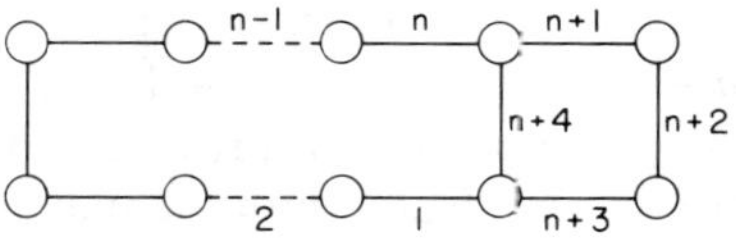

FIG. 6. A simple two-cycle network.

tained simply by making sure that this branch is included only once in the set of C independent cycles. But, in general, this will not be the case and the problem becomes combinatorial. The various cycle selection algorithms are aimed at minimizing the burden of computation and storage without an excessive expenditure of efforts.

With reference to Fig. 7, Epp and Fowler (E2) gave the following description of their "loop defining algorithm":

> ... to determine the pipes which are not in loops; start at a node of degree 1, work through nodes of degree 2, and stop when a node of degree higher than 2 is reached. Temporarily "remove" the tail (pipes 1 and 2) from the network. (Note that with these pipes removed, node 1 now is of degree 2.) Define nodes of degree 2 to be "key" nodes. To define a loop any key node is selected (e.g. node 1). The two nodes connected by pipes to the key node are then known (i.e. nodes 4 and 7). The shortest path between these two nodes which does not pass through the key node is then determined. In the example the path will be pipes 4, 7 and 6 and so pipes 3, 4, 7, 6, 5 become a defined loop. Pipes 3 and 5 are now temporarily "removed" from the network and thus node 4 now has degree 1 and node 7 has degree 2 and is hence a key node. Hence the pipe 4, being a tail is "removed" and the algorithm is repeated at a new key node, e.g. node 7. If at any time only nodes of degree higher than 2 are present, arbitrarily choose a node of least degree and proceed as before. In this manner the "natural set" of loops is found.

Notice that no claim was made that the cycles so obtained would be minimal nor indeed that the algorithm would always work.

We shall now offer two simple examples to illustrate these points. Example 1 shown in Fig. 8a consists of 2 independent cycles. Starting from vertex **1**, we delete edges **1** and **2** to locate (**1, 2, 3, 4, 5**) as cycle **1**. Similarly, starting from either vertex **2** or vertex **5**, the algorithm traces (**3, 4, 5, 6, 7, 8, 9,**

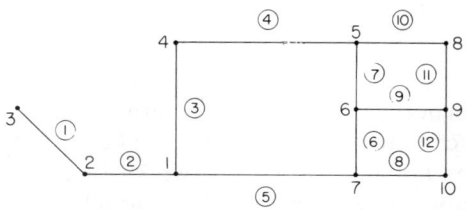

FIG. 7. A simple network. Reproduced from Epp and Fowler (E2).

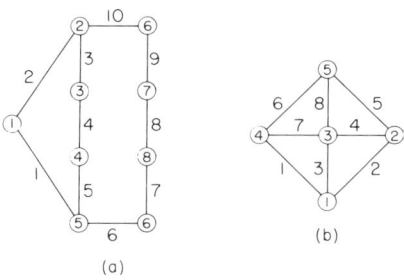

Fig. 8. Two counter-examples.

10) as cycle 2. But this choice is clearly nonminimal, because the alternative (**2, 1, 6, 7, 8, 9, 10**) contains fewer edges.

Example 2 shown in Fig. 8b contains 4 independent cycles. Again starting from vertex **1** and deleting edges **1** and **2**, we obtain (**1, 2, 4, 7**) as cycle 1. Similarly, with vertices **2** and **3** as "key nodes" the Epp–Fowler algorithm yields (**4, 5, 8**) and (**6, 7, 8**) as the two remaining cycles. In this case it would appear that only 3 of the 4 independent cycles are located by the algorithm.

We offer these two counter examples not to castigate the method, for indeed the Epp–Fowler algorithm has worked well for a wide variety of networks ranging up to 307 edges, 170 vertices, and 135 cycles (E2), but as an illustration of the difficulties of constructing a completely reliable algorithm that will work for all cases.

In an independent investigation on pipeline network calculations using sparse computation techniques, Mah (M1) devised the following cycle construction algorithm (algorithm MLC):

(i) Eliminate all vertices of degree 2, merge the corresponding edges and increment the length counts on these merged edges.

(ii) Pick the longest edge (highest length count), e_1, in the network, between vertices v_A and v_B, say.

(iii) Find the shortest path between v_A and v_B in the subgraph in which e_1 has been eliminated.

(iv) Let C_1 be the cycle containing all the edges in the shortest path and the selected edge, e_1.

(v) Pick the next edge to be the longest edge in the subgraph which does not include the edges in the previously constructed cycles, C_i. Replace e_1, v_{A_1}, v_{B_1} and C_1 in steps (ii)–(iv) and repeat these steps until no more fresh edge is available.

(vi) Trace the cycles in the subgraph that exclude the selected edges and add these cycles to complete the cycle set, $\{C\}$.

It can be shown (M1) that if all C independent cycles are found in steps (i)–(v), the algorithm yields a minimal length cycle set (in the sense that the total number of length counts in the cycle set is minimal). Otherwise, a near-minimal length cycle set is obtained.

Mah (M1) pointed out that step (iii) is the longest step in algorithm MLC and gave an estimate of $3CN^2$ operations (additions or comparisons) assuming that the shortest paths would be determined using Dijkstra's algorithm (D5). Both the Epp–Fowler algorithm and algorithm MLC are one-pass algorithms.

3. Sparse Computation Techniques

For medium and large networks, the occurrence matrix that is of the same structure (isomorphic) as the coefficient matrix of the governing equations is usually quite sparse. For example, Stoner (S5) showed a 155-vertex network with a density of 3.2% for the occurrence matrix (i.e., 775 nonzeros out of a total of 155^2 entries) using formulation C. Still lower densities have been observed on larger networks. In these applications it is of paramount importance that the data structure and data manipulations take full advantage of the sparsity of the governing equations. Sparse computation techniques are also needed in order to capture the full benefit of cycle selection and row and column reordering.

In ordinary matrix computation, the matrices are stored in 2-dimensional arrays and arithmetic operations are carried out row by row or column by column. In sparse matrix computation only the nonzero elements are stored and operated upon. The reduction in computer storage and computing time is achieved at the expense of more complicated data structures and data manipulation techniques. For example, one scheme of storing the sparse matrix in Fig. 9a is shown in Fig. 9b. In this scheme the N nonzero elements of the matrix are stored in list A. The corresponding column indices are stored in list JA. A third list IA contains the row pointers that point to the starting positions of each row in lists A and JA. Thus, the number of nonzero elements in row i is given by $IA(i + 1) - IA(i)$, the values of these elements are given by $A(IA(i)), \ldots, A(IA(i + 1) - 1)$ and the column indices of these elements are given by $JA(IA(i)), \ldots, JA(IA(i + 1) - 1)$. For a matrix of order n list IA contains $n + 1$ elements the last of which points to the end of lists A and JA. This scheme was originally devised by Chang (C5) and has been applied to natural gas networks by Stoner (S5). We shall refer to it as scheme I. Notice that Scheme I requires $2N + n + 1$ locations to store N nonzero elements in a matrix of order n. It provides ready access to elements in a given row, but not to elements in a given column. It also requires significant revamping of lists in order to insert or delete a nonzero element.

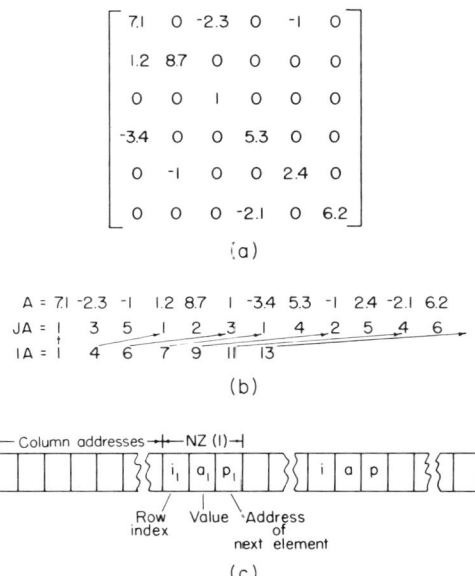

FIG. 9. Sparse matrix storage schemes: (a) a square matrix, (b) scheme I, (c) scheme II (linked lists).

By contrast, Fig. 9c shows an alternative scheme using linked list. In this scheme (scheme II) the information associated with a nonzero element is stored in a triplet containing the row index, the value of the nonzero element, and a pointer to the address of the next element in the same column. The starting addresses of each column are stored in another n locations. Notice that in this scheme the successive elements need not be stored in consecutive locations. To insert or delete an element requires only the change of one or two pointers; no rearrangement of the list is necessary. On the other hand, the storage requirement for the same matrix is now $3N + n$ and, as it stands, to find a specific nonzero element requires a linear search through the chain.

We offer these two examples of data storage scheme in order to illustrate the interrelationship between data structure, storage requirement, and the types of operations to be performed. The specific data structure and data manipulation techniques to be used should always be tailored to the structure of the matrix and the requirement of the application. In point of fact both schemes I and II can be modified to overcome some of the stated deficiencies. Gustavson (G9) discussed modifications of scheme I to permit both row- and column-oriented operations and to accommodate "fill-ins" generated during Gaussian elimination, and Tewarson (T1) proposed the use of an ordered linked list to reduce the search required by scheme II. Specific storage schemes for band and symmetric matrices have also been devised (T1). Finally, a sparse computation scheme which takes advantage of the specific features of networks with continuous piecewise-linear elements has been presented by Fujisawa, et al. (F7).

Although sparse computation techniques have been successfully applied to pipeline networks by Stoner (S5) and Bending and Hutchison (B5) without row and column reordering, it is well known that the sparsity of the original coefficient matrix may be partially destroyed by fill-ins. In a study using randomly generated sparse matrices Brayton, et al. (B11) showed up to a 10-fold increase in density using Gaussian elimination and up to a 20-fold increase in density using Gauss–Jordan elimination. The generation of fill-ins may be controlled by row and column ordering (selection of pivotal sequence). A review of methods for selecting pivotal sequences in Gaussian elimination may be found in a recent paper by Duff and Reid (D9) and a new method for selecting pivotal sequences in Gauss–Jordan elimination is given by Lin and Mah (L7). Row and column reordering algorithms specifically for formulation B have been presented by Mah (M1).

C. Networks with Regulators and Other Nonlinear Elements

In the treatment of steady-state pipeline network problems so far we have tacitly assumed that there is a unique solution for each problem. For certain types of networks the existence of a unique solution can indeed be rigorously established. The existence and uniqueness theorems for formulation C were proved by Duffin (D10) and later extended by Warga (W1). In Warga's derivation the governing relation for each network element assumes the form,

$$q_{ij} = f_{ij}(p_i - p_j) \tag{87}$$

where the function f_{ij} must satisfy the following conditions in order to ensure the existence of a unique solution:
  (i) $f_{ij}(x) = -f_{ji}(-x)$.
  (ii) $f_{ij}(x)$ is continuous for all x.
  (iii) for all edges $\{i, j\}$, $f_{ij}(x)$ is nondecreasing as x increases.
  (iv) Between every pair of vertices there exists a path of network elements, $\{i, j\}$, for which $f_{ij}(x)$ increases as x increases and assumes all values between $-\infty$ and $+\infty$.

For all practical purposes these conditions can only be met by a network of

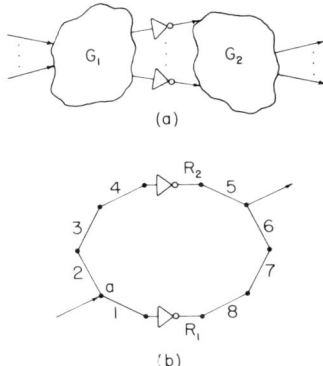

Fig. 10. Networks with pressure regulators: (a) a cut-set formed by regulators, (b) regulators in a cycle.

pipes and flow control valves, and then, only if the valves do not completely surround any vertex. Fujisawa *et al.* (F7) proved the existence of at least one solution for any arbitrary input to a piecewise-linear resistive network, if the determinants of the Jacobian in successive iterations are nonzero and of the same sign. For networks with pumps, compressors, check valves or regulators, neither the existence, nor the uniqueness of a solution is guaranteed.

Cheng (C7) recently made some interesting observations on networks with regulators. Since the purpose of the regulators is either to protect the downstream installation by regulating pressures below a certain limit, or to maintain the upstream installation pressures above a certain threshold, a regulator should be put on every path between the "source" and the "sink." Hence, the edges corresponding to the regulators should always form a cut-set (see Fig. 10a). For a given set of conditions, each regulator is either throttling or it is fully open. If it is throttling, then either the downstream pressure (for a downstream regulator) or the upstream pressure (for an upstream regulator) is known. For simplicity we shall refer to the downstream regulator in subsequent discussions.

Using the specifications [Eqs. (45)–(48)] at least one vertex pressure in each cycle is known. If the cycle contains one throttling regulator (for instance, R_1 in Fig. 10b), then the pressure difference between the reference vertex (vertex **a**) and the regulator is fixed and Eq. (38) may be simplified to

$$\sum_{k=2}^{8} \sigma_k(q_k) - p_a + p_{R_1} = 0 \qquad (88)$$

Similarly, if both regulators R_1 and R_2 in Fig. 10b are throttling, then the pressure difference along the connecting path is fixed and the simplified cycle equation is now

$$\sum_{k=5}^{8} \sigma_k(q_k) - p_{R_2} + p_{R_1} = 0 \tag{89}$$

For any given simulation, it is generally not possible to determine the state of each regulator beforehand. Hence, the states of all regulators must be included in the initial guesses. These initial assumptions must then be validated by subsequent computation. If the computed pressures at the regulators are in agreement with the regulator constraints, then the assumptions are valid. Otherwise, a different set of states must be assumed.

Since the regulators form a cut-set, one decomposition strategy is to tear the original network along the cut-set. Referring to Fig. 10a the decomposition creates two separate networks, G_1 and G_2, and $(2m + 1)$ new specifications where m is the number of edges in the cut-set. For instance, the pressures downstream of the regulators, p_D, and the overall material balance, Eq. (46) may be introduced as the specifications for the downstream network G_2, which can then be solved for the flows across the regulators. These flow rates form the additional specifications for the upstream network G_1, which can now be solved for pressures upstream of the regulators, and so on. The new specifications are in effect tear variables in the overall iterative scheme. It should be emphasized that there is no guarantee that such an iterative scheme will not oscillate, nor that it will indeed converge. Very little has been published on networks with "flip-flop" elements. There is definitely room for some definitive work.

IV. Design Optimization and Synthesis

In the discussions so far we have focused on the formulation, structural features, and numerical solution of pipeline network problems. In actual application a single steady-state solution rarely provides all the information required in either design or operation analysis. Given a set of requirements and specifications the ultimate design goal is to produce an optimal network that will meet all the constraints at minimal cost or maximal profit. Similarly for an operating network we may be interested in minimizing operating costs while meeting the external demands or in the optimal expansion of the existing facilities to meet anticipated future demands. In this section we shall review published methods and results pertaining to these problems. But before we do so, a brief discussion of the morphology of the subject area will help to clarify the considerations in our selection of topics.

TABLE V

APPLICATIONS OF PIPELINE NETWORK DESIGN AND SYNTHESIS

Applications	References
Agricultural engineering	Karmeli et al. (K2), Liang (L6), Yang et al. (Y1)
Control and actuation mechanism	Jacoby (J1)
Gas distribution	O'Callaghan (O1), Brameller et al. (B10), Flanigan (F4)
Gathering networks and off-shore production	Rothfarb et al. (R5), Babayev and Karayeva (B1)
Oil and gas transmission	Larson (L3), Wong and Larson (W10), Martch and McCall (M3), Cheesman (C6), Shamir (S1), Stoner (S5), Flanigan (F4), Zimmer (Z1), Bickel et al. (B7)
Pressure-relieving network	Murtagh (M9), Cheng and Mah (C8)
Sanitary engineering	Dajani et al. (D1), Cembrowicz and Harrington (C4)
Water distribution	Tong et al. (T2), Jacoby (J1), Lam and Wolla (L1, L2), Kally (K1), Swamee et al. (S8), Watanatada (W2), Donachie (D8), Gerlt and Haddix (G3), Nakajima (N2), Bonansinga (B8)

For the most part the material that we shall cover has appeared in publications since 1970. Researches on pipeline network design and synthesis appear to have germinated spontaneously in several application areas and have proceeded on a somewhat uneven keel, often without apparent awareness of similar efforts in other disciplines and applications. Table V shows the major areas to which these techniques have been applied. Although the field of application is quite diverse, the methodologies developed and employed are clearly related. Indeed, one of the purposes of this review is to bring these related developments to the attention of all engineers engaged in pipeline network design and synthesis. But since space limitations do not permit us to treat all these applications in equal detail, in our treatment we shall attempt to give prominence to those applications of greatest interest to chemical engineers.

The various contributions can also be classified in accordance with the optimization techniques used. However, this method of organization gives rise to an even more diverse classification, since the techniques used range all the way from rules of thumb (A3–A5, M6–M8, O1, T2) and analytical solution (S8) to the more recent developments in mathematical programming. Most of the techniques reported are continuous, but some are discrete (C8, R5) and still others are of mixed integer types (G3). Table VI shows such a classification for the papers reviewed. It is clearly beyond the scope of this review to delve into the mathematical bases of these methods. We shall

TABLE VI

OPTIMIZATION METHODS IN PIPELINE NETWORK DESIGN

Methods	References
Conjugate gradient	Watanatada (W2)
Discrete merging	Rothfarb et al. (R5), Cheng and Mah (C8)
Dynamic programming	Liang (L6), Wong and Larson (W10), Larson (L3), Shamir (S1), Zimmer (Z1)[a]
Generalized reduced gradient with branch and bound	Bickel et al. (B7)
Geometric programming	Cheng and Mah (C8)
Gradient projection	Murtagh (M9)
Linear programming	Goda and Ogura (G5), Karmeli et al. (K2), Brameller et al. (B10), Dajani et al. (D1), Kally (K1)
LP with branch and bound	Gerlt and Haddix (G3)
Steepest descent	Flanigan (F4)
Zoutendijk's method of feasible direction	Cembrowicz and Harrington (C4)

[a] The description given by Zimmer seems to indicate a grid search based on local optimization rather than dynamic programming.

refer interested readers to a previous review in these Advances (W3) and to standard references in this area (H1, H4, W5). Some space, however, will be devoted to those methods that appear to show the greatest merit and that pertain especially to pipeline network problems.

From the structural viewpoint there is much to commend the classification of problems based on the topology of the pipeline network— single branch pipelines, tree networks, and cyclic networks. However, since some methods are applicable to more than one category, rigorous adherence to this classification will lead to unnecessary duplication and overlaps.

The material under discussion may also be classified according to the problem requirements. In sensitivity analysis we are primarily concerned with the behavior of the network in the neighborhood of the design conditions. Such behaviors can often be studied using linearization techniques. In design optimization we broaden our viewpoint to include the whole feasible space. Depending on the problem context, the decision variables may either be continuous (e.g., nodal pressure) or discrete (e.g., pipe diameter), but the network configuration remains fixed. In network synthesis, on the other hand, the network configuration is not specified, only the flows and pressures are prescribed.

It should be pointed out that while the different classifications discussed above are very useful in imparting a perspective of the subject matter, they are not without ambiguities. Design optimization based on sensitivity information could be discussed in the context of either sensitivity analysis or design optimization. Likewise, optimal pipeline routing may be classified as either design optimization or synthesis with some justification in either case. In selecting our topics we were guided by the relevance of the subject matter, the utility of the method and the clarity and brevity of presentation.

A. Sensitivity Analysis

1. *Computation of the Sensitivity Matrix*

In analyzing the design or operation of a network, it is frequently of interest to know how the system would behave in the event that certain specifications such as delivery pressures or nodal flow rates were changed. If the changes are large and numerous; it would probably be expeditious to recompute the network conditions. But for many situations it may be sufficient to determine the approximate behavior from sensitivity information based on linearized approximation in the neighborhood of the original design solution. Such an analysis is most readily carried out when the Newton–Raphson method is used in the steady-state solution. Sensitivity analysis has been applied to pipeline network problems by Shamir and Howard (S2), Stoner (S5), and Donachie (D8). The application of similar analysis to distillation and process flowsheet calculations has been reported by Goldstein and Stanfield (G6) and Mah and Rafal (M2).

Let the network specifications and parameters be collectively denoted by u and the state variables be denoted by x as before. Then we may represent the steady-state pipeline network equations as

$$f(x, u) = 0 \qquad (90)$$

As we have shown in Sections II and III, Eq. (90) may be solved to determine the state variables x given a consistent set of values of u. Since the u's are specified by mathematical relations external to Eq. (90), they are sometimes referred to as the external variables (M2). The effect of varying the external variables and at the same time satisfying Eq. (90) is given by

$$df = \left(\frac{\partial f}{\partial x}\right)_u dx + \left(\frac{\partial f}{\partial u}\right)_x du = f_x\, dx + f_u\, du \qquad (91)$$

or

$$\left(\frac{\partial f}{\partial x}\right)_u \left(\frac{\partial x}{\partial u}\right) = -\left(\frac{\partial f}{\partial u}\right)_x \qquad (92)$$

where $(\partial x/\partial u)$ is an $(n \times m)$ matrix of partial derivatives of the n state variables with respect to the m external variables. We shall refer to it as the sensitivity matrix.

If the Newton–Raphson method is used to solve Eq. (1), the Jacobian matrix $(\partial f/\partial x)_u$ is already available. The computation of the sensitivity matrix amounts to solving the same Eq. (59) with m different right-hand side vectors which form the columns $-(\partial f/\partial u)_x$. Notice that only the partial derivatives with respect to those external variables subject to actual changes in values need be included in the m right-hand sides.

2. Parameter Estimation and Optimization Using Constrained Derivatives

A special adaptation of this analysis was applied by Donachie (D8) to estimate the pipe resistances using field measurements of pressures. In that application the u's correspond to pipe resistances and the x's correspond to the nodal pressures. Using linearized sensitivity information adjustments of u's are made, one at a time, to reduce the discrepancies between calculated and measured pressures according to the least-squares criterion. The Δu which causes the largest change in the sum of squares of discrepancies is given precedence over others.

An extension of the linearization technique discussed above may be used as a basis for design optimization. Such an application to natural gas pipeline systems was reported by Flanigan (F4) using the so-called constrained derivatives (W4) and the method of steepest descent. We offer a more concise derivation of this method following a development by Bryson and Ho (B14).

Let the objective function to be minimized be denoted by $\phi(x, u)$. Then the constrained optimization is governed by Eq. (91) together with

$$d\phi = \sum_{i=1}^{n} \frac{\partial \phi}{\partial x_i} dx_i + \sum_{j=1}^{m} \frac{\partial \phi}{\partial u_j} du_j = \phi_x \, dx + \phi_u \, du \tag{93}$$

or

$$\phi_u = \phi_x f_x^{-1} f_u \tag{94}$$

where ϕ_u is a vector of partial derivatives of ϕ with respect to u, the so-called constrained derivatives. These derivatives could be directly incorporated in the method of steepest descent. In Flanigan's derivation the constrained derivatives are given as ratios of two determinants. Flanigan (F4) illustrated the application with three sample problems from natural gas transmission and distribution but gave no computation strategy nor performance data.

A generalization of this method, known as the generalized reduced gradient (GRG) method, is treated by Himmelblau (H4) and discussed in Section IV,B,3.

B. Design Optimization

In many respects the tree network occupies a much more important role than the morphology of the network structure may suggest. Many large scale networks occur in this form. Long distance oil and gas transmission, pressure-relieving networks in refineries and chemical plants, gathering networks in oil and gas production as well as municipal sewers all fall within this category. Although it has never been rigorously established, many investigators (C4, F4, K1, M6–M8, N2, R5, W2) have observed that the minimum cost network for meeting stipulated flows and pressures at a set of vertices is a tree network, if reliability and vulnerability do not enter significantly into the design considerations. In some instances the tree network also appears to be a natural stepping stone in designing a cyclic network. For these reasons three of the following four sections will be devoted to applications to tree networks. Optimization of cyclic networks will be treated collectively in the last section.

Publications on optimal design of tree networks are further divided into single-branch trees or pipelines (C6, F4, L3, L6, S8) and many-branch trees (B7, C7, F4, K1, K2, M3, M9, N1, R5, W10, Y1, Z1). For our purposes, since the pipeline problems can always be solved using the optimization methods developed for the many-branch tree networks, we need to dwell no further on this special case. On the other hand, it is important to note that the form of the objective function could influence the applicability of a given optimization method. For the sake of concreteness, problem formulations and optimization techniques will be discussed in the context of applications.

1. *Pressure-Relieving Piping Networks*

The networks that interconnect various process units and vessels to the discharge zones or flares occur widely in refineries and chemical plants. Figure 11 shows a typical configuration in which the root represents the flare, the terminal vertices represent the relief valves, and the edge (each labeled with an arabic numeral) represents a pipe section between two physical junctions (valves, flare, or pipe joints). The configuration of such a network is dictated by the layout of the process unit. In this discussion both the lengths of the pipe sections and the interconnections will be treated as specified variables.

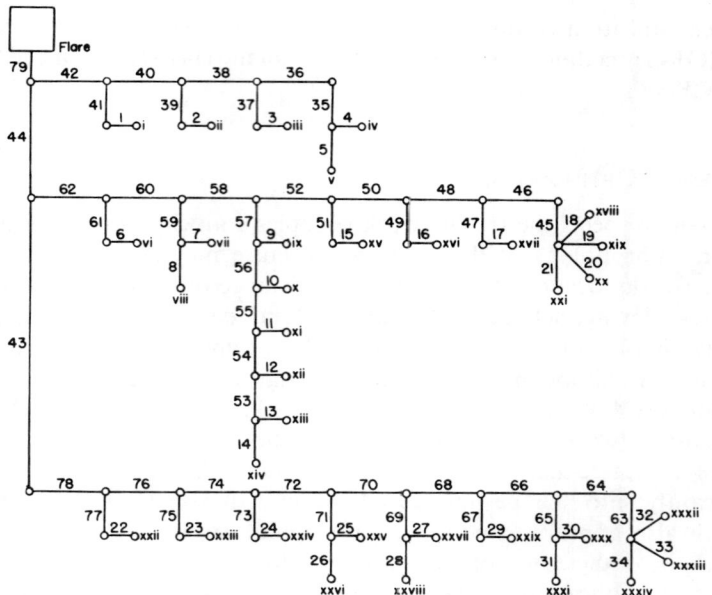

FIG. 11. A relief header network. Reproduced from Cheng and Mah (C8).

The design of the network calls for the selection of pipe diameters such that the discharge through each valve attains the maximum (sonic) velocity for an initial transitory period. Since the flare pressure and the process unit pressures are specified, this requirement amounts to the stipulation of a maximum allowable pressure drop over each path S_j (labeled with a roman numeral) from the valve to the flare. The optimal design in this case may be formulated as the following constrained minimization problem:

$$\min_{d} \sum_{i=1}^{P} l_i(\alpha_0 + \alpha_1 d_i) \tag{95}$$

where l_i and d_i are the length and the diameter of the ith pipe section, α_0 and α_1 are the cost coefficients, and P is the total number of pipe sections. Notice that in this case since there is no operating cost to speak of, the objective function is simply the capital investment which is assumed to be a linear function of diameters and lengths. If the customary API standard is followed, it can be shown (C8, M9) that the constraints may be simplified to assume the form,

$$b_j \geq \sum_{i \in S_j} (p_{u_i}^2 - p_{d_i}^2) = \sum_{i \in S_j} K_i/d_i^{4.814} \qquad j = 1, 2, \ldots, S \tag{96}$$

where p_{u_i} and p_{d_i} are the upstream and downstream pressures of the ith pipe section and K_i is a constant which includes the flow rate in the ith pipe.

The constrained minimization problem stated above may be transformed into a form well-suited to gradient projection methods of nonlinear programming by making the following substitution:

$$x_i = 1/d_i^{4.814}, \quad i = 1, \ldots, P \tag{97}$$

whereupon we obtain a linearly constrained minimization problem:

$$\min_x \sum_{i=1}^{P} l_i(\alpha_0 + \alpha_1 x_i^{-0.2077}) \tag{98}$$

subject to

$$\sum_{i \in S_j} K_i x_i \leq b_j \tag{99}$$

$$0 \leq x_i \leq \hat{x}_i \tag{100}$$

Murtagh (M9) pointed out that rounding errors and storage limitations restrict the applicability of such techniques to networks of approximately 100 pipe sections or less. As an alternative he proposed to solve the following dual problem:

$$\max_{\lambda \leq 0} \min_{x \in E^n} \psi(x, \lambda) \tag{101}$$

where the Lagrangian function,

$$\psi(x, \lambda) = \sum_{i=1}^{P} l_i(\alpha_0 + \alpha_1 x_i^{-0.2077}) - \sum_{j=1}^{S} \lambda_j \left(\sum_{i \in S_j} K_i x_i - b_j \right) \tag{102}$$

The advantage of this approach stems from the fact that the minimization over x can be carried out analytically. Consequently, the dimensionality of the optimization problem is reduced from P, the number of pipe sections to S, the number of paths. Murtagh (M9) reported computer storage reduction of more than 50% and computing time reduction of up to 80% using the dual instead of primal formulation.

The success of the dual approach and the form of the objective function and constraints suggest geometric (posynomial) programming as an alternative optimization technique. In the absence of the so-called reverse constraints, the posynomial program takes the following form:

$$\min_u g_0(u) \tag{103}$$

subject to

$$g_j(u) \leq 1, \quad j = 1, 2, \ldots, S \qquad (104)$$

$$u \geq 0 \qquad (105)$$

where

$$g_j(u) = \sum_i \gamma_i \prod_k u_k^{\alpha_{ik}}, \quad j = 0, 1, 2, \ldots, S \qquad (106)$$

the coefficients γ_i are positive and the exponents α_{ik} are arbitrary real constants. For the optimal design of pressure-relieving networks,

$$g_0(d) = \sum_{i=1}^{P} l_i(\alpha_0 + \alpha_1 d_i) \qquad (107)$$

$$g_j(d) = \frac{1}{b_j} \sum_{i \in S_j} k_i d_i^{-4.814}, \quad j = 1, 2, \ldots, S \qquad (108)$$

This formulation contains P primal variables and $2P$ to $(S+1)P - (S-1)S$ terms. Cheng and Mah (C8) evaluated this method and found both computing time and storage requirements to be quite excessive. A 14-branch and 5-path problem, for instance, took 11.9 CPU seconds and occupied 9728 words of storage on a CDC 6400 computer.

The applicability of all aforementioned continuous optimization techniques depends very heavily on the simplified form of the objective function and constraints. They also suffer the common drawback that the solution must be rounded to the nearest standard pipe sizes, a procedure whose pitfalls must then be eliminated by a partially enumerative search. By contrast, the discrete merge method (C8) which is an enhancement of an earlier development by Rothfarb et al. (R5), is fast, direct, compact, and flexible.

Consider a 100-branch network with 5 permissible pipe sizes for each branch. There are $5^{100} = 8 \times 10^{69}$ possible combinations of pipe diameters. Direct enumeration is clearly not a practical strategy even for a moderately sized network. In the discrete merge method only a small fraction of possible combinations is examined: Optimal partial assignments are created using information inherent in the branches and their interconnections. Since the flows and the pipe lengths are specified, each branch in the network is characterized by a pressure drop, a cost, and a size index. Numerical values of these three quantities arranged in the order of increasing pipe sizes for a given pipe section form a "branch list," which is the basic information used in the discrete merge method. The optimal partial assignments are created by merging the appropriate branch lists beginning with the pendant

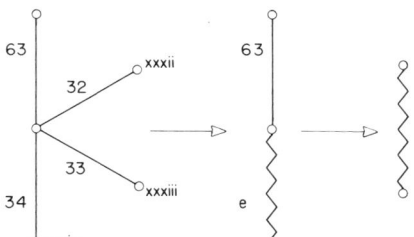

FIG. 12. Discrete merging.

branches (branches incident to vertices of degree one). The merge makes use of the following two properties: (1) For a given pipe length and mass flow rate, the pressure drop is a monotonically decreasing function of the pipe diameter, (2) but the piping cost, on the other hand, is a monotonically increasing function of the pipe diameter. As the merge proceeds, the nonoptimal partial assignments are progressively eliminated from further consideration. The merges are carried out in stages; the optimal partial assignments created in one merge becomes an "equivalent branch list" in the next merge.

There are two kinds of merges. Parallel merge is the procedure used to assign optimal diameters to the branches (or equivalent branches) which directly connect vertices of degree one to a common vertex. Serial merge, on the other hand, is used to create optimal partial assignments for two branches linked through a common vertex of degree two, if at least one of the two branches is a pendant branch or an equivalent branch. The two kinds of merges are illustrated schematically in Fig. 12. Through repeated applications of parallel and serial merges, the discrete merge method eventually transforms a tree network to a single equivalent branch list which gives the optimal diameter assignment and minimal cost not only for the desired b_j's, but also for all the parametric cases covered by the information inherent in the branch lists. This parametric capability is particularly useful when we analyze the trade-off between pressure drop constraints and the network cost.

A detailed description of the discrete merge algorithm and an analysis of its performance characteristics may be found elsewhere (C8). It suffices to point out that experience with this algorithm to date has been most encouraging. Cheng and Mah (C8) reported that the discrete merge method is about 2–4 times faster than the best previously known procedure (M9) and requires only about $\frac{1}{3}$ as much storage. Table VII summarizes the computation performance data based on the CDC 6400 computer.

The discrete merge method is a special case of dynamic programming applied to branch list processing. It is applicable to a wide class of network

TABLE VII

DISCRETE MERGE METHOD: COMPUTING TIME AND STORAGE REQUIREMENTS[a]

Number of pipes P	Number of paths S	Computing time (CPU seconds)	Storage requirement (words)
14	5	0.759	11,456
30	13	1.693	12,160
35	16	1.733	12,544
79	34	3.998	15,680

[a] Reproduced from Cheng and Mah (C8).

design problems of which the pressure-relieving and natural gas gathering networks are but two special cases. The basic requirements are that the network be acyclic and that the objective function and the constraints be monotonically increasing and decreasing functions, respectively. In the next section we shall treat a similar problem using an alternative approach which linearizes the problem at the expense of increasing the problem dimension.

2. Water Distribution Networks

Although typical municipal water distribution networks contain loops and bypasses, tree networks appear to be favored in agricultural applications. The single-source water distribution network is almost an exact analog of the pressure-relieving system with the flow directions reversed. We shall assign label **0** to the source node and label the other nodes and their corresponding branches numerically from **1** to **N** in an $(N + 1)$-node tree. As before, the length l_i and the flow rate q_i in each of the N branches are specified. To obtain a slightly more general formulation we shall allow each branch **i** to be made up of pipes selected from v_i different sizes, each of length ξ_{ij}. It follows that

$$\sum_{j=1}^{v_i} \xi_{ij} = l_i \qquad i = 1, 2, \ldots, N \tag{109}$$

and

$$\xi_{ij} \geq 0 \qquad j = 1, 2, \ldots, v_i \quad \text{and} \quad i = 1, 2, \ldots, N \tag{110}$$

Let α_{ij} and σ_{ij} be the cost and friction head loss per unit length of size j pipe in branch **i**. Then the actual hydraulic head at any node **k** must not be less than a specified minimum head $h(k)$, that is

$$H(0) - \sum_{i \in S_k} \sum_{j=1}^{v_i} \sigma_{ij}\xi_{ij} \geq h(k), \qquad k = 1, 2, \ldots, N \tag{111}$$

where S_k is the path from node 0 to node k, $H(0)$ is the pump head at the source node and

$$H(0) \geq h(0) \tag{112}$$

If the pumping cost is α_0 per unit head, then the cost to be minimized is given by

$$\alpha_0 H(0) + \sum_{i=1}^{N} \sum_{j=1}^{v_i} \alpha_{ij}\xi_{ij} \tag{113}$$

Equations (113) and (109)–(112) constitute the objective function and constraints of a linear programming problem. Notice that in this formulation the minimization is carried out with respect to both $H(0)$ and ξ_{ij}. Linearization is effected at the expense of increasing the number of independent (decision) variables to $1 + \sum_{i=1}^{N} v_i$. However, it can be shown that each branch is comprised of no more than two pipe sections in an optimal solution (K2). One merit of the linear programming formulation is the existence of highly developed and highly efficient computer codes. Unfortunately, no computation performance data on network optimization using this approach have been reported to allow a comparison with alternative methods in spite of the fact that the method has been used by a number of investigators (see Table VI). However, Kally (K1) claimed a cost reduction of 11% in one application as a result of design optimization.

3. Long Distance Gas Transmission

Optimal design of long distance natural gas pipeline is a subject of considerable commercial importance because of the very large capital investment involved. The subject has attracted a number of investigations since 1968 (see Table V). From a technical standpoint the problem is interesting because it is a transition between design optimization and network synthesis. Figure 13 shows a transmission tree with one branch point in which the nodes denote the compressors and edges the pipeline segments.

In the formulation of Bickel et al. (B7) which appears to be the most general formulation on this problem to date, both the feed and delivery conditions (temperature, pressure, flow rate, and composition) are specified. As before, the decision variables include the pipe diameters. But in addition, the number, placement, suction, and delivery pressures of compressors may also be varied within the constraints of overall pipeline lengths and network

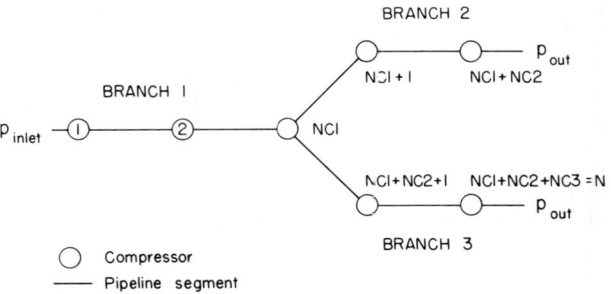

FIG. 13. A transmission tree.

configuration. In other words, the design optimization is to be carried out not on a fixed underlying graph, but over a family of homeomorphic graphs.[4]

The objective function is comprised of the sum of the annual operating and maintenance costs of the compressors and the sum of the discounted capital costs of the pipeline segments and compressors. The annualized capital cost for each pipe section is assumed to be proportional to its length and its diameter. Similarly the annual operation and maintenance charges of a compressor is assumed to be proportional to the horsepower which is given by

$$0.08531\omega q\theta_1[(p_d/p_s)^{Z/\omega} - 1] \qquad (114)$$

where $q =$ flow rate into the compressor, $\theta_1 =$ suction temperature, $p_s =$ suction pressure, $p_d =$ discharge pressure, $Z =$ compressibility factor of inlet gas, and $\omega = C_p/(C_p - C_v)$, C_p and C_v being specific heats at constant pressure and volume. If the capital cost of a compressor is also assumed to be proportional to its horsepower, then the minimum cost formulation of the design problem may be stated mathematically as

$$\min_{p_d, p_s, l, d} \left\{ \sum_{i=1}^{N} 0.08531(\alpha_0 + \alpha_c)\omega q_1 \theta_1[(p_{d_{i+1}}/p_{s_i})^{Z/\omega} - 1] + \sum_{j=1}^{N+1} \alpha_s l_j d_j \right\} \qquad (115)$$

where N is the number of compressors and α_0, α_c and α_s are the annual operating cost, compressor capital cost and pipe capital cost, respectively. The constraints in this minimization are the overall path length L_j,

$$\sum_{i \in S_j} l_i = L_j \qquad (116)$$

---

[4] Two graphs are said to be homeomorphic if one graph can be obtained from the other by the creation of edges in series or by the merger of edges in series (i.e. by insertion or deletion of vertices of degree two).

the pipeline element equation [the Weymouth formula was used by Bickel *et al.* (B7)], and the inequality constraints on compression ratio, pressures, pipe lengths, and diameters.

The minimal cost design problem formulated above was solved by Bickel *et al.* (B7) using the generalized reduced gradient (GRG) method of Abadie and Guigou (H4). If x and u are vectors of state and decision (independent) variables and $\phi(x, u)$ is the objective function in a minimization subject to constraints [Eq. (90)], then the reduced gradient $d\phi/du$ is given by

$$d\phi = [\nabla_u^T \phi - \nabla_x^T \phi (\nabla_x^T f)^{-1} \nabla_u^T f] \, du \qquad (117)$$

Initial search is in the direction of steepest descent given by the reduced gradient, z_0 say. Subsequent search directions s_{k+1} are generated by a conjugate direction formula (F6),

$$s_{k+1} = -z_{k+1} + s_k \frac{z_{k+1}^T z_k}{z_k^T z_k} \qquad (118)$$

In order to maintain feasibility, periodic updating of state variables and relinearization of active constraints are carried out.

For a 10-compressor, 2-path problem Bickel *et al.* (B7) gave a solution time of 353 CPU seconds on a CDC 6600 computer. As a comparison, the same problem was later solved by Chen and Fan using a direct search procedure (F2). The solution was obtained in less than 3 minutes on an IBM 370/158 computer.

In the formulation above, the discrete optimization on the number of compressors has been transformed into a continuous optimization on suction and delivery pressures. This transformation was made possible by the form of the compressor cost function which vanishes when $p_d = p_s$. However, if the compressor costs include a fixed capital outlay, i.e., the cost function is a linear function of horsepower with a nonzero constant term, then a branch and bound procedure must be used in conjunction with the GRG method.

4. Cyclic Networks

The first publications (A3–A5, M6–M8, T2) on cyclic network optimization used rules that appear to be based entirely on empirical observations. The best known of these rules formulated in terms of "equivalent pipe lengths" was enunciated by Tong *et al.* (T2) in 1961. The equivalent length is defined as the length of 8-in. diameter pipe with a roughness coefficient of 100, which gives rise to the same head loss at the same flow rate as the pipe under consideration. According to this design rule, an optimal design (pipe diameter assignments) is attained when the algebraic sum of the equivalent lengths of pipes in each and every loop in the network is equal to zero. This method has apparently been used in designing water distribution networks. More recently Deb and Sarkar (D3) proposed a method based on "equivalent diameters." While references to these two methods in published literature (N2, S8, W2) have been generally quite critical, to the best of our knowledge no comparative studies have been made nor counter examples presented to allow an impartial evaluation of their merits.

Optimization of cyclic networks using mathematical programming techniques has been reported by several investigators. Jacoby (J1) formulated his objective (merit) function to include penalty functions on nodal imbalance and nonzero loop pressure drop. He proposed a heuristic search procedure to solve the unconstrained minimization problem. Watanatada (W2) formulated the water distribution network design as a nonlinear programming problem. His objective function, which takes into account the capital costs of the pipes and pumps and the operating costs of pumps, is a nonlinear function of the pipe diameters and pump heads. After suitably transforming the inequality constraints into equality constraints and incorporating the latter into his objective function by means of Lagrange multipliers, he used the Davidson–Fletcher–Powell method (F5) to carry out the unconstrained minimization. Finally, Cembrowicz and Harrington (C4) proposed a procedure that makes use of chord flows in a cyclic network. This procedure requires a minimal flow to be specified for the network. By assigning the minimal flows to a chord set, the flows in the remaining arcs may be uniquely determined. This set of flows is termed a "minimal flow solution." Cost minimization based on this flow distribution is carried out with respect to head losses to yield a "local minimum." The global minimum is found by comparing all such local minima based on admissible minimal flow solutions. Since the examples used by these authors are illustrative rather than evaluative and no computation data were given, the computational merits of these three methods are somewhat debatable. Published literature gave no evidence that any of the above three methods has been applied to a problem of realistic dimension.

The linear programming approach outlined in Section IV,B,1,b has also been applied to cyclic networks (K1), the lack of theoretical validity notwithstanding. On an operational level, linear programming has been used to determine the most efficient means of supplying the water requirements of a major metropolitan area (G3) and to guide the allocation of production and supply of gas for the northwestern counties of England (B10). In the latter application the results of the grid optimization are used to determine (i) optimal allocation of natural gas supply, (ii) production plant to be operated during daytime (16 hours) and nightime (8 hours) and the method of operation at each station, (iii) holder stocks at all sites during maximum and minimum stock periods, (iv) jet boosters and compression required at all sites during day and night periods, (v) gas movement at different pressure levels during these periods, and (vi) marginal cost of production and distribution. The main grid optimization involving over 1000 variables was under development at the time of publication (B10) in 1971.

C. Synthesis

The ultimate goal of engineering design is to synthesize an optimal system directly from its functional requirements and specifications. In the context of pipeline networks neither the network elements nor the network configuration are given at the start of the synthesis; only the input and output conditions are specified. The simplest form of network synthesis involves a single input and a single output—the problem of optimal routing of a pipeline. The next simplest form involves a tree—the synthesis of a gathering or distribution network. Although very little has been written on network synthesis, two recent papers (R5, S1) have provided a most auspicious beginning. We shall summarize these developments in the following two sections.

1. *Optimal Pipeline Routing*

Although the topology of a single pipeline is trivial, the routing must take into consideration many topographical and environmental factors which may drastically affect the cost of the pipeline. These factors include the terrain in the corridor through which the pipeline passes, soil types, tree cover, water courses and rivers, roads and railroads, and property rights as well as labor and materials. In long distance transmission the capital requirement can vary significantly on the routes selected.

The optimal routing of a two-phase flow pipeline was investigated by Shamir (S1). Such pipelines are commonly used to convey both oil and gas from producing wells to collecting facilities and plants. They obviate the necessity for oil–gas separation facilities at the well head, which are sometimes uneconomical or impractical. In this context it is reasonable to assume that the pipeline will operate under the pressure differential naturally available between the source (well) and the point of delivery (refinery) and that the desired flow rate is specified. Hence one constraint on the optimal route is given by

$$\sigma_l l + \sigma_h h \leq \sigma \tag{119}$$

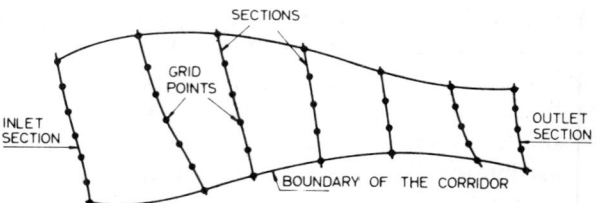

FIG. 14. Establishing a grid. Reproduced from Shamir (S1).

where σ = available pressure difference between the source and the delivery point, σ_l = pressure loss per unit length of pipe due to friction, σ_h = pressure loss per unit length of the vertical rises of all uphill sections along the pipe, l = length, and h = sum of the "ups." Both σ_l and σ_h are computed from the two-phase flow equation [the Flanigan method (S1)]. They do not depend on l or h. Notice that the uphill pressure loss term would not be present in the case of a single phase flow.

The first step in the route selection is to define a suitable corridor which connects the inlet and outlet sections. The choice of this corridor, its location and width, is based on engineering judgment. To discretize the problem the topographical characteristics of the corridor are defined over a two-dimensional grid as shown in Fig. 14. The grid is divided into sections ($i = 1, \ldots, n_i$) in the longitudinal direction of the corridor. Each section is divided transversely by grid points [$j = 1, \ldots, n_j(i)$]. We shall refer to the length of pipe connecting two grid points on adjacent sections as a segment. Clearly there is a trade-off between the resolution and the computational requirements in selecting the grid spacing. Rough guidelines might be $\frac{1}{2}$ to 2 miles between sections and a few hundred to a thousand feet between grid points.

Let $\phi[(i - 1, m); (i, j)]$ be the cost of the segment from point $(i - 1, m)$ to point (i, j). Then the route selection may be formulated as a cost minimization problem,

$$\min \sum_{i=2}^{n_i} \phi[(i - 1, m); (i, j)] \qquad (120)$$

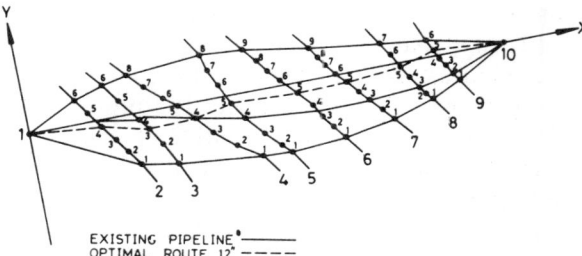

FIG. 15. Grid, existing and optimal routes. Reproduced from Shamir (S1).

subject to the constraint [Eq. (119)] in which l and h are functions of the point (i, j) and the route taken. The minimization may be readily carried out using dynamic programming (B4). Starting from the inlet, the suboptimal routes from the inlet to point (i, j) are generated successively for $\mathbf{i} = 1, 2, \ldots, \mathbf{n}_i$ and $\mathbf{j} = 1, 2, \ldots, \mathbf{n}_j(i)$ for different feasible values of σ. The cost of such a suboptimal route is given by

$$\phi_{ij}^*[\mathbf{i}, \mathbf{j} | \sigma_k] = \min_m \{\phi[(\mathbf{i} - 1, \mathbf{m}); (\mathbf{i}, \mathbf{j})] + \phi_{i-1, m}^*[(\mathbf{i} - 1, \mathbf{m}) | \sigma_k - (\sigma_l l_{mj} + \sigma_h h_{mj})]\} \quad (121)$$

where σ_k, $k = 1, 2, \ldots, K$ ranges over all the feasible values of σ at point $(\mathbf{i}, \mathbf{j})$. At each point $(\mathbf{i}, \mathbf{j})$ the cost associated with the suboptimal route $\phi_{ij}^*[(\mathbf{i}, \mathbf{j}) | \sigma_k(\mathbf{i}, \mathbf{j})]$ and the grid point $m_{ij}^*[\sigma_k(\mathbf{i}, \mathbf{j})]$ are stored. The latter is used to trace out the minimal cost route at the conclusion of the optimization. In the algorithm implemented by Shamir (S1), an "available length of pipe" and "available hills" were used in place of σ_k. Shamir also discussed interpolation procedures and the optimization procedure starting from the outlet.

It may be noted in passing that, unlike l and h, the influence of the pipe diameter is deeply imbedded in the two-phase flow model which is used to compute σ_l and σ_h. Consequently the design optimization with reference to pipe diameters cannot be effected within the same framework. However, in two-phase flow pipelines, the same pipe diameter is usually maintained throughout the sections in which the discharges do not change. This design practice is dictated by "pigging," an operating procedure in which "pigs"— cylindrical or spherical bodies—are sent with the flow down the line to clean the pipe walls from deposits, as well as to move fluid slugs over the hills. Hence, the diameter optimization can be adequately handled by case studies.

The procedure outlined above was evaluated using the data on an existing 5.8 mile pipeline (S1) as illustrated in Figs. 15 and 16. The optimal pipeline

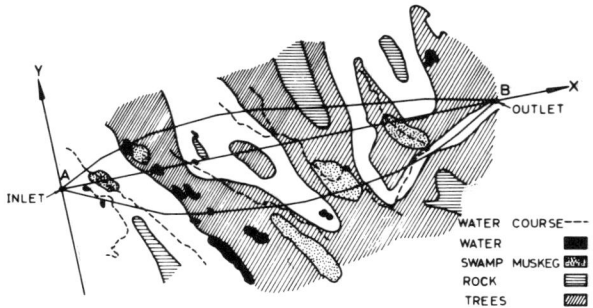

FIG. 16. Soil types and tree cover. Reproduced from Shamir (S1).

was found to be about 5% cheaper than the existing pipeline based on the same unit costs. The cost of a computer run was reported to be about $20 (1971).

2. Offshore Natural Gas Gathering Networks

For many combinatorial problems there is no known algorithmic procedure that will lead unerringly to the desired solution. However, it may be possible to devise heuristic procedures that will yield acceptable solutions most of the time and occasionally even the desired solutions to these problems. Although such procedures may lack the rigor and analytical foundation commonly associated with algorithmic procedures, from a practical viewpoint the shortcomings may not be nearly as overwhelming as they appear. A near-optimal solution that is significantly better than other solutions may be almost as good as an optimal solution, especially if the heuristic procedure used requires relatively little computing time and storage. The procedures, devised by Rothfarb et al. (R5), for synthesizing tree networks appear to belong to this category. These procedures have been applied to the design of offshore natural gas gathering networks.

The synthesis is carried out in two phases. In the first phase a heuristic starting routine is used to generate an initial tree. It was found through experience that the pipes connected to the separation plant (the root of the tree) play a special role in the development of a tree. These pipes will be referred to as "arms" and the pipes connected to each arm its "subtree." It is assumed that the number and location of the arms are specified as program input by the engineer. The two heuristic rules used by the starting routine are (i) efficient trees have low total pipe mileage; (ii) efficient trees have nearly equal flow in their arms. Notice that the application of rule (i) by itself

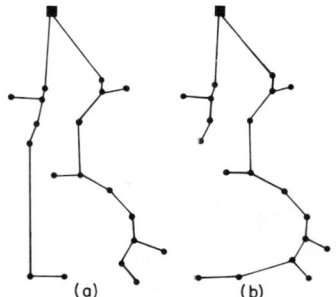

FIG. 17. Examples of two-arm network: (a) total 20-year cost $198,572,000, (b) total 20-year cost $210,100,000. Reproduced from Rothfarb et al. (R5).

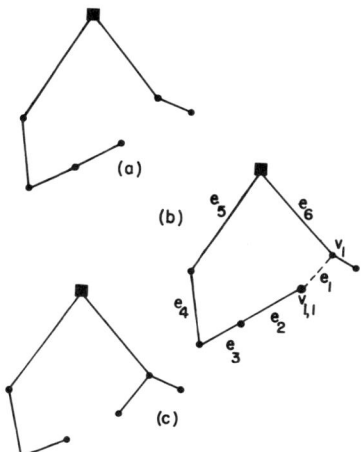

FIG. 18. Illustration of a Δ-change: (a) original tree, (b) original tree with added branch e_1 and circuit e_1, e_2, e_3, e_4, e_5, e_6, (c) new tree obtained by omitting e_2. Reproduced from Rothfarb et al. (R5).

leads to a shortest spanning tree, which is not always the lowest cost tree. The tree shown in Fig. 17b is closer to the shortest spanning tree than that shown in Fig. 17a, but its cost is actually 5.8% greater.

The second phase of synthesis is carried out using the so-called "Δ-opt" routine which improves on the initial tree by "Δ-changes." Specifically, for a given vertex v_1, find the vertex closest to v_1 but not adjacent to it. Add an edge from v_1 to this vertex $v_{1,1}$ and determine the circuit formed. Let e_1 be the new edge and $e_1, e_2, \ldots, e_m$ be the edges in this circuit. Then new trees can be formed by deleting each of the edges $e_2, e_3, \ldots, e_m$ in turn. The process of generating each new tree is termed a Δ-change and is illustrated in Fig. 18a–c. The Δ-changes can now be repeated with the nonadjacent vertex $v_{1,2}$ next closest to v_1. In the Δ-opt routine the three nearest nonadjacent neighbors of each vertex are used to generate Δ-changes. If there are k edges in each circuit on the average, approximately $3kN$ trees must be evaluated to certify an optimal network evolving from a given initial tree. Global optimality is approximated by repeating the procedure with different initial trees.

Since it is highly likely that a vertex will be connected to one of its nearest neighbors in the global optimum, Δ-changes will uncover a large fraction of local improvements without resorting to exhaustive enumerations. Restricting the choice to the three nearest nonadjacent neighbors helps to keep down the computing time, which is substantially more than network simulation or design. A typical running time for a 14-vertex problem is 6.5 minutes on a Univac 1108. However, it is encouraging to note the application experience reported by Rothfarb et al. (R5) that the design produced by their network synthesis procedure were better than those generated by "many trial-and-error human iterations."

V. Transient and Compressible Flows in Pipeline Networks

So far we have been concerned primarily with the design and synthesis of a pipeline network to meet steady demands; however, in reality steady-state operation at design loads is rarely attained. Deviations from design conditions are of particularly great economic significance in gas transmission systems in which transient states are often encountered. The loss of a compressor, the addition or loss of supply or sales points, replacement of equipment, and a customer who demands variable sale rates can each contribute to the creation of line transients. Cost-effective design must go beyond the engineering of facilities for nominal steady-state operation. It should consider the alternative means of meeting fluctuating demands. For instance, the extent to which such demands can and should be met by line-pack as opposed to peak-shaving facilities. It should also consider the sizing and location of standby equipment in the context of the system capacity and the penalty for failing to meet deliveries. A realistic assessment of such contingencies requires the capability to predict the time-dependent response of the system in addition to its steady-state behavior.

A. Governing Equations

For flow through pipes the governing equations are the equations of state, motion, continuity (material conservation), and energy conservation, and the state (dependent) variables are density, pressure, flow rate, and temperature. For engineering purposes, the following assumptions are usually made: (i) the flow is isothermal, (ii) steady-state friction is valid, (iii) expansion of pipe walls due to pressure changes may be neglected, and (iv) the pipe is assumed to have a constant slope over any particular reach. The first assumption is very reasonable for long pipelines, and in any case, the principal effect is on sonic velocity for which the isothermal sonic velocity is a lower limit. It allows us to dispense with the energy conservation equation.

Since the derivation of governing equations may be found elsewhere (S6, W6), we need only state them briefly in order to establish our notation and to point out some further simplifications. Let c be the isothermal sonic velocity, then the equation of state in terms of compressibility factor Z is

$$p/\rho = ZR\theta = c^2 \tag{122}$$

Making use of this relationship allows us to express the continuity equation as

$$\frac{c^2}{a}\frac{\partial W}{\partial x} + \frac{\partial p}{\partial t} = 0 \tag{123}$$

where t is the time, x the distance along the pipe, W the mass flow rate, and a the cross-sectional area. To state the equation of motion explicitly we shall choose a specific form of the steady-state friction equation:

$$\frac{\tau_0}{\rho} = \frac{fc^4 W|W|}{8p^2 a^2} \tag{124}$$

where τ_0 is the wall shear stress and f the Darcy–Weisbach friction factor. Using Eqs. (122)–(124), the equation of motion may be written as

$$\frac{\partial p}{\partial x}\left(1 - \frac{c^2 W^2}{a^2 p^2}\right) + \frac{pg \sin \delta}{c^2} + \frac{fc^2 W|W|}{2a^2 pd} + \frac{1}{a}\left(\frac{\partial W}{\partial t} + \frac{2c^2 W}{ap}\frac{\partial W}{\partial x}\right) = 0 \tag{125}$$

where δ is the angle between the pipeline and the horizontal. As a rule,

$$(cW/ap)^2 \ll 1 \tag{126}$$

$$\frac{2c^2 W}{ap}\frac{\partial W}{\partial x} \ll \frac{\partial W}{\partial t} \tag{127}$$

Hence Eq. (125) may be further simplified to

$$\frac{\partial p}{\partial x} + \frac{1}{a}\frac{\partial W}{\partial t} + \frac{pg \sin \delta}{c^2} + \frac{fc^2 W|W|}{2a^2 pd} = 0 \tag{128}$$

The physical significance of the first term is obvious. The second term is the inertia term. The third term accounts for gravitational effects; it vanishes for horizontal pipelines. The last term represents the frictional losses.

In terms of transient behaviors, the most important parameters are the fluid compressibility and the viscous losses. In most field problems the inertia term is small compared with other terms in Eq. (128), and it is sometimes omitted in the analysis of natural gas transient flows (G4). Equations (123) and (128) constitute a pair of partial differential equations with p and W as dependent variables and t and x as independent variables. The equations are hyperbolic as shown, but become parabolic if the inertia term is omitted from Eq. (128). As we shall see later, the hyperbolic form must be retained if the method of characteristics (Section V,B,1) is to be used in the solution.

Before we move on to a discussion of the methods of solution, we may note the two special cases of this general formulation. The special case of steady-state compressible flow ($\partial p/\partial t = \partial W/\partial t = 0$) was discussed by Daniel (D2). For incompressible transient flows, Eq. (123) is replaced by

$$\partial W/\partial x = \rho(\partial q/\partial x) = 0 \tag{129}$$

and Eq. (128) becomes

$$\frac{\partial p}{\partial x} + \frac{\rho}{a}\frac{\partial q}{\partial t} + \frac{pg \sin \delta}{c^2} + \frac{f\rho q|q|}{2a^2 d} = 0 \tag{130}$$

This case was studied by Nahavandi and Catanzaro (N1) using an explicit forward difference formula with Δx set equal to the pipe length and the time step dictated by stability considerations.

In the discussion above only a single pipe section was considered. For a network, nodal continuity equations in flows and pressures are also required (N1, W12). If a pipe section is too long to be treated as one cell, it may be divided into as many cells as necessary. Intermediate nodes are introduced between cells and the equations are augmented accordingly.

B. Methods of Solution

The transient network flow equations have been solved using implicit and explicit finite-difference procedures (I2). The chief advantage of the implicit method is its stability. Large time steps may be used for long-duration transients, while more detailed simulation of fast transients may be obtained with shorter time steps. The disadvantage of the method is the storage and computing time required in solving large sets of simultaneous algebraic equations which are typically nonlinear, if a second-order correct-in-time procedure is used. The implicit approach was used in implementing CAP (H3) and investigated by Streeter and associates (W12, S6).

The explicit methods avoid the need of solving large sets of equations and can therefore be used on smaller computers. However, these methods tend to be unstable unless the step sizes are kept small and an artificial constraint is introduced on the variables. In most formulations the time step must be less than the reach length divided by the isothermal sonic velocity (see Section V,B,1). Explicit methods are used in PIPETRAN (D7) and SATAN (G4).

Probably the least flexible of all methods with respect to the time-step and distance relationship is the method of characteristics (MOC). It requires the pipe lengths in a network to be adjusted to satisfy the condition of a common time interval, but provides an accurate solution of the differential equations. MOC has been successfully implemented by Goacher (G4), Streeter and associates (S6), and Masliyah and Shook (M5). More recently, attempts have been made to overcome the step-size restrictions using the so-called inertia multiplier (Y3, W13). However the accuracy and reliability of this procedure have been challenged (R1).

Since 1972 there has been a revival of interest in the classical variational methods of treating partial differential equations. Specifically, the Galerkin method has been applied to natural gas transmission problems with considerable success by Rachford and Dupont (R1, R2). In these methods a family of approximating functions are selected such that the initial and boundary conditions are automatically satisfied regardless of the values of the undetermined parameters. The undetermined constants or undetermined functions of a single variable are then chosen to approximate the solution behavior within the domain of interest. Different criteria may be employed in this approximation. In the Galerkin method weighted averages of residuals are used for this purpose. These methods are sometimes also referred to as the methods of weighted residuals. The success of these methods depends heavily on the selection of the trial functions. In general the desirable attributes of trial functions are that they should be simple and that they give wide variations of possible solutions, all of which satisfy the boundary and initial conditions. Hermite cubic polynomials were used by Rachford and Dupont (R1).

In the following sections we shall discuss two of these methods, MOC and implicit finite-differences, in slightly more detail.

1. *The Method of Characteristics*

Briefly the idea behind this method is to delineate families of curves in the x-t plane, called characteristic curves, along which the partial differential equations [(123) and (128)] become a system of ordinary differential equations which could then be integrated with greater ease. However, only hyperbolic partial differential equations possess two families of characteristics curves required by the method.

For the sake of simplicity the elevation (or gravitational effects) term will be omitted in the following development. Let Eq. (123) be multiplied by an unknown multiplier λ and then added to Eq. (128),

$$\frac{1}{a}\left(\lambda c^2 \frac{\partial W}{\partial x} + \frac{\partial W}{\partial t}\right) + \lambda\left(\frac{\partial p}{\partial x}\frac{1}{\lambda} + \frac{\partial p}{\partial t}\right) + \frac{fc^2 W|W|}{2a^2 pd} = 0 \qquad (131)$$

It is clear that with any two real and distinct values of λ, Eq. (131) will be equivalent to Eqs. (123) and (128) in every way. Let us choose λ so that

$$dx/dt = \lambda c^2 = 1/\lambda \qquad (132)$$

Then the quantities in brackets in Eq. (131) become total differentials and Eq. (131) becomes

$$\frac{1}{a}\frac{dW}{dt} + \lambda\frac{dp}{dt} + \frac{fc^2 W|W|}{2a^2 pd} = 0 \quad (133)$$

Solving Eq. (132) for the two values of λ and making explicit substitutions in Eqs. (132) and (133), we obtain

$$dx/dt = c \quad (134)$$

$$\frac{1}{a}\frac{dW}{dt} + \frac{1}{c}\frac{dp}{dt} + \frac{fc^2 W|W|}{2a^2 pd} = 0 \quad (135)$$

$$dx/dt = -c \quad (136)$$

$$\frac{1}{a}\frac{dW}{dt} - \frac{1}{c}\frac{dp}{dt} + \frac{fc^2 W|W|}{2a^2 pd} = 0 \quad (137)$$

Equations (135) and (137) are valid only along the characteristic directions defined by Eqs. (134) and (136), respectively. Let us refer to these directions as c^+ and c^-, respectively.

The computational procedure can now be explained with reference to Fig. 19. Starting from points P_1 and P_2, Eqs. (134) and (135) hold true along the c^+ characteristic curve and Eqs. (136) and (137) hold true along the c^- characteristic curve. At the intersection P_3 both sets of equations apply and hence they may be solved simultaneously to yield p and W for the new point. To determine the conditions at the boundary, Eq. (135) is applied with the downstream boundary condition, and Eq. (137) is applied with the upstream boundary condition. It goes without saying that in the numerical procedure Eqs. (135) and (137) will be replaced by finite difference equations. The Newton–Raphson method is recommended by Streeter and Wylie (S6) for solving the nonlinear simultaneous equations. In the specified-time-

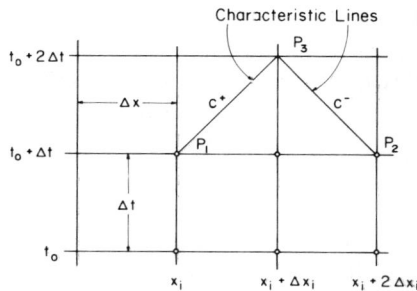

FIG. 19. Specified time interval method.

interval method equations for equally spaced sections along the pipe are solved. By virtue of Eq. (132)

$$\Delta t = \Delta x/c \qquad (138)$$

where Δx is referred to as the "reach length."

In the modified MOC due to Yow (Y3, W13), the contribution of the inertia term is deliberately enlarged by a constant factor α^2, and Eq. (128) is modified to read

$$\frac{\partial p}{\partial x} + \frac{\alpha^2}{a}\frac{\partial W}{\partial t} + \frac{pg\sin\delta}{c^2} + \frac{fc^2 W|W|}{2a^2 pd} = 0 \qquad (139)$$

With this modification and again neglecting the elevation term, we obtain

$$\frac{dx}{dt} = \frac{c}{\alpha} \qquad (140)$$

$$\frac{\alpha^2}{a}\frac{dW}{dt} + \frac{\alpha}{c}\frac{dp}{dt} + \frac{fc^2 W|W|}{2a^2 pd} = 0 \qquad (141)$$

$$\frac{dx}{dt} = -\frac{c}{\alpha} \qquad (142)$$

$$\frac{\alpha^2}{a}\frac{dW}{dt} - \frac{\alpha}{c}\frac{dp}{dt} + \frac{fc^2 W|W|}{2a^2 pd} = 0 \qquad (143)$$

in place of Eqs. (134)–(137). For a given reach length the time step is now given by

$$\Delta t = \alpha\,\Delta x/c \qquad (144)$$

The net effect of introducing the inertia multiplier is to increase the time step by a factor equal to α. The crucial question is how large a value can α be without substantially distorting the true physical behavior of the system. Yow (Y3) and Wylie et al. (W13) developed correlations for this purpose, but the procedures are apparently somewhat ambiguous and unreliable for pipeline networks. Rachford and Dupont (R1) gave a counter example for which an apparently reasonable choice of $\alpha(=5)$ miss the pressure prediction by as much as 100 psi.

2. *Implicit Finite Difference Methods*

In contrast to the method of characteristics, which gives faithful simulation of transient flows but which is very restrictive in time step sizes, the stability of the implicit methods permit large time steps and drastic reduc-

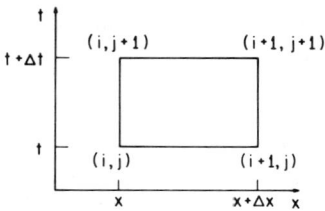

Fig. 20. A finite difference cell in the x-t plane.

tion of computer time for slow transients. The combined use of these two types of methods for pipeline network transients has in fact been proposed by Streeter and Wylie (S6). They suggested the use of MOC to break a complex system into regions so that the application of the implicit scheme does not require the solution of a large number of simultaneous nonlinear equations. At the same time the time steps of the combined method may be greatly increased over that of using MOC alone. In this section we shall present one such implicit scheme given by Wylie et al. (W12) as an illustration of this class of methods.

Referring to Eqs. (123) and (128) and Fig. 20 the partial derivatives may be expressed as

$$\frac{\partial p}{\partial x} = \frac{p_{i+1,j+1} + p_{i+1,j} - p_{i,j+1} - p_{i,j}}{2\,\Delta x} \tag{145}$$

$$\frac{\partial p}{\partial t} = \frac{p_{i,j+1} + p_{i+1,j+1} - p_{i,j} - p_{i+1,j}}{2\,\Delta t} \tag{146}$$

$$\frac{\partial W}{\partial x} = \frac{W_{i+1,j+1} + W_{i+1,j} - W_{i,j+1} - W_{i,j}}{2\,\Delta x} \tag{147}$$

$$\frac{\partial W}{\partial t} = \frac{W_{i,j+1} + W_{i+1,j+1} - W_{i,j} - W_{i+1,j}}{2\,\Delta t} \tag{148}$$

Omitting the elevation term and substituting the centered finite differences for the partial derivatives, Eqs. (123) and (128) become

$$\begin{aligned}f_1(p_{i,j+1}, p_{i+1,j+1}, W_{i,j+1}, W_{i+1,j+1}) \\ = (p_{i,j+1} + p_{i+1,j+1} - p_{i,j} - p_{i+1,j}) \\ + \frac{\Delta t}{\Delta x}\frac{c^2}{a}(W_{i+1,j+1} + W_{i+1,j} - W_{i,j+1} - W_{i,j}) \\ = 0\end{aligned} \tag{149}$$

$$f_2(p_{i,j+1}, p_{i+1,j+1}, W_{i,j+1}, W_{i+1,j+1})$$
$$= (p_{i+1,j+1} + p_{i+1,j} - p_{i,j+1} - p_{i,j})$$
$$+ \frac{\Delta x}{a\,\Delta t}(W_{i,j+1} + W_{i+1,j+1} - W_{i,j} - W_{i+1,j})$$
$$+ \frac{\Delta x f c^2}{4a^2\,d} \frac{(W_{i,j} + W_{i+1,j} + W_{i,j+1} + W_{i+1,j+1})}{(p_{i,j} + p_{i+1,j} + p_{i,j+1} + p_{i+1,j+1})}$$
$$\times |W_{i,j} + W_{i+1,j} + W_{i,j+1} + W_{i+1,j+1}|$$
$$= 0 \qquad (150)$$

Note that Eqs. (149) and (150) each contain four unknown quantities and that there are as many pairs of such equations as there are cells.

Ordinarily each pipe section is treated as a cell. But if the section is very long, it may be divided into many cells with intermediate nodes between them. Hence there are as many cells as the number of edges in the network. For a network with P edges and N vertices, there are N pressures, N nodal inflows or outflows, and $2P$ internal flows (bearing in mind that the flows at the two extremities of an edge are distinct). On the other hand, there are $2P$ governing equations of the type [(149) and (150)] and N nodal balances. Consequently, N specifications are needed to complete the description of the problem. Some caution is again necessary in picking admissible specifications. As a rule, it is safe to specify one variable at each node. The $2P + N$ equations along with the N specifications may be solved to yield the new conditions for the next time increment. The sparse computation techniques described in Section III,B,3 can again be used to reduce the storage and computing time requirements.

3. Computational Performance

In simulating transient flows in pipeline networks, the importance of accuracy cannot be over-emphasized. Because the transient behaviors are less well-understood, they are often rich in surprises. Physical intuition affords less guidance in these situations than in steady-state phenomena. Rachford and Dupont (R2) provided two instructive and deceptively simple examples to illustrate the interaction between regulators and compressors and the oscillatory response which can produce pressures higher than the supply pressure through reinforcement.

The unmodified MOC appears to give excellent accuracy in simulation when the time and distance increments are properly chosen. Streeter and Wylie (S6) showed one example for which MOC gave the same results for a reach length of 2.5 miles or less and a time increment of 12 seconds or less. The solution was accurate to within 1% using a reach length of 5 miles and a time increment of 24 seconds. At increments of 12.5 miles and 60 seconds a stable but incorrect solution was obtained. As a comparison a solution with 1% error was obtained with the implicit method using increments of 6.25 miles and 240 seconds. As a framework of reference, long transmission transients of the order of 1 day are not uncommon. At 12 second increments many thousands of time steps are needed for a complete simulation.

Wylie et al. (W13) presented a case study of a network with 16 vertices and 21 edges using the modified MOC. Satisfactory results were obtained using reach lengths of approximately 11.4 miles and time steps between 2 and 28 minutes.

Computational performance data on the Galerkin method were reported by Rachford and Dupont (R1). Time increments of the order of 5 minutes are typical. According to these authors an average of 120 pipe steps per CPU second on a CDC 6600 computer was achieved. Approximately 100 words of storage per pipe were required for array storage and 12,000 words of storage for the program and subroutines. On that basis the authors estimated that a 1000-pipe network can be simulated on a full-size CDC 6600 computer at about 100 CPU seconds per simulated hour.

VI. Concluding Remarks

In this review we have deliberately steered clear of any evaluation of available computer programs because an adequate discussion of this subject would require considerably more space than we have at our disposal. It is of course true that the input format and program flexibility can make a big difference to the utility of a computer program apart from any considerations of computational methods and strategy. Some discussions on these subjects for steady-state network simulation programs may be found in papers by Shamir and Howard (S2), Rosenhan (R4), O'Callaghan (O1), Boyer (B9), and Williams and Pestkowski (W8). For many applications, particularly steady-state network design or simulation, commercial programs are readily available. The relative merits of licensing these programs, using them on commercial time-share computer networks, or developing one's own programs are discussed by Bonansinga (B8) with reference to water distribution applications.

There is little doubt that computer-aided design techniques can give rise to drastic reductions of time and cost in designing pipeline networks. For small to medium applications Bonansinga (B8) estimated that a job which would have cost $500 by manual solution could now be accomplished at a cost of $7 using a computer. The comparison is even more favorable to the latter approach when multiple case studies are required. For large applications Neufville et al. (N3) estimated that the time required is about 1/100 of that required by manual calculations.

Equally impressive are the benefits of applying computer-aided design and synthesis. Typical overland gas transmission facility costs are $4000/in-mile for pipelines and $300/hp for compressors according to the figures quoted by Flanigan in 1972 (F4). Offshore facilities are even more costly. Table VIII lists cost data for the Gulf of Mexico pipeline around 1970 (R5). Based on investigations reported in the literature (F4, R5, S1) cost reductions on the order of 5–10% may be expected as a result of design optimization. Hence the savings on each application could range up to several million dollars on capital investment alone. These estimates are probably conservative, since Neufville et al. (N3) projected a cost reduction of 30% or $100 million in their study of design alternatives for the proposed additions to the New York City primary water supply network. Up to 20% savings in construction material was claimed by Smith (S3) as a result of using transient simulation in design. Although we know of no comparable figures for process applications, it is well known that process plant piping

TABLE VIII

APPROXIMATE GULF OF MEXICO PIPELINE COSTS PER MILE[a]

Nominal external pipe diameter (inches)	Onshore-marsh	Water depth (feet)	
		0–90	90–200
30	$260,000	$295,000	$325,000
26	195,000	205,000	235,000
24		188,000	215,000
20		153,000	175,000
16		131,000	150,000
$12\frac{3}{4}$		83,000	95,000
$10\frac{3}{4}$		65,000	75,000

[a] Reproduced from Rothfarb et al. (R5).

can run as high as 80% of process equipment cost or 15% of the installed plant cost (P2). There appears to be a growing awareness of this potential for design improvements among practicing chemical engineers (K4-K6).

In this review we have attempted to present a unified treatment of this emerging area. By relating the diverse developments we hope that we have succeeded in pointing out some new opportunities for research and innovation and reducing the likelihood of overlaps and rediscoveries. The subject area of this review has hitherto fallen outside the scope of traditional chemical engineering instruction. We hope that this review will stimulate efforts to introduce such material into the classroom.

Appendix: Description of Test Problems

Sample problem A is a small gravity-fed water distribution network taken from Carnahan and Wilkes (C2). The network is shown schematically in Fig. 21 with an arbitrary assignment of flow directions. The numerical data on this network including the initial guesses and the final solution are given in Tables IX and X. In addition, the following data are used in the computation: $\rho = 62.4 \text{ lb}_m/\text{ft}^3$, $\mu = 1 \text{ cP}$, $\epsilon = 0.01$ in. for all pipes.

TABLE IX

SAMPLE PROBLEM A: PIPE LENGTHS, DIAMETERS, AND FLOW RATES

Pipe number	Length l (ft)	Diameter d (in.)	Estimated flow rate $q^{(0)}$ (gpm)	Converged flow rate q^* (gpm)
1	40	4	1063.2	615.8923
2	40	3	−312.2	151.9614
3	40	3	−40.0	134.3869
4	80	3	−119.0	98.3495
5	40	3	−359.0	32.7364
6	60	3	509.0	117.2636
7	40	4	390.0	215.6131
8	60	3	252.5	86.1657
9	40	3	201.5	18.3119
10	40	3	−70.7	35.8865
11	40	3	220.7	114.1135
12	120	2	51.0	67.8537

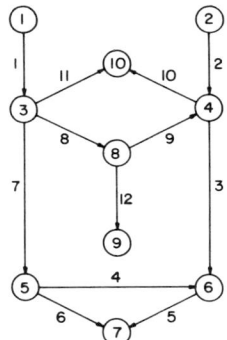

FIG. 21. Sample problem A.

Sample problem B shown in Fig. 22 is taken from Gay and Middleton (G1). All nodes in this network are at the same elevation and all pipes are 100 ft long and 6 in. in diameter. The values of fluid density and viscosity, and pipe roughness factor are taken to be the same as in the previous problem. Tables XI and XII summarize the numerical description of this network including initial guesses and final solution.

TABLE X

SAMPLE PROBLEM A: ELEVATION, PRESSURES AND WITHDRAWAL RATES

Node number	Elevation (ft)	Pressure p (psig)	Withdrawal rate $-w$ (gpm)
1	50	0	—
2	40	0	—
3	20	—	200
4	20	—	—
5	0	—	—
6	0	—	200
7	0	—	150
8	20	—	—
9	20	—	—
10	20	—	150

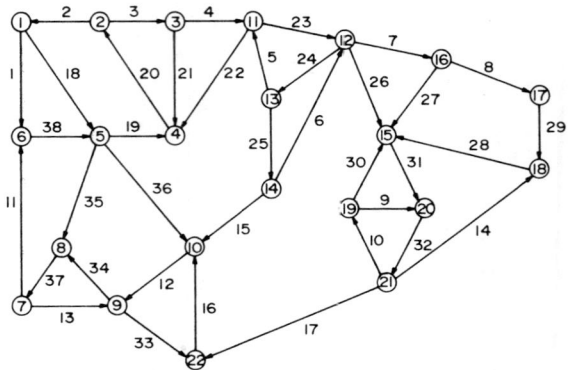

Fig. 22. Sample problem B.

TABLE XI

Sample Problem B: Initial and Final Flow Rates

Pipe number	Estimated flow rate $q^{(0)}$ (gpm)	Converged flow rate q^* (gpm)	Pipe number	Estimated flow rate $q^{(0)}$ (gpm)	Converged flow rate q^* (gpm)
1	−48.5	−111.1	20	−247.5	−285.0
2	−640.7	−664.8	21	−145.7	−246.1
3	393.2	379.7	22	−684.6	−674.8
4	538.9	625.8	23	293.6	360.2
5	−32.3	−42.7	24	−261.3	−357.0
6	490.2	478.9	25	−229.0	−314.4
7	649.9	633.3	26	−395.1	562.9
8	904.7	917.4	27	−254.8	−284.1
9	112.2	132.1	28	−268.9	−387.8
10	232.5	262.0	29	−890.6	−877.9
11	402.3	474.5	30	120.3	129.8
12	−420.4	−403.4	31	−8.1	20.9
13	687.3	573.9	32	−344.7	−295.8
14	621.7	490.2	33	473.1	402.1
15	−719.2	−793.3	34	−206.2	−231.5
16	52.7	−27.42	35	−275.0	−290.9
17	1199.0	1047.9	36	351.5	362.5
18	305.4	344.0	37	−481.2	−522.5
19	582.8	635.8	38	353.9	363.3

TABLE XII

Sample Problem B: Pressures and Inflow/Outflow Rates

Node number	Pressure p (psig)	Inflow rate w (gpm)
1	—	897.6
7	—	1570.9
11	—	−897.6
17	0	−1795.3
20	—	−448.8
22	—	673.2

Nomenclatures[5]

A	An occurrence matrix in Section II (40). A band matrix in Section III (86)	f	A friction factor in Sections II and V (20) and (124), and a function in Sections III and IV (53) and (90)
A	A data list in Fig. 9b	f_u	Partial derivative of function f with respect to u (91)
a	Cross sectional area (27)		
a_{ij}	An element of A	f_x	Partial derivative of function f with respect to x (91)
B	An incidence matrix containing the chords only. B refers to a digraph (17), and $\tilde{B}$ refers to an undirected graph (14)	G	A graph (3)
		G'	A subgraph of G
		g	Gravitational acceleration
B	Bandwidth of a matrix	g_c	Dimensionless gravitational conversion factor (20)
b	The number of network parameters β		
		H	Approximation of the negative inverse of the Jacobian (62)
b_j	An upper bound of optimization constraint (96)		
		$H(0)$	Pump head at the source node (111)
C	Cycle matrix. C refers to a digraph (8b) and (8d), and $\tilde{C}$ refers to an undirected graph (8a) and (8c)	h	Minimum pump head (112), and sum of "ups" (119)
		h_p	Compressor horsepower (32)
C	Cyclomatic number (5)	I	An identity matrix (13)
C_i	The ith fundamental cycle (38)	IA	Row pointers in Fig. 9b
c	Isothermal sonic velocity (122)	JA	Column index of nonzero elements in Fig. 9b
d	Pipe diameter (18)		
E	A set of edges (20)	K	Cut-set matrix. K refers to a digraph, $\hat{K}$ refers to a digraph containing only internal edges (42), and $\tilde{K}$ refers to an undirected graph (14)
E'	A subset of E		
E^n	Euclidean n-space (101)		
e_i	The ith edge (2)		

---

[5] Numbers in parentheses refer to the equation in which the symbol first appears. Letters and numerals that serve as labels of points and lines are shown in boldface when they appear in the text and are not given in the list of Nomenclature.

K_i	A constant (96)		refers to a digraph containing only internal edges (42), and $\tilde{T}$ refers to an undirected graph (13)
L_j	Length of path j (116)		
l_i	Length of ith pipe section (95)		
M	Reduced incidence matrix. M refers to a digraph (7), $\tilde{M}$ refers to an undirected graph (12), and $\hat{M}$ refers to a digraph containing only internal edges (34)	T	Indicates the transpose of a matrix, when used as a superscript
		T_i	Tree i
		t	Time
		u	Decision variable (90)
M'	Incidence matrix. M' refers to a digraph (6b), and $\tilde{M}'$ refers to an undirected graph (6a)	V	Vertex set (1)
		V'	Subset of V
		v_i	Vertex i (1)
M	Number of external flows (45)	W	Mass flow rate (123)
m	Number of decision variables u (93)	w	A vector of w_i (34)
N	Number of nodes (1). Number of nonzero elements in Section III,B	w(s)	Laplace transform of w(t) (49)
		w_i	Rate of inflow or outflow at node i (35)
n	Number of equations		
P	Number of edges (2)	w_i°	Specified rate of inflow or outflow at node i (45)
p	A vector of p_k		
p(s)	Laplace transform of $p(t)$ (51)	x*	Solution vector (54)
p_d	Discharge pressure (114)	x	State variable in general (53) and distance along the pipe in Section V (123)
p_k	Pressure at node k (18)		
p_s	Set pressure in Section II (28) and suction pressure in Section IV (114)		
		$\hat{x}_i$	An upper bound of state variable x_i (100)
p°	Reference pressure (23)	Y(s)	Diagonal edge admittance matrix (50)
p_k°	Specified pressure at node k (47)		
q	Vector of stream flow rates (7). $\hat{q}$ refers to the subset associated with the internal edges (34)	Y	A quantity defined by Eq. (25)
		y_k	Vector of changes of residuals in successive iterations (64)
$\hat{q}(s)$	Laplace transform of $\hat{q}(t)$ (50)	Z	Compressibility factor (114)
q_C	Vector of flow rates associated with chords or mesh flow rates (43)	z_k	Step direction vector in Section III (63) and reduced gradient in Section IV (118)
q_T	Vector of flow rates associated with tree arcs (43)		
		z	Elevation or height
q_{ij}	Stream flow rate from node i to node j (18)		
		Greek Letters	
R	Gas constant (122)		
R'	Mean convergence rate (79)	$\alpha, \boldsymbol{\alpha}$	Constants differently defined in different equations
R_i	Regulator i		
Re	Reynolds number (21)	β_i	ith network element parameter (48)
r	Number of function evaluations (79)	Γ	Fundamental cycle matrix. Γ refers to a digraph, $\hat{\Gamma}$ refers to a digraph containing only internal edges (44), and $\tilde{\Gamma}$ refers to an undirected graph (13)
S	Number of paths (96)		
S_j	The jth path (96)		
s	Search direction (118)		
s	Laplace transform variable (49)	γ_i	Nonnegative constant (106)
T	A partition of a fundamental cycle matrix, containing the columns corresponding to the tree branches only. T refers to a digraph (17). $\hat{T}$	δ	Angle between pipeline and the horizon (125)
		ϵ	Roughness factor in Section II (18);

	error tolerance in Section III (55)	$\dot{\sigma}$	Edge voltage vector referring to internal edges only.
ζ	An expansion factor (27)		
η	Pipe efficiency (24)	$\dot{\sigma}(s)$	Laplace transform of $\dot{\sigma}$ (50)
θ	Temperature (23)	σ_{ij}	Pressure drop across an edge $\{i, j\}$ (36)
$\theta°$	Reference temperature (23)		
λ	Lagrange multiplier in general (82) and a weighting factor in the method of characteristics in Section V (131)	τ	A damping factor (60)
		τ_0	Wall shear stress (124)
		ϕ	Objective function (80)
		ψ	A Lagrangian function (82)
μ	Absolute viscosity	ω	$= C_p/(C_p - C_v)$ (114)
ν_i	Number of different pipe sizes in branch i (109)		
ξ_{ij}	Length of the jth section of branch i (109)		OTHER SYMBOLS
		$\in$	Belongs to
π_i	Dimensionless weights (67)	$\subseteq$	Contained in or equal to
ρ	Fluid density (20)	$\|\cdot\|$	Norm of
ρ_i	Fluid density at node i (27)	∇	"Nabla" or "del" operator
σ	Pressure drop vector (9)	$\cong$	Is congruent with

ACKNOWLEDGMENTS

Acknowledgment is made to the Donors of the Petroleum Research Fund, administered by the American Chemical Society, for the support of this research, and to Wai-Biu Cheng and Iren Suhami for their invaluable assistance in assembling and editing this manuscript.

References

A1. Alexander, S. M., Glenn, N. L., and Bird, D. W., *J. Am. Water Works Assoc.* **67**, 343 (1975).
A2. Anonymous, "Engineering Data Book," 9th Ed. Gas Processors Suppliers Assoc., Tulsa, Oklahoma, 1972.
A3. Aoki, Y., *J. Jpn. Water Works Assoc.* **310**, 43 (1960); cf. Nakajima (N2).
A4. Aoki, Y., *J. Jpn. Water Works Assoc.* **314**, 32 (1960); cf. Nakajima (N2).
A5. Aoki, Y., *J. Jpn. Water Works Assoc.* **320**, 28 (1960); cf. Nakajima (N2).
B1. Babayev, D. A., and Karayeva, E. M., *Proc. Azerbaidzhan Acad. Sci. Ser. Physico-Tech. Math. Sci.* **3**, 42 (1971).
B2. Barlow, J. F., and Markland, E., *Proc. Inst. Civ. Eng.* **43**, 429 (1969).
B3. Bauer, W. J., Loui, D. S., and Voorduin, W. L., *in* "Handbook of Applied Hydraulics" (Davis, C. V., and Sorensen, K. E., ed.), p. 2-1. McGraw-Hill, New York, 1969.
B4. Bellman, R., "Dynamic Programming." Princeton Univ. Press, Princeton, New Jersey, 1957.
B5. Bending, M. J., and Hutchison, H. P., *Chem. Eng. Sci.* **28**, 1857 (1973).
B6. Berge, C., "The Theory of Graphs and Its Applications." Wiley, New York, 1962.
B7. Bickel, T. C., Himmelblau, D. M., and Edgar, T. F., "The Optimal Design of a Long Gas Transmission Line with Compressors in Series," presented at the 81st AIChE National Meeting, Kansas City, Missouri, April 12, 1976.

B8. Bonansinga, P., *J. Am. Water Works Assoc.* **67**, 347 (1975).
B9. Boyer, H. M., "A User Oriented System Design Program," presented at the Pacific Coast Gas Assoc. Distribution Conf., Los Angeles, California, 1970.
B10. Brameller, A., Chancellor, V. E., Hamam, Y., and Yalcindag, C., *Inst. Gas Eng. J.* **11**, 188 (1971).
B11. Brayton, R. K., Gustavson, F. G., and Willoughby, R. A., *Math. Comp.* **24**, 937 (1970).
B12. Brown, K. M., *SIAM J. Numer. Anal.* **6**, 560 (1969).
B13. Broyden, C. G., *Math. Comp.* **19**, 577 (1965).
B14. Bryson, A. E., and Ho, Y-C., "Applied Optimal Control; Optimization, Estimation and Control." Blaisdell, Waltham, Massachusetts, 1969.
C1. Carnahan, B., Luther, H. A., and Wilkes, J. O., "Applied Numerical Methods." Wiley, New York, 1969.
C2. Carnahan, B., and Wilkes, J. O., in CACHE "Computer Programs for Chemical Engineering Education Vol. VI Design." (R. Jelinek, ed.), p. 71. Sterling Swift, Manchaca, Texas, 1972.
C3. Carnahan, N. F., and Christensen, J. H., "Chemical Engineering Computing," Vol. 2. Am. Inst. Chem. Eng., New York, 1972.
C4. Cembrowicz, R. G., and Harrington, J. J., *J. Hydraul. Div., Am. Soc. Civ. Eng.* **99**, 431 (1973).
C5. Chang, A., in "Sparse Matrix Proceedings" (R. A. Willoughby, ed.), p. 113. Report RA1 (#11707) IBM Thomas J. Watson Research Center, 1968.
C6. Cheesman, A. P., *Oil Gas J.* **69**, 64 (1971).
C7. Cheng, W. B., M.S. Thesis, Northwestern University, Evanston, Illinois, 1976.
C8. Cheng, W. B., and Mah, R. S. H., *AIChE J.* **22**, 471 (1976).
C9. Christensen, J. H., and Rudd, D. F., *AIChE J.* **15**, 94 (1969).
C10. Collins, A. G., and Johnson, R. L., *J. Am. Water Works Assoc.* **67**, 385 (1975).
C11. Cross, H., *Ill. Univ. Eng. Exp. Sta. Bull.* **286** (1936).
D1. Dajani, J. S., Gemmell, R. S., and Morlok, E. D., *J. Sanit. Eng. Div., Am. Soc. Civ. Eng.* **98**, 853(1972).
D2. Daniel, P. T., *Trans. Inst. Chem. Eng.* **44**, T77 (1966).
D3. Deb, A. K., and Sarkar, A. K., *J. Sanit. Eng. Div., Am. Soc. Civ. Eng.* **97**, 141 (1971).
D4. Deo, N., "Graph Theory with Applications to Engineering and Computer Science." Prentice-Hall, Englewood Cliffs, New Jersey, 1974.
D5. Dijkstra, E. W., *Numer. Math.* **1**, 269 (1959).
D6. Dillingham, J. H., *Water Sewage Works*, 43 (1967); cf. Shamir and Howard (S2).
D7. Distefano, G. P., "A Digital Computer Program for the Simulation of Gas Pipeline Network Dynamics." PIPETRAN, Version IV, Am. Gas Assoc., New York, 1970.
D8. Donachie, R. P., *J. Hydraul. Div., Am. Soc. Civ. Eng.* **100**, 393 (1974).
D9. Duff, J. S., and Reid, J. K., *J. Inst. Math. Its. Appl.* **14**, 281 (1974).
D10. Duffin, R. J., *Bull. Am. Math. Soc.* **53**, 963 (1947).
E1. Enger, T., and Feng, C. C., *J. Hydraul. Div., Am. Soc. Civ. Eng.* **97**, 1607 (1971).
E2. Epp, R., and Fowler, A. G., *J. Hydraul. Div., Am. Soc. Civ. Eng.* **96**, 43 (1970).
F1. Fair, G. M., and Geyer, J. C., "Elements of Water Supply and Waste Water Disposal." Wiley, New York, 1958.
F2. Fan, L. T., Chen, T. C., and Aldis, D., *Proc. Symp. Computers Design Erection Chem. Plants*, p. 239. Czechoslovakia, 1975.
F3. Fietz, T. R., *J. Hydraul. Div., Am. Soc. Civ. Eng.* **99**, 1165 (1973).
F4. Flanigan, O., *J. Pet. Technol.* **24**, 549 (1972).
F5. Fletcher, R., and Powell, M. J. D., *Comput. J.* **6**, 163 (1963).

F6. Fletcher, R., and Reeves, C. M., *Comput. J.* **7**, 149 (1964).
F7. Fujisawa, T., Kuh, S. E., and Ohtusuki, T., *IEEE Trans. Circuit Theory* **19**, 571 (1972).
F8. Fulkerson, D. R., *J. Soc. Ind. Appl. Math.* **9**, 18 (1961).
F9. Fanning, J. T., "Treatise on Hydraulics and Water-Supply Engineering," Van Nostrand, New York, 1877 (cf. also 13th Ed., 1896).
G1. Gay, B., and Middleton, P., *Chem. Eng. Sci.* **26**, 109 (1971).
G2. Gay, B., and Preece, P. E., *Trans. Inst. Chem. Eng.* **53**, 12 (1975).
G3. Gerlt, J. L., and Haddix, G. F., *J. Am. Water Works Assoc.* **67**, 381 (1975).
G4. Goacher, P. S., *Inst. Gas. Eng. J.* **10**, 242 (1970).
G5. Goda, T., and Ogura, Y., *Jpn. Soc. Civ. Eng. Trans.* **138**, 21 (1967).
G6. Goldstein, R. P., and Stanfield, R. B., *Ind. Eng. Chem. Proc. Des. Dev.* **9**, 78 (1970).
G7. Graves, Q. B., and Branscome, D., *J. Sanit. Eng. Div., Am. Soc. Civ. Eng.* **84**, 1608 (1958).
G8. Gupta, P. K., Hendry, J. E., Hughes, R. R., and Westerberg, A. W., *AIChE J.* **20**, 397 (1974).
G9. Gustavson, F. G., in "Sparse Matrices and their Applications" (D. J. Rose and R. A. Willoughby, eds.). Plenum, New York, 1972.
H1. Hadley, H. G., "Non-Linear and Dynamic Programming." Princeton Univ. Press, Princeton, New Jersey, 1957.
H2. Harary, F., "Graph Theory." Addison-Wesley, Reading, Massachusetts, 1969.
H3. Heath, M. J., and Blunt, J. C., *Inst. Gas. Eng. J.* **9**, 261 (1969).
H4. Himmelblau, D. M., "Applied Nonlinear Programming." McGraw-Hill, New York, 1972.
H5. Hoag, L. N., and Weinberg, G., *J. Am. Water Works Assoc.* **49**, 517 (1957).
H6. Hsing, H. Y., M.S. Thesis, University of Texas at Austin, 1974.
H7. Hyman, S. I., and Jones, R. I., *Am. Gas Assoc. Workshop on Comput. Appl. Distribut. Des. Problems*, Washington D.C. February 14–16, 1967.
I1. Ingels, D. M., and Powers, J. E., *Chem. Eng. Prog.* **60**, No. 2, 65 (1964).
I2. Isaacson, E., and Keller, H. B., "Analysis of Numerical Methods." Wiley, New York, 1966.
J1. Jacoby, S. L. S., *J. Hydraul. Div., Am. Soc. Civ. Eng.* **94**, 641 (1968).
J2. Jeppson, R. W., and Tavallaee, A., *J. Hydraul. Div., Am. Soc. Civ. Eng.* **101**, 576 (1975).
K1. Kally, E., *Water Sewage Works*, **119**, R-121 (1972).
K2. Karmeli, D., Gadish, Y., and Meyers, S., *J. Pipeline Div., Am. Soc. Civ. Eng.* **94**, 1 (1968).
K3. Katz, D. L., Cornell, D., Kobayashi, R., Poettmann, F. H., Vary, J. H., Elenbaas, J. R., and Weinaug, C. H., "Handbook of Natural Gas Engineering." McGraw-Hill, 1959.
K4. Kern, R., *Chem. Eng.* **82**, No. 16, 107 (1975).
K5. Kern, R., *Chem. Eng.* **82**, No. 19, 129 (1975).
K6. Kern, R., *Chem. Eng.* **82**, No. 24, 209 (1975).
K7. King, I. P., *Int. J. Num. Methods Eng.* **2**, 523 (1970).
K8. Kniebes, D. V., and Wilson, G. G., *Chem. Eng. Prog. Symp. Ser.* No. 31, **56**, 49 (1960).
K9. Knights, I. A., and Allen, J. W., *Midland* Jun. Gas Assoc. (1965); cf. Gay and Middleton (G1).
K10. Kron, G., "Diakoptics." Macdonald, London, 1963.
L1. Lam, C. F., and Wolla, M. L., *J. Hydraul. Div., Am. Soc. Civ. Eng.* **98**, 335 (1972).
L2. Lam, C. F., and Wolla, M. L., *J. Hydraul. Div., Am. Soc. Civ. Eng.* **98**, 447 (1972).
L3. Larson, R. E., *IEEE Trans. Autom. Control* **12**, 767 (1967).
L4. Ledet, W. P., and Himmelblau, D. M., *Adv. Chem. Eng.* **8**, 186 (1970).
L5. Lemieux, P. F., *J. Hydraul. Div., Am. Soc. Civ. Eng.* **98**, 1911 (1972).
L6. Liang, T., *J. Hydraul. Div., Am. Soc. Civ. Eng.* **97**, 383 (1971).
L7. Lin, T. D., and Mah, R. S. H., *Math. Programming* **12**, 260 (1977).

L8. Liu, K. T. H., *Proc. Cong., Int. Assoc. Hydraul. Res. 13th* **1**, 136 (1969).
M1. Mah, R. S. H., *Chem. Eng. Sci.* **29**, 1629 (1974).
M2. Mah, R. S. H., and Rafal, M., *Trans. Inst. Chem. Eng.* **49**, 101 (1971).
M3. Martch, H. B., and McCall, N. J., "Optimization of the Design and Operation of Natural Gas Pipeline Systems," Paper No. SPE 4006, Soc. Pet. Eng., AIME, 1972; cf. T. C. Bickel, *et al.* (B7).
M4. Martin, D. W., and Peters, G., *J. Inst. Water Eng.* **17**, 115 (1963).
M5. Masliyah, J. H., and Shook, C. A., *Can. J. Chem. Eng.* **53**, 469 (1975).
M6. Matsuda, N., *J. Jpn. Water Works Assoc.* **328**, 31 (1962).
M7. Matsuda, N., *J. Jpn. Water Works Assoc.* **329**, 25 (1962).
M8. Matsuda, N., *J. Jpn. Water Works Assoc.* **330**, 21 (1962).
M9. Murtagh, B. A., *Chem. Eng. Sci.* **27**, 1131 (1972).
M10. Mah, R. S. H., Stanley, G. M., and Downing, D. M., *Ind. Eng. Chem. Prod. Res. Dev.* **15**, 175 (1976).
N1. Nahavandi, A. N., and Catanzaro, G. V., *J. Hydraul. Div., Am. Soc. Civ. Eng.* **99**, 47 (1973).
N2. Nakajima, S., *J. Am. Water Works Assoc.* **67**, 390 (1975).
N3. Neufville, R., Shaake, J., and Stafford, J. H., *J. Sanit. Eng. Div., Am. Soc. Civ. Eng.* **97**, 825 (1971).
O1. O'Callaghan, R. T., *Am. Gas Assoc. Oper. Sect. Distribution Conf., Philadelphia, 1969*.
O2. Ore, O., "Graphs and Their Uses." Random House, New York, 1963.
P1. Perry, R. H., and Chilton, C. H., "Chemical Engineers' Handbook," 5th ed., p. 5-11. McGraw-Hill, New York, 1973.
P2. Ibid. p. 25-18.
R1. Rachford, H. H., and Dupont, T., *Soc. Pet. Eng. J.* **14**, 165 (1974).
R2. Rachford, H. H., and Dupont, T., *Soc. Pet. Eng. J.* **14**, 179 (1974).
R3. Raman, V., *J. Sanit. Eng. Div., Am. Soc. Civ. Eng.* **96**, 1249 (1970).
R4. Rosenhan, A. K., ASME Publ. No. 69-WA/PVP-9 (1969).
R5. Rothfarb, R., Frank, H., Rosenbaum, D. M., Steiglitz, K., and Kleitman, D. J., *Oper. Res.* **18**, 992 (1970).
S1. Shamir, U., *Soc. Pet. Eng. J.* **11**, 215 (1971).
S2. Shamir, U., and Howard, D. D., *J. Hydraul. Div., Am. Soc. Civ. Eng.* **94**, 219 (1968).
S3. Smith, B., *Process Eng.* **90**, October (1974).
S4. Steward, D. V., *SIAM Rev.* **4**, 321 (1962).
S5. Stoner, M. A., *Soc. Pet. Eng. J.* **12**, 115 (1972).
S6. Streeter, V. L., and Wylie, E. B., *Soc. Pet. Eng. J.* **10**, 357 (1970).
S7. Stuckey, A. T., *Water Water Eng.* **73**, 104 (1969).
S8. Swamee, P. K., Kumar, V., and Khanna, P., *J. Environ. Eng. Div., Am. Soc. Civ. Eng.* **99**, 123 (1973).
T1. Tewarson, R. P., "Sparse Matrices." Academic Press, New York, 1973.
T2. Tong, A. L., O'Connor, T. F., Stearns, D. E., and Lynch, W. O., *J. Am. Water Works Assoc.* **53**, 192 (1961).
V1. Varga, R. S., "Matrix Iterative Analysis," p. 9, Prentice-Hall, Englewood Cliffs, New Jersey, 1962.
V2. Voyles, C. F., and Wilke, R., *J. Am. Water Works Assoc.* **54**, 285 (1962).
W1. Warga, J., *Proc. Instrum. Soc. Am.* **9**, Pt. 5, Paper 54-43-4 (1954).
W2. Watanatada, T., *J. Hydraul. Div., Am. Soc. Civ. Eng.* **99**, 1497 (1973).
W3. Wilde, D. J., *Adv. Chem. Eng.* **3**, 273 (1962).
W4. Wilde, D. J., *Oper. Res.* **13**, 848 (1965).

W5. Wilde, D. J., and Beightler, C. S., "Foundations of Optimization." Prentice-Hall, Englewood Cliffs, New Jersey, 1967.
W6. Wilkinson, J. F., Holliday, D. V., Batey, E. H., and Hannah, K. W., "Transient Flow in Natural Gas Transmission Systems." Tracor, Am. Gas Assoc., New York, 1965.
W7. Williams, G. N., *J. Hydraul. Div., Am. Soc. Civ. Eng.* **99**, 1057 (1973).
W8. Williams, J. D., and Pestkowski, T., *J. Inst. Water Eng.* **26**, 381 (1972).
W9. Wolfe, P., *Commun. ACM* **2**, 12 (1959).
W10. Wong, P. J., and Larson, R. E., *IEEE Trans. Autom. Control* **13**, 475 (1968).
W11. Wood, D. J., and Charles, C. O. A., *J. Hydraul. Div., Am. Soc. Civ. Eng.* **98**, 1157 (1972).
W12. Wylie, E. B., Stoner, M. A., and Streeter, V. L., *Soc. Pet. Eng. J.* **11**, 356 (1971).
W13. Wylie, E. B., Streeter, V. L., and Stoner, M. A., *Soc. Pet. Eng. J.* **14**, 35 (1974).
Y1. Yang, K-P., Liang, T., and Wu, I-P., *J. Hydraul. Div., Am. Soc. Civ. Eng.* **101**, 167 (1975).
Y2. Yao, K. M., *J. Am. Water Works Assoc.* **56**, 703 (1964).
Y3. Yow, W., *Trans. ASME* **94**, 422 (1972).
Z1. Zimmer, H. I., *Oil Gas J.* **73**, 204 (1975).

MASS-TRANSFER MEASUREMENTS BY THE LIMITING-CURRENT TECHNIQUE

J. Robert Selman and Charles W. Tobias

Department of Chemical Engineering
Illinois Institute of Technology
Chicago, Illinois

Materials and Molecular Research Division
Lawrence Berkeley Laboratory and
Department of Chemical Engineering
University of California
Berkeley, California

I. Introduction	212
II. Basic Theory	213
A. Limiting Current and Diffusion Layer	213
B. Measurement of Mass-Transfer Rates	215
III. Limiting-Current Measurement	217
A. Historical Survey	217
B. Electrochemical Reactions Used in Limiting-Current Measurements	219
C. Measurement of Electrode Potential	223
D. Boundary Conditions at the Electrode	227
E. Approach to the Limiting Current	228
IV. Interpretation of Results	229
A. Definition of the Limiting-Current Plateau	229
B. Migration Effects	231
C. Diffusion Coefficients	233
D. Unsteady-State Effects	235
E. Interaction with the Potential Distribution at Elongated Electrodes	243
F. Effect of Surface Roughness	247
V. Conditions for Valid Measurement and Interpretation of Limiting Currents	252
VI. Review of Applications	253
A. Simple Laminar Flows	254
B. Laminar Flows with Complications	259
C. Free Convection and Mixed Convection	263
D. Turbulent Forced Convection	268
E. Oscillating Flows	273
F. Stirred Cells	274
G. Convection by Gas Evolution	275
H. Particulate Electrodes	276

VII. Concluding Remarks 279
　　Nomenclature 309
　　References 310

I. Introduction

Electrochemical measurements of mass-transfer rates by the limiting-current technique have been employed with increasing frequency in the last 20 years. This chapter offers a discussion of the underlying principles, conditions of validity, and selected applications.

The basic features of the limiting-current phenomenon are described in Section II. In Section III a synopsis is given of electrochemical mass-transfer theory, to the extent required for analysis of limiting-current conditions. Although *surface overpotential* is introduced to allow consideration of the effect of rate limitations at the electrode–solution interface, the kinetics of surface reaction are not discussed as competent treatments of electrode kinetics and reaction mechanisms are available in numerous textbooks and monographs (e.g., G4b, N8a, V4). In Section IV various complicating factors in the interpretation of limiting-current measurements are considered: migration effects, the choice of appropriate diffusivity values, unsteady-state conditions, current distribution below and at the limiting current, and the effect of formation of rough metallic deposits. Consideration of the effect of a magnetic field on the limiting current is not included; initial results have been reported only recently (M13b, M13c). Section V is a summary of the criteria for obtaining valid limiting-current data and for their correct interpretation.

A selective review of mass-transfer studies based on limiting-current measurements is given in Section VI. To assist the reader, a table (Table VII) of mass-transfer correlations obtained by means of the limiting-current technique in over 200 separate investigations is provided. Part of these studies were aimed specifically at developing a more rational basis for the design of electrochemical reactors. The greater part of the experimental work, however, was undertaken by various investigators to exploit the unique advantages of the limiting-current technique for obtaining local and average mass-transfer coefficients under a large variety of hydrodynamic conditions. The primary interest here was to gain insight into the nature of convective mass transport and also to provide practical correlations for the design and operation of process equipment. This chapter therefore should be of interest not only to engineers directly concerned with electrochemical processes but

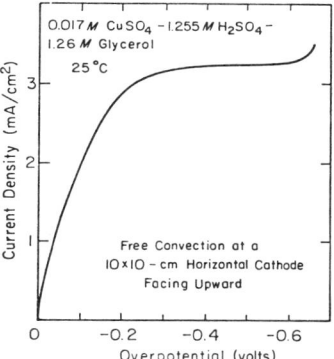

FIG. 1. Experimental limiting-current curve for copper deposition from an acidified cupric solution. [From Hickman (H3).]

also to chemical and mechanical engineers involved in research in fluid mechanics and heat and mass transport.

II. Basic Theory

A. Limiting Current and Diffusion Layer

When copper ions, for instance, in an acidified aqueous solution are discharged at an electrode by applying to the latter a negative potential with respect to a copper anode, for example, it is found that, upon increasing the applied potential slowly, the current at first increases rapidly and then reaches a saturation level, as evidenced by a current " plateau " (Fig. 1). Only upon relatively great further increase of the applied potential will the current rise appreciably again. The appearance of gas bubbles at the cathode indicates that the electrode reaction is no longer exclusively the deposition of copper; hydrogen ions are also being discharged, resulting in the evolution of hydrogen gas.

This simple experiment illustrates the basic features of the limiting-current method. A particular electrode reaction proceeds at the highest possible rate, indicated by a current plateau. From the limiting current thus recorded the mass-transfer rate and the mass-transfer coefficient at the electrode in question may be determined.

The term *limiting-current density* is used to describe the maximum rate at 100% current efficiency, at which a particular electrode reaction can proceed in the steady state. This rate is determined by the composition and transport properties of the electrolytic solution and by the hydrodynamic condition at the electrode surface.

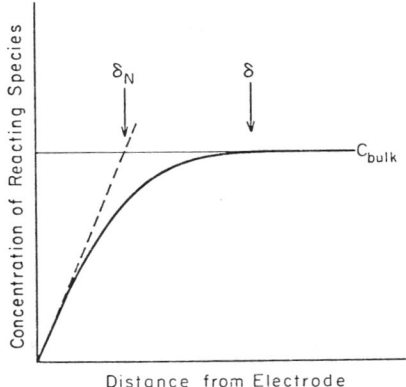

FIG. 2. Concentration profile of the reacting ion at an electrode. The so-called Nernst diffusion layer thickness is indicated by δ_N.

When a limiting current is encountered, it is almost always caused by the slowness of transport of charged (ionic) or uncharged (molecular) species through the solution.[1] These species move toward the appropriate electrodes, where they are consumed in the electrode reaction, or in a reaction coupled with it. Whenever the supply of a dissolved species from the solution to the electrode surface becomes a rate-limiting factor, limiting-current phenomena may be observed.

Electrode reactions are intrinsically surface reactions, and cause changes in the electrolyte composition near the surface. A thin layer, impoverished in the reacting species, develops at the electrode surface, and the ions or molecules move across this layer by diffusion down the concentration gradients. Ions move also under the influence of the applied electric field; that is, they migrate.

Early investigators assumed that this so-called diffusion layer was stagnant (Nernst–Whitman model), and that the concentration profile of the reacting ion was linear, with the film thickness δ_N chosen to give the actual concentration gradient at the electrode. In reality, however, the thin diffusion layer is not stagnant, and the fictitious δ_N is always smaller than the real mass-transfer boundary-layer thickness (Fig. 2). However, since the actual concentration profile tapers off gradually to the bulk value of the concentration, the well-defined Nernst diffusion layer thickness has retained a certain convenience in practical calculations.

---

[1] Limiting-current phenomena may also result from a slow step in the reaction mechanism at the surface of the electrode (V4, p. 233 ff). Electrode reactions of this type are not suitable for mass-transport studies.

In a stagnant solution, free convection usually sets in as a density gradient develops at the electrode upon passing current. The resulting convective velocity, which is zero at the wall, enhances the transfer of ions toward the electrode. At a fixed applied current, the concentration difference between bulk and interface is reduced. For a given concentration difference, the concentration gradient of the reacting species at the electrode becomes steeper (equivalent to a decrease of the Nernst layer thickness), and the current is thereby increased.

For a given hydrodynamic condition near the electrode in steady state, the maximum gradient is obtained when the concentration at the electrode is zero, or virtually zero. From the definition of "limiting-current density," this situation corresponds to the limiting-current condition.

Limiting currents are usually associated with cathodic reactions (e.g., in metal deposition), although anodic reactions are by no means excluded. Whenever the supply of a dissolved species from the solution to the electrode surface becomes the rate-limiting factor, limiting-current phenomena may be observed. Anodic limiting currents can be obtained, for example, in the oxidation of ferrous to ferric ion, or ferro- to ferricyanide ion (E1). Diffusion of H_2O limits O_2 evolution in fused NaOH (A2). In these examples the limiting current is caused by depletion of the reactant species at the anode.

B. Measurement of Mass-Transfer Rates

If the rate of the reaction is not restricted by the kinetics of the surface reaction, the reaction is called "reversible" and its rate is transport controlled. By Faraday's law the current density i is proportional to the reacting ion (or molecule) flux N_i:

$$s_i i = -nFN_i \tag{1a}$$

Here, s_i is the number of molecules or ions of species i participating in the transfer of n electrons to or from the electrode:[2]

$$s_1 M_1^{z_1+} + s_2 M_2^{z_2+} + \cdots + s_i M_i^{z_i+} \to ne^- \tag{1b}$$

The driving force of the mass-transfer process now can be related to the concentration gradient of the reacting species, or to the concentration difference between electrode and bulk solution. Mass-transfer rates then can be related in a general way to the concentration driving force. For example, if

---

[2] $s_i > 0$ for reactant species in anodic reactions and product species in cathodic reactions; $s_i < 0$ for product species in anodic reactions and reactant species in cathodic reactions.

the concentration gradient is known, an effective diffusivity can be determined:

$$N_i = -D_i^{\text{eff}} \nabla c_i. \tag{2}$$

In the majority of cases, however, only the difference of concentration between bulk and surface is known; for these situations the "mass-transfer coefficient" can be conveniently defined by

$$N_i = k(c_{i,0} - c_{i,b}) = -s_i i/nF \tag{3}$$

In the limiting-current method of measuring mass-transfer coefficients, the reacting-ion concentration at the electrode is made vanishingly small by applying a sufficiently large potential. In this case, therefore, only the current density and the bulk concentration need to be known.

Optical techniques, in particular interferometry, may be used to measure a nonzero concentration of the reactant at the electrode. However, such measurements are restricted to (a) dilute solutions, because refraction occurs in addition to interference (B4a), and (b) solutions in which only the concentration of the reacting species varies, that is, to solutions of a single salt. If the solution contains two electrolytes with dissimilar concentration profiles in the diffusion layer, then a second independent measurement is needed to establish the reactant concentration at the electrode. Interferometric methods are considered in detail by Muller (M14).

In binary solutions, for example, $CuSO_4$ in H_2O, the limiting current exceeds that due to convective diffusion alone by a factor of about two. The excess mass transfer is caused by *migration* of the reacting ion in the electric field. In both forced and free convection it is important to know the ion flux contributed by migration, which can never be suppressed completely.

The migration flux can be very much decreased by increasing the conductivity of the solution, thereby lowering the electric field strength. Thus, in studies involving electrochemical mass-transfer measurement, a large amount of "supporting" or "inert" electrolyte is usually added to the solution. Examples are: (1) H_2SO_4 added to $CuSO_4$ for the copper-deposition reaction; and (2) NaOH or KOH added to $K_3Fe(CN)_6$ for the cathodic reduction of ferricyanide. Therefore, in practice, multicomponent electrolytic solutions are used more frequently than single-salt solutions.

The complications associated with the presence of nonreacting ions in the diffusion layer are discussed extensively by Selman and Newman (S9a, S9b). In free convection, where the driving force, that is, the density difference, is affected by the accumulation or depletion of nonreacting ions at the electrode, consideration of the concentration profiles of these species assumes special importance (Section IV,B).

III. Limiting-Current Measurement

A. HISTORICAL SURVEY

The dependence of the limiting current density on the rate of stirring was first established in 1904 by Nernst (N2) and Brunner (B11a). They interpreted this dependence using the stagnant layer concept first proposed by Noyes and Whitney. The thickness of this layer ("Nernst diffusion layer thickness") was correlated simply with the speed of the stirring impeller or rotated electrode tip.

The lack of hydrodynamic definition was recognized by Eucken (E7), who considered convective diffusion transverse to a parallel flow, and obtained an expression analogous to the Lévêque equation of heat transfer (L5b, B4c, p. 404). Experiments with Couette flow between a rotating inner cylinder and a stationary outer cylinder did not confirm his predictions (see also Section VI,D). At very low rotation rates laminar flow is stable, and does not contribute to the diffusion process since there is no velocity component in the radial direction. At higher rotation rates, secondary flow patterns form (Taylor vortices), and finally the flow becomes turbulent. Neither of the two flow regimes satisfies the conditions of the Lévêque equation.

However, flow generated by a cylinder rotating at high speed was subsequently used by others, and in particular by King and co-workers (K3, K4a), to demonstrate that dissolution and electrochemical corrosion may both be transport limited. The dependence of the mass-transfer coefficient on the rotation rate and on the diffusivity of the dissolving species was established by correlation of experimental data (see Table VII, System 43).

The conceptual development of limiting-current measurement was advanced substantially by Agar and Bowden (A2), who investigated the current–overpotential relationship for oxygen evolution at nickel electrodes in fused sodium hydroxide. Here water transport is the limiting step:

$$2OH^- \to \tfrac{1}{2}O_2 + H_2O + 2e \tag{4}$$

A more detailed review of this and other early work in ionic mass transfer is given by Tobias *et al.* (T3).

Full development of electrochemical techniques in mass-transfer measurement had to await a quantitative formulation of the role played by convective diffusion in electrode processes, which was successfully provided in

1942–1944 by the groundbreaking work of Levich on the rotating-disk electrode (L6, L7). By 1947, conditions were fulfilled for a matching of theory and experiment (S14). In the same year, Agar (A1b) pointed out that electrochemical mass-transfer correlations should resemble closely those established for heat transfer. The first experimental test of the theory of convective diffusion in electrochemically generated free convection was made by Wagner (W1). From then on, limiting-current measurements have been used with increasing frequency and confidence to establish mass-transfer rates at reactive surfaces in flow situations, either where the rates cannot be predicted on theoretical grounds or where theory needs to be confirmed.

Only a few reviews have appeared in which application of the limiting-current method is discussed from a chemical engineering viewpoint. In the review of Tobias et al. (T3) mentioned earlier, the authors examined the knowledge available on electrochemical mass transport during the early stages of its application in 1952. Ibl (I1) reviewed early work on free convection, to which he and his co-workers contributed notably by development of optical methods for study of the diffusion layer. A discussion of the application of optical techniques for the study of phase boundaries has been given by Muller (M14).

Another survey by Ibl (I3) in 1963 listed 13 mass-transfer correlations established by the limiting-current method, only four of which were derived from quantitative considerations. At the time of writing the total number of publications is more than 200. The majority of these concern flow conditions under which theoretical predictions are, at best, qualitative. More recently, an increasing number of publications deal with model hydrodynamic studies of more complex situations, for example, packed and fluidized beds.

In 1971 Mizushina (M9) reviewed the limiting-current method with particular emphasis upon shear–stress and fluid-velocity measurements. Mass-transfer measurements, that is, limiting-current measurements in the original more restricted sense, are documented fairly extensively. The electrochemical analysis of limiting-current measurements is touched upon, but not elaborated.

A concise instructive review of limiting-current densities in several simple flow conditions during electroplating was presented by Jahn (J2).

A more recent review by Fahidy (F1) concerns the chemical engineering approach to electrochemical processes, such as fluidized-bed reactors, bipolar particulate reactors, pulsed electrochemical reactors, gas-phase electrochemical reactors, electrocrystallization and electrodissolution, and the enhancement of heat and mass transfer in electric fields. In this review, the author also discusses dimensionless mass-transfer equations applied in cell design. Such equations are reviewed in greater detail in Section VI.

B. Electrochemical Reactions Used in Limiting-Current Measurements

The choice of electrochemical systems used for mass-transport studies may appear to the reader to be narrowly restricted. Certain criteria to be met should be rather obvious: (1) chemical stability; (2) high solubility; (3) electrode potential sufficiently different from that of hydrogen (or oxygen) to give long, well-defined plateaus; (4) low cost. For the interpretation of limiting currents, physical properties of the electrolyte must be known accurately over large ranges of composition. Therefore it is not at all surprising that relatively few systems are employed by the various investigators. Of the systems used so far, listed in Table I, two in particular have earned a well-deserved popularity: (1) deposition of copper from acidified copper sulfate solution, and (2) reduction of ferri- to ferrocyanide from solutions containing a large excess of NaOH or KOH.

TABLE I

MODEL REACTIONS USED IN ELECTROCHEMICAL MASS-TRANSFER STUDIES

	$E°(V)$	Electrode metal	Supporting electrolyte	Ref.
Cathodic				
1. $Cd^{2+} + 2e \rightarrow Cd$	-0.352	Cd (Hg)	None	L10a
2. $Cu^{2+} + 2e \rightarrow Cu$	$+0.337$	Cu	H_2SO_4	W1, F3
			None	I5b, I6, O2
3. $Fe(CN)_6^{3-} + e \rightarrow Fe(CN)_6^{4-}$	$+0.360$	Ni, Pt	NaOH	E1, L9
			KOH	A6a
4. $O_2 + 2H_2O + 4e \rightarrow 4OH^-$	$+0.401$	Pt, Ag	NaOH	L9, S15a
			NaCl	E4
5. $I_3^- + 2e \rightarrow 3I^-$	$+0.536$	Pt	KI	D2, N9a
6. benzoquinone + $2H^+ + 2e \rightarrow$ hydroquinone	$+0.699$	Ag	Phosphate buffer	L9
7. $Fe^{3+} + e \rightarrow Fe^{2+}$	$+0.771$	Pt, Cu	H_2SO_4	V3
8. $Ag^+ + e \rightarrow Ag$	$+0.799$	Ag	$HClO_4$	W5, R21a
9. $Ce^{4+} + e \rightarrow Ce^{3+}$	$+1.61$	Pt, Cu	H_2SO_4	J3, V3
Anodic				
10. $CuCl_4^{3-} \rightarrow CuCl_4^{2-} + e$	$+0.153$	Pt	$CaCl_2$	D14a
11. $Fe(CN)_6^{4-} \rightarrow Fe(CN)_6^{3-} + e$	$+0.360$	Ni, Pt	NaOH	E1, L9
			KOH	A6a
12. $Ce^{3+} \rightarrow Ce^{4+} + e$	$+1.61$	Pt	H_2SO_4	J3

For studies of free-convection mass transfer, or combinations of forced and free convection, the copper deposition reaction is preferred, in spite of the precautions necessary to prevent excessive surface roughening. Copper deposition has inherent advantages: (1) cupric salts, in particular $CuSO_4$, have a relatively high solubility at room temperature; and (2) the electrode reaction, being a deposition reaction, does not produce a soluble product species. Deposition of copper from moderately concentrated solutions results in large density differences between bulk and catholyte solution and hence in large driving forces for free convection. In contrast, the ferri–ferrocyanide system offers a much smaller range of density differences because the reaction product remains in solution for reaction in either direction. Consequently, the opposing effects of reactant depletion and product accumulation at a given electrode nearly cancel, depending on the exact magnitudes of the relevant diffusivities and densification coefficients (see Table II, which gives data for nearly saturated solutions). Since the densification coefficient of ferrocyanide is larger than that of ferricyanide, the electrolyte in the vicinity of the cathode becomes more dense than the bulk solution and, in the vicinity of the anode, less dense.

TABLE II

COMPARISON OF INTERFACIAL DENSITY DIFFERENCES RESULTING FROM THE DEPOSITION OF COPPER, THE CATHODIC REDUCTION OF FERRICYANIDE, AND THE ANODIC OXIDATION OF FERROCYANIDE AT THE LIMITING CURRENT DENSITY ($T = 25°C$)

Properties[a]:	Composition:	1.4 M $CuSO_4$	0.9 M $CuSO_4$ +1.5 M H_2SO_4	0.2 M $K_3Fe(CN)_6$ +0.2 M $K_4Fe(CN)_6$ + 2.0 M NaOH	
				Reduction	Oxidation
Bulk electrolyte	ρ (gm ml^{-1})	1.214	1.217	1.150	
	μ (cP)	1.824	1.914	1.647	
	ν (cSt)	1.502	1.573	1.432	
Differences between bulk and surface	α_R (liter M^{-1})[b]	0.140	0.140	0.167	0.226
	Δc_R (M)	1.4	0.9	0.2	0.2
	$\Delta\rho$ (gm ml^{-1})	0.196	0.137	−0.0812	+0.0957
	$\Delta\rho/\rho$	0.161	0.125	−0.0706	0.0832

[a] Values taken from Selman and Newman (S9b) and Hsueh (H7).
[b] The densification coefficient α_i is defined as $(1/\rho)(d\rho/dc_i)$.

Densification is also influenced by the presence of supporting electrolyte. As shown in the last line of Table II, the relative densification in acidified cupric sulfate is less than that in binary cupric sulfate solution. In the case of the supported redox reaction, that is, in the presence of KOH or NaOH, the migration effect makes the density difference larger than that expected from overall reaction stoichiometry.

For forced-convection studies, the cathodic reaction of copper deposition has been largely supplanted by the cathodic reduction of ferricyanide at a nickel or platinum surface. An alkaline-supported equimolar mixture of ferri- and ferrocyanide is normally used. If the anolyte and the catholyte in the electrochemical cell are not separated by a diaphragm, oxidation of ferrocyanide at the anode compensates for cathodic depletion of ferricyanide.[3]

Oxidation of ferrocyanide, although used occasionally, offers no advantages relative to reduction of ferricyanide. Because the potential for oxygen liberation in alkaline solutions is close to the oxidation potential of the ferrocyanide couple, the limiting-current plateaus obtained in this case are quite narrow (E1).

Nickel, and to a lesser degree platinum, are most commonly employed in ferricyanide reduction. Both are sensitive to the presence of cyanide ion, and exhibit "poisoning" effects. A poisoned electrode has lost its catalytic activity, and on such a surface the redox reaction, which otherwise occurs close to the reversible potential, requires large excess driving force (overpotential). High overpotentials caused by poisoning may lead to poor definition, or even loss, of the limiting current plateau because the potential required for the consecutive electrode process (in the specific case of ferricyanide reduction, hydrogen evolution) may be reached before mass transfer becomes limiting.

Cyanide results from the reaction of ferricyanide in base:

$$Fe(CN)_6^{3-} + 3OH^- \rightarrow Fe(OH)_3 + 6CN^- \tag{5a}$$

Ferrocyanide when exposed to light slowly hydrolyses (E1):

$$Fe(CN)_6^{4-} + H_2O \rightarrow Fe(CN)_5 \cdot H_2O^{3-} + CN^-$$
$$CN^- + H_2O \rightarrow HCN + OH^- \tag{5b}$$

The poisoning of the nickel surface is manifested in a sharp increase of electrode potential relative to the value in the active state. Since at most only part of the surface remains active, the actual current density can be much

---

[3] A change in overall composition will occur only if one or the other electrode operates above the limiting current.

higher than is assumed on the basis of the superficial electrode area. This type of behavior leads to errors, the observed limiting current being lower than if the entire surface were in the active state. Upon progressive poisoning under otherwise identical conditions, limiting currents measured from the same solution show a progressive decline.

Ferricyanide is a strong oxidizing agent, and reacts rapidly with organic materials. Although many of the polymeric materials commonly used for equipment components (e.g., Lucite or Plexiglas) are attacked only very slowly, the ferricyanide solutions undergo gradual change in composition. Because of this effect, as well as the spontaneous degradation of experimental solutions according to Eqs. (5a) and (5b), storage of solutions in the dark, frequent reanalysis for composition, and replacement after a few days are recommended.

When a redox reaction is used for limiting-current measurements, it is generally necessary to ensure that the observed limiting current is due to depletion at the working electrode, not to a limitation caused by the counterelectrode. There are several ways to ensure this: (1) the counterelectrode surface is made much larger than that of the working electrode; (2) the concentration of the species reacting at the counterelectrode is made much larger than that of the reactant at the working electrode; (3) electrode compartments are separated by a diaphragm, and the flow rate (or stirring rate) at the counterelectrode is kept higher than that at the working electrode.

The reactions listed in Table I other than ferricyanide reduction and copper deposition have not attained the same popularity for experimental mass-transfer investigations. Oxygen and iodine reduction reactions satisfy the requirement of a reasonably long plateau in the plot of current versus electrode overpotential (> 250 mV, depending on the pH of the supporting electrolyte); moreover, the electrode surface remains unchanged. However, reducible impurities and gases (e.g., O_2, if iodine is the reactant) have to be rigorously excluded. In the case of oxygen reduction, it is difficult to maintain a known oxygen concentration in the bulk solution because: (1) oxygen has a tendency to supersaturate, and (2) its solubility is extremely sensitive to the salt concentration of the solution.

Silver deposition has been used only occasionally. Dendritic deposits form very readily unless small quantities ($\sim 0.01\%$) of organic additives are present, usually colloids such as gelatine, dextrin, or glue. These substances generally improve the adhesion and smoothness of the metal deposits. Although their exact function is not yet fully understood (I2, I4), additives usually increase the surface overpotential (see Section III,C). Thus, these substances can complicate the electrode kinetics by obscuring the limiting current through the shortening or skewing of the limiting-current plateau. In addition, many additives undergo chemical change during electrolysis so that their concentration is neither known nor controlled. Rousar *et al.* (R21a) used deposition of radioactive silver from $NaClO_4$ solutions very dilute in $AgClO_4$ to determine the current distribution in channel flow under mass-transfer limited conditions.

C. Measurement of Electrode Potential

The potential of the working electrode is measured most conveniently as the *overpotential* with respect to a reference electrode immersed in the bulk solution; by definition the overpotential (η) is the deviation, upon current passage, of the electrode potential from the value measured at equilibrium, that is, at zero net current. For convenience, the reference electrode is usually chosen to be identical with the working electrode, that is, the electrode at which the limiting current is to be measured. This is only inadvisable if the reaction is markedly "irreversible," that is, if its electrode kinetics are very slow (exchange current density $\ll 0.1$ mA/cm^2) as for example in oxygen reduction; generally, these and other conditions that favor the development of a mixed potential at the reference electrode should be avoided (N8a, pp. 110, 184; G4b, p. 207). When one does not want to disturb the flow pattern in the main cell, a capillary liquid bridge connecting it to an external reference electrolyte compartment is commonly used. The essential condition is that no appreciable current should flow between the reference electrode and the working electrode.

It is convenient to distinguish three components of the overpotential, η. Two of these are associated respectively with mass-transfer restrictions in the electrolyte near the electrode (concentration overpotential, η_c), and with kinetic limitations of the reaction taking place at the electrode surface (surface overpotential, η_s); the third one is related to ohmic resistance.

The concentration overpotential η_c is the component of the overpotential due to concentration gradients in the electrolyte solution near the electrode, not including the electric double layer. The concentration overpotential is usually identified with the "Nernst potential" of the working electrode with respect to the reference electrode; that is, the thermodynamic electromotive force (*emf*) of a concentration cell formed between the working electrode (immersed in electrolyte depleted of reacting species) and the reference electrode (of the same kind but immersed in bulk electrolyte solution):

$$\eta_c = (RT/nF) \ln(c_{R,\text{ electrode}}/c_{R,\text{ bulk}}) \tag{6}$$

In rigorous treatments, e.g., Newman (N8a, Ch. 20), the concentration overpotential is defined as the potential difference (excluding ohmic potential drop) between the reference electrode and a second, imaginary reference electrode situated in the electrolyte adjacent to the working electrode, just outside the double layer. It can be shown that this potential difference includes, in addition to the *emf* defined by Eq. (6), a diffusion potential caused by the unequal diffusivities of ions present in the diffusion layer. This contribution, however, is negligible if a large excess of inert electrolyte is present.

The second component of the overpotential, η_s, is associated with the passage of reacting species and electrons across the electric double layer, discharge of the reacting species, and changes in the electrode surface structure. Following Newman (N8a), this component is called the *surface overpotential*. It depends on the reaction rate, the species concentrations in the double layer, and the kinetic characteristics of the electrode reaction at the surface in question.

The potential observed between the reference electrode and the working electrode comprises not only the surface and concentration overpotential (η_s and η_c), but also an ohmic contribution $\Delta\phi_\Omega$, which constitutes a part of the total ohmic drop between the two electrodes between which current is passed. For a given current, $\Delta\phi_\Omega$ depends on the resistivity of the electrolyte solution and on the relative location of the reference electrode. It may be made very small by placing the interconnecting capillary tip very close to the working electrode; however, by doing so, the flow pattern near the electrode may be disturbed, or the current distribution at the electrode may be altered (B1) since the capillary has a shielding effect, which, in turn, will affect the overpotential. These distortions can be avoided by the use of "backside capillaries" (E2) connected from behind the electrode to an insulated orifice in the electrode surface, which is small enough to cause a minimal disturbance of the mass-transfer boundary layer. In most limiting-current measurements, the ohmic component of overpotential is negligible anyway, since excess supporting electrolyte is added to suppress the migration flux of the reacting ion (see Section IV,B). Moreover, accurate measurement of the electrode potential is not essential for limiting-current measurements.

The total overpotential in limiting-current measurements is therefore represented as the sum of these three components:

$$\eta = \eta_c + \eta_s + \Delta\phi_\Omega \tag{7}$$

The concentration overpotential η_c is the component directly responsible for the steep increase in potential observed as the current approaches the limiting current, since the "Nernst potential difference" (Eq. 6) becomes very large as the concentration of the reacting ion at the electrode approaches zero. However, the surface overpotential η_s and the ohmic contribution $\Delta\phi_\Omega$ also increase with increasing current. If η_s at the limiting current is of the same magnitude as the concentration overpotential, this may obscure what would otherwise be a limiting-current plateau. A high ohmic contribution shifts the total overpotential to higher negative values without affecting the plateau.

The ohmic contribution to the overpotential can be minimized by suitable placement of the reference electrode, but the surface overpotential cannot be reduced similarly. In making limiting-current measurements, the surface overpotential, or rather its rate of increase with current density, should be low enough to permit observation of a long, clearly defined limiting-current plateau.

Therefore, criteria in the selection of an electrode reaction for mass-transfer studies are: (1) sufficient difference between the standard electrode potential of the reaction that serves as a source or sink for mass transport and that of the succeeding reaction (e.g., hydrogen evolution following copper deposition in acidified solution), and (2) a sufficiently low surface overpotential and rate of increase of surface overpotential with current density, so that, as the current is increased, the potential will not reach the level required by the succeeding electrode process (e.g., H_2 evolution) before the development of the limiting-current plateau is complete.

For most of the reactions frequently employed in limiting-current studies, the surface overpotential is not negligible. A criterion for assessing its magnitude is the exchange-current density i_0, which is a measure of the reaction rate at the equilibrium potential of the electrode (i.e., when anodic and cathodic rates are equal).

Figure 3a is an illustration of the effect of surface overpotential on the limiting-current plateau, in the case of copper deposition from an acidified solution at a rotating-disk electrode. The solid curves are calculated limiting currents for various values of the exchange current density, expressed as ratios to the limiting-current density. Here the surface overpotential is related to the current density by the Erdey–Gruz–Volmer–Butler equation (V4):

$$i = i_0[\exp(\alpha_a \eta_s) - \exp(-\alpha_c \eta_s)] \tag{8}$$

with the transfer coefficients $\alpha_a = 1.5$, $\alpha_c = 0.5$. The dependence of the exchange-current density on $CuSO_4$ concentration is assumed to be

$$i_0 \sim c^\gamma \tag{9}$$

where $\gamma = 0.41$ in accordance with data reported by Mattson and Bockris (M4e), and Brown and Thirsk (B10).

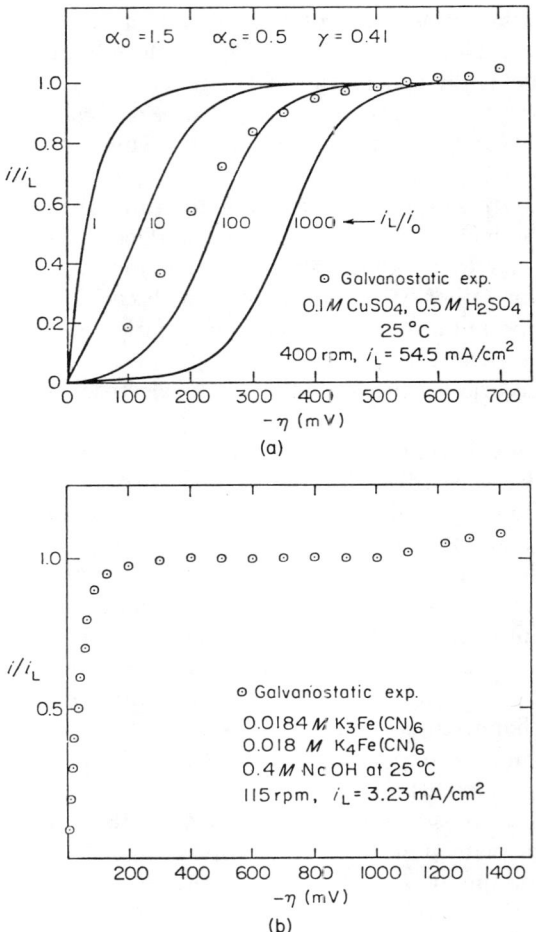

FIG. 3. (a) Typical galvanostatic limiting-current curve for copper deposition at a copper disk in acidified $CuSO_4$ solution. The circles indicate the experimental curve. The solid curves were calculated using kinetic parameters as indicated. (b) Typical galvanostatic limiting current curve for ferricyanide reduction at a nickel electrode in equimolar ferri–ferrocyanide solution with excess NaOH. [From Selman (S8).]

The experimental limiting current curve of Fig. 3a suggests that the exchange-current density is 0.5–1.5 mA cm^{-2}, which is slightly lower than the values report by Mattson and Bockris (M4e) for highly purified solutions of $CuSO_4$ in 0.5 M H_2SO_4. In general, one would not expect to find good agreement since the large volumes of solutions employed in mass-transfer studies make it impossible to keep them free of organic contaminants that may affect the electrode-kinetic characteristics strongly.

It is clear from the calculated limiting-current curves in Fig. 3a that the plateau of the copper deposition reaction at a moderate limiting-current level like 50 mA cm^{-2} is narrowed drastically by the surface overpotential. On the other hand, the surface overpotential is small for reduction of ferricyanide ion at a nickel or platinum electrode (Fig. 3b). At noble-metal electrodes in well-supported solutions, the exchange current density appears to be well above 0.5 A/cm^2 (T1a, S20b, D6b, A3e). At various types of carbon, the exchange current density is appreciably smaller (T1a, S17a, S17b).

In limiting-current measurements, the counterelectrode is sometimes used as a reference electrode. In that case, the surface overpotential of the counterelectrode contributes to the recorded overpotential; that is, the potential of the "reference electrode" is now current dependent. Unless precautions are taken (e.g., the area of the counterelectrode is much larger than that of the working electrode), a properly defined limiting-current plateau may not be obtained.

D. Boundary Conditions at the Electrode

Potentiostatic current sources, which allow application of a controlled overpotential to the working electrode, are used widely by electrochemists in surface kinetic studies and find increasing use in limiting-current measurements. A decrease in the reactant concentration at the electrode is directly related to the concentration overpotential, η_0 (Eq. 6), which, in principle, can be established directly by means of a potentiostat. However, the controlled overpotential is made up of several contributions, as indicated in Section III,C, and hence, the concentration overpotential is by no means defined when a given overpotential is applied; its fraction of the total overpotential varies with the current in a complicated way. Only if the surface overpotential and ohmic potential drop are known to be negligible at the limiting current density can one assume that the reactant concentration at the electrode is controlled by the applied potential according to Eq. (6).

Potentiostatic measurements are analogous to heat-transfer experiments in which the wall temperature is controlled, whereas galvanostatic measurements are similar in character to those in which the heat flux is controlled. However, whereas heat transfer may be measured readily with a uniform flux generated at the surface, there is no assurance that a known current applied to an extended electrode will yield a uniform current distribution over the surface, unless the surface is divided into electrically insulated segments and identical current densities are imposed externally on these segments. The nonuniform current distribution (which is naturally established at an unsegmented electrode) is a consequence of the interplay of concentration overpotentials and surface overpotentials at the two electrodes and the ohmic potential drop between them.

Even if a uniform current distribution were to be successfully imposed on a segmented electrode, yet another difficulty would arise. No simple relationship would exist between the overpotential measured at the electrode and the concentration at the electrode surface, unless the overpotential were predominantly a concentration overpotential, that is, the current were effectively at its limiting value. Under these circumstances, the reactant concentrations at the various electrode segments would be effectively zero. Flow geometries in which the limiting-current condition coincides with a uniform limiting-current distribution are indeed rare. (The rotating-disk electrode in the laminar flow regime and the rotating-cylinder electrode are examples.)

In most flow geometries the limiting-current condition at an extended electrode implies a nonuniform current distribution, the nature of which is determined by the character of the flow. The analogy with the uniform wall temperature condition in heat transfer is appropriate in this case. However, generating the limiting-current condition at the entire electrode simultaneously is not always possible, and this effect may lead to complications in the interpretation of the limiting-current curve (see Section IV,E). In practice it is advisable to use a segmented electrode, and to record sectional limiting-current curves.

In many limiting-current measurements the expected current distribution is only moderately nonuniform, and a single unsegmented electrode will yield well-defined limiting-current plateaus. The various techniques by which the limiting current at a single electrode can be generated are discussed in the next section.

E. Approach to the Limiting Current

To obtain the limiting current for a particular electrode reaction, one needs to generate the condition in which the concentration of the reacting species at the electrode becomes vanishingly small. In practice the simplest way to accomplish this is to increase the current up to the point at which the electrode potential shows a steep increase for a very small increase in the current. In this galvanodynamic approach, the current is controlled and may be increased in steps or continuously, while the electrode potential is being monitored as the indicator of the limiting-current condition. Of course, it is essential that the electrode potential be measured with an instrument of sufficiently high input impedance so as to pass only a negligible current between the working electrode and the reference electrode.

Alternatively, one may control the electrode potential and monitor the current. This potentiodynamic approach is relatively easy to accomplish by use of a constant-voltage source if the counterelectrode also functions as the reference electrode. As indicated in the previous section, this may lead to various undesirable effects if a sizable ohmic potential drop exists between the electrodes, or if the overpotential of the counterelectrode is strongly dependent on current. The potential of the working electrode can be controlled instead with respect to a separate reference electrode by using a potentiostat. The electrode potential may be varied in small increments or continuously. It is also possible to impose the limiting-current condition instantaneously by applying a potential step.

A third, rather unusual way of generating the limiting-current condition is possible in situations controlled by forced convection. The current is held at a known level while the convective velocity is diminished in a controlled manner, for example, by decreasing the rate of rotation of the electrode, until the electrode potential increases steeply (H9).

Which of these three methods is preferred in a particular limiting-current measurement may depend on special circumstances. The method of controlled convection, for example, has the advantage of a constant ohmic potential drop, and would be suitable where a low-conductivity electrolyte is used (e.g., seawater in corrosion studies). The copper deposition reaction frequently used in limiting-current measurements poses special problems because of the formation of rough deposits near the limiting current; thus, the galvanodynamic method is preferable in this case (see Section IV,F). Unsteady-state effects caused by too rapid an approach to the limiting current in either the galvanodynamic or the potentiodynamic mode are discussed in Section IV,D.

IV. Interpretation of Results

Since current can be measured with ease and precision, the limiting-current technique provides a convenient and, under certain conditions, accurate method for measuring mass-transfer rates. The conditions for valid measurement and correct interpretation of limiting currents are discussed in the following sections.

A. Definition of the Limiting-Current Plateau

In principle, the accuracy with which mass-transfer rates may be measured is limited by the precision with which the limiting-current plateau or inflection point can be read. Furthermore, the electrode area, the current distribution, and the bulk concentration of the reactant, as well as all relevant transport properties, should be known accurately.

The type of electrode reaction employed, the cell geometry, and the manner in which the limiting-current measurement is carried out determine the shape of the current versus electrode-potential curve. Often the ideal horizontal inflection in such curves is absent, making the determination of true limiting current problematical if not impossible. Characteristics of satisfactory limiting current plateaus are as follows:

1. *A steep rise of current to the plateau.* This rise would be complete within approximately 200 mV, if surface overpotential were negligible. (See, e.g., Fig. 3a, illustrating cathodic deposition of copper in which the experimental current-voltage curve indicates an appreciable surface overpotential, and Fig. 3b, illustrating cathodic reduction of ferricyanide in which the surface overpotential is negligible.)

2. *Termination of the plateau at a sufficiently high overpotential.* The potential at which a consecutive electrode reaction sets in (e.g., hydrogen evolution in cathodic reactions) is determined by the composition of the electrolyte (specifically, the pH) and by the nature and state of the electrode surface (hydrogen overpotential). The reduction of ferricyanide in alkaline solution on nickel also provides a better-defined plateau in this respect than the deposition of copper in acid solution.

3. *Zero inclination of the plateau with respect to the current axis.* In practice this condition is almost never met. Possible causes are:

(a) The surface area of the electrode increases due to deposition, for example, because rough copper deposits form at the limiting current (see Section IV,F);

(b) The steady-state limiting current is attained very slowly, for example, in free convection under the influence of a very small density difference (see Section IV,D);

(c) The bulk concentration of the reactant changes, for example, due to prolonged electrolysis in cases where the initial bulk concentration was low.

Although the reduction of ferricyanide has distinct advantages over the copper deposition reaction, in terms of obtaining a well-defined plateau, its usefulness in free-convection measurements is limited by a combination of Effects (3b) and (3c), as well as its chemical instability (see Section III,B).

(d) Finally, if an electrode of extended surface area is used, the current distribution may not be uniform below or at the limiting current (see Section IV,E).

B. MIGRATION EFFECTS

Migration of the reacting ion in the electric field, briefly referred to in Section II,B, is usually suppressed by the addition of excess inert electrolyte. Incorrect values for mass-transfer rates are obtained if migration contributes more than a negligible fraction of the total limiting current.

To estimate the contribution of migration to the limiting current, early investigators used the transference number of the reacting ion (defined as the ratio of reacting-ion mobility to total ionic mobility, on a concentration-proportional basis). The migration contribution could thus be subtracted from the limiting current. The same stratagem of estimating ion fluxes by means of an ad hoc transference number was first applied by Wilke et al. (W4, W5) to arrive at an estimate of the interfacial concentration of the supporting ion, which is essential for free-convection correlations. This method has since been used extensively, since it requires only simple calculations. However, complications may arise when the number of electrons transferred is not equal to the reacting-ion charge number, as in reduction of ferricyanide (B9, F5, I5a, T1b). There is no theoretical basis for the manipulation of ionic mobilities to estimate fluxes; as pointed out by Newman (N4), the concept of a transference number in solutions with concentration gradients is meaningless.

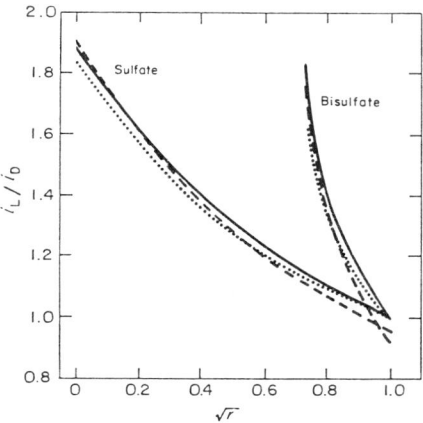

FIG. 4. Migration contribution to the limiting current in acidified $CuSO_4$ solutions, expressed as the ratio of limiting current (i_L) to limiting diffusion current (i_D); $r = c_{H_2SO_4}/(c_{H_2SO_4} + c_{CuSO_4})$. "Sulfate" refers to complete dissociation of HSO_4^- ions, "bisulfate" to undissociated HSO_4^- ions. "Forced convection" refers to steady-state laminar boundary layers, as at a rotating disk or flat plate; "free convection" refers to laminar free convection at a vertical electrode; "penetration" to unsteady-state diffusion in a stagnant solution. [From Selman (S8).]

Recently, the contribution of migration to the limiting current has been computed by numerical solution of the complete set of transport equations for several types of diffusion layer and for specific combinations of reacting and supporting ions in common use (N5, N6, S9a). The value of the migration contributions for a given composition of the bulk solution may be read directly from graphs (S9a).

In the case of the supported ferri–ferrocyanide system the migration contribution is relatively unimportant; for example, 1% for cathodic reduction of 0.01 M $K_3Fe(CN)_6$ + 0.01 M $K_4Fe(CN)_6$ in 1 M KOH, and 4% for 0.1 M $K_3Fe(CN)_6$ + 0.1 M $K_4Fe(CN)_6$, both in laminar forced convection. Figure 4 is an illustration of the migration effect in acidified $CuSO_4$ solution. The effect is more dependent on the degree of dissociation of the HSO_4^- ion than on the type of diffusion layer. Although the experimental evidence is not as conclusive as one would wish, it appears fairly certain (H10) that dissociation of the HSO_4^- ion is largely complete in solutions of the ionic strength typically used in limiting-current measurements.

Note in Fig. 4 that the free-convection limiting current in the presence of excess inert electrolyte (H_2SO_4) is not equal to the diffusion current of Cu^{2+} ions, as in forced convection and stagnant diffusion, but slightly less. This results from accumulation of H_2SO_4 at the cathode, which decreases the driving force for free convection. Migration thus has an indirect, as well as a direct, effect on free-convection-limiting currents (S9a). Figure 5 illus-

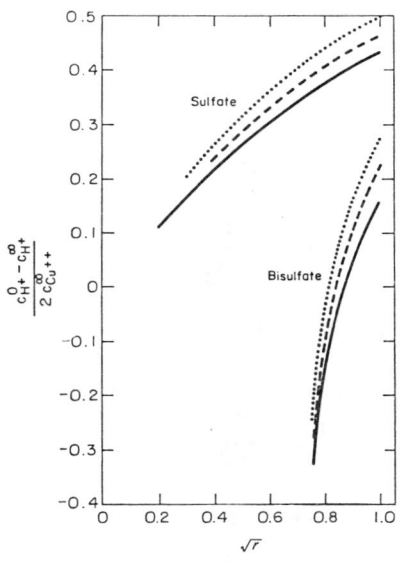

Fig. 5. Accumulation of H_2SO_4 at the cathode, due to migration in acidified $CuSO_4$ solutions, at the limiting current. See Fig. 4 for meaning of legends. [From Selman (S8).]

trates that the accumulation of H_2SO_4 at the cathode is again more dependent on the degree of dissociation of HSO_4^- ion than on the type of diffusion layer.

C. Diffusion Coefficients

From an analysis of the electrochemical mass-transfer process in well-supported solutions (N8a), it becomes evident that the use of the molecular diffusivity, for example, of $CuSO_4$, is not appropriate in investigations of mass transfer by the limiting-current method if use is made of the copper deposition reaction in acidified solution. To correlate the results in terms of the dimensionless numbers, Sc, Gr, and Sh, the diffusivity of the reacting ion must be used.

On the other hand, the diffusivity of an ion, for example, Cu^{2+}, is only known in the limit of infinite dilution where the Nernst–Einstein equation is valid:

$$D_i = RTu_i = RT\lambda_i/|z_i| \tag{10}$$

The quantity D_i cannot be derived from molecular diffusivities; at infinite dilution the calculated ionic diffusivity of Cu^{2+} is approximately 20% lower than the molecular diffusivity of $CuSO_4$.

Nevertheless, many investigators have used the molecular diffusivities measured by Cole and Gordon (C12a) for $CuSO_4$ in the ternary system $CuSO_4$–H_2SO_4–H_2O. The validity of these diffusivities for the purpose of limiting-current correlations has been questioned repeatedly (see Section IV,C,1).

Mass-transfer rates from limiting-current measurements in well-supported solutions should invariably be correlated with ionic and not with molecular diffusivities. The former can be calculated from limiting-current measurements, for example, at a rotating-disk electrode.

The diffusivities thus obtained are necessarily *effective diffusivities* since: (1) they reflect a migration contribution that is not always negligible; and (2) they contain the effect of variable properties in the diffusion layer that are neglected in the well-known solutions to constant-property equations. It has been shown, however, that the limiting current at a rotating disk in the laminar range is still proportional to the square root of the rotation rate if the variation of physical properties in the diffusion layer is accounted for (D3e, H8). Similar invariant relationships hold for the laminar diffusion layer at a flat plate in forced convection (D4), in which case the mass-transfer rate is proportional to the square root of velocity, and in free convection at a vertical plate (D1), where it is proportional to the three-fourths power of plate height.

The effective diffusivities determined from limiting-current measurements appear at first applicable only to the particular flow cell in which they were measured. However, it can be argued plausibly that, for example, rotating-disk effective diffusivities are also applicable to laminar forced-convection mass transfer in general, provided the same bulk electrolyte composition is used (H8). Furthermore, the effective diffusivities characteristic for laminar free convection at vertical or inclined electrodes are presumably not significantly different from the forced-convection diffusivities.

Effective ionic diffusivities at a rotating-disk electrode are calculated from the Levich equation as derived for constant physical properties, used here in inverted form:

$$D = v(0.6205 c_\infty nF \sqrt{v\omega}/i_{\text{lim}})^{-3/2} \tag{11}$$

1. *Effective Diffusivities of* Cu^{2+} *in* $CuSO_4-H_2SO_4-H_2O$

Measurements of $CuSO_4$ molecular diffusivity by Cole and Gordon (C12a), referred to above, were carried out in diaphragm cells, mostly at 18°C. Their results were correlated by Fenech and Tobias (F3) using the Stokes–Einstein equation

$$\mu D/T = (2.495 + 0.0173 c_{H_2SO_4} + 0.0692 c_{CuSO_4}) \times 10^{-10} \quad \text{dyn}/°K \tag{12}$$

This expression, originally given for 22°C, has been used extensively in later work.

The validity of Eq. (12) for correlations of limiting-current measurements was first questioned by Arvía et al. (A5), and later by Wragg and Ross (W13a). The latter found that limiting currents in an annular flow cell could be correlated in better agreement with the Leveque mass-transfer theory if a lower mobility (Stokes–Einstein) product were employed, such as

$$\mu D/T = 2.09 \times 10^{-10} \quad \text{dyn}/°K \tag{13}$$

This value is based on Cu^{2+} diffusivities calculated Arvía et al. (A5) from limiting-current measurements at a rotating-disk electrode by, with $CuSO_4$ concentrations below 0.1 M. In practical applications (e.g., copper refining or electrowinning) higher Cu^{2+} concentrations are often required, as is also the case in free-convection limiting-current measurements.

Recently, effective diffusivities at a rotating-disk electrode were determined for a wide range of $CuSO_4$ and H_2SO_4 concentrations (H7, S8, L1a). These data, together with the data of Arvía et al. (A5), yield for the mobility product, at or near 25°C:

$$\mu D/T = (1.98 + 2.34 c_{Cu}) \times 10^{-10} \quad \text{dyn}/°K \tag{14a}$$

The mobility product is not significantly dependent on the concentration of H_2SO_4, but is strongly dependent on the concentration of $CuSO_4$. The latter dependence may include the effect of surface area increase due to roughness of the deposit formed near the limiting current (S8). The extent of this effect depends also on the method of generating the limiting current (e.g., potential scan or current ramp), as discussed in Section IV,F. The roughness effect is probably responsible for the appreciable scatter in the effective diffusivity values; Eq. (14a), based on 56 data, has a standard error of 8%.

Diffusivity values obtained from Eq. (14a) for solutions that are dilute or moderately concentrated in $CuSO_4$, differ appreciably from the molecular diffusivities hitherto used.

Glycerol is often added to $CuSO_4$–H_2SO_4–H_2O solutions to vary the Schmidt number. The addition of glycerol to solutions dilute in $CuSO_4$ ($< 0.1\ M$) apparently increases the mobility product. From data of Arvía et al. (A5) and of Landau (L1a) one can deduce that for solutions of $CuSO_4$ ($< 0.1\ M$), H_2SO_4 (1.5 M) and glycerol ($c_G < 6\ M$), at or near 25°C:

$$\mu D/T = 2.075 + 0.06 c_G \tag{14b}$$

with a standard error of 0.05 in $\mu D/T$.

2. *Effective Diffusivities of* $Fe(CN)_6^{4-}$ *and* $Fe(CN)_6^{3-}$

Effective diffusivities for these ions in equimolar concentration ratio and with various inert electrolytes, have been determined by several methods (see Table III). The mobility products obtained from capillary cell (stagnant diffusion) and rotating-disk measurements are in fairly good agreement.

3. *Effective Diffusivities of Oxygen in* KOH

Values of the oxygen diffusivity in KOH solutions of various concentrations are also included in Table III.

D. Unsteady-State Effects

1. *Free Convection*

Investigations on mass-transfer rates along planar electrodes (F2, H3) in which the rate of increase of current, or of cell voltage, was varied systematically from one measurement to the other revealed that the time taken for attaining the limiting current influenced the limiting-current curve. This unsteady-state effect was noticeable both in the quality of definition of the

TABLE III

VALUES OF $\mu D/T$ FOR THREE REACTANT SPECIES
(in 10^{-10} dyn/°K^{-1})

	\multicolumn{4}{c}{Reactant}					
	K$_3$Fe(CN)$_6$ equimolar + Excess				K$_4$Fe(CN)$_6$ equimolar + Excess				
Ref.	KOH	NaOH	KCl	KNO$_3$	KOH	NaOH	KCl	KNO$_3$	Method[a]
Lin et al. (L9)	2.67	2.67	2.67	2.67	2.31	2.31	2.31	2.31	LM
Eisenberg et al. (E3)	—	2.50	—	—	—	2.15	—	—	CC
Arvia et al. (A6a)	2.78	2.52	2.00	—	2.52	2.09	1.75	—	RDE
Gordon et al. (G5)	(2.34 + 0.0141)[b]	—	—	—	(1.87 + 0.0341)[b]	—	—	—	RDE
Van Shaw et al. (V2)	—	2.30[c]	—	—	—	—	—	—	LFA
Noordsij and Rotte (N10)	—	2.27[c]	—	—	—	—	—	—	DS
Sih (S11c)	—	—	—	1.49[c]	—	—	—	—	RDE
Smyrl and Newman (S16)	—	—	—	—	—	—	—	1.81	RDE

Reactant O$_2$

	0.015M KOH	2M KOH	
Lin et al. (L9)	7.37	—	ICT
Davis et al. (D7)	5.40	4.36	RDE, CC
Gubbins and Walker (G8)	5.46	5.60	POL

[a] CC, Capillary cell (stagnant diffusion); DS, diffusion to spherical electrode; ICT, from mobility measurements (International Critical Tables); LFA, laminar-flow annular cell (Lévêque relation); LM, from limiting mobility at infinite dilution; POL, polarographic cell; RDE, rotating-disk electrode.

[b] I, ionic strength = $\frac{1}{2} \sum c_i z_i^2$.

[c] Based on measurements with one composition only.

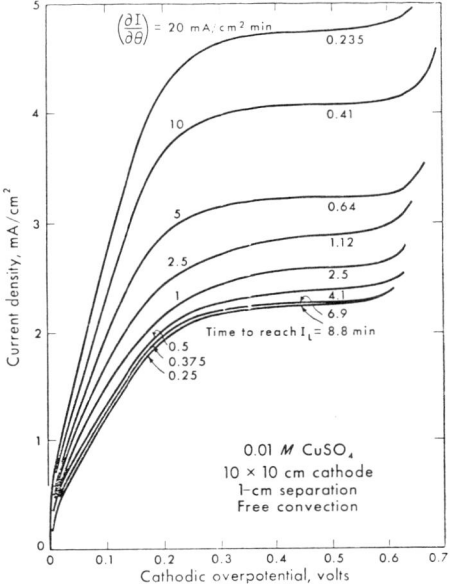

FIG. 6. Limiting-current curves recorded for various current application rates in free convection at a horizontal electrode. [From Hickman (H3).]

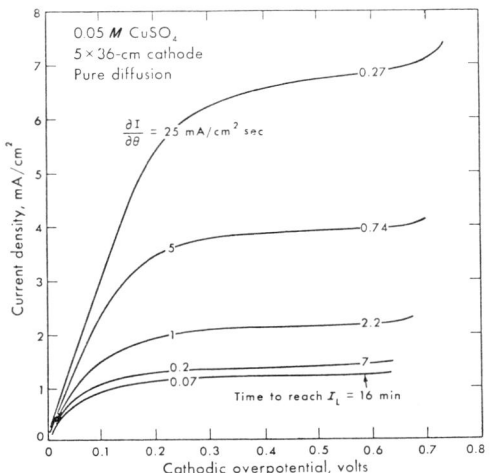

FIG. 7. Limiting-current curves recorded for various current application rates in pure diffusion at a horizontal cathode facing downward. [From Hickman (H3).]

plateau, and in the current reading at the plateau or inflection point in the curve.

The effect is most prominent in free convection. Limiting-current curves recorded by Hickman (H3) at a horizontal cathode facing upward in free convection are shown in Fig. 6. The apparent limiting-current value is definitely dependent on the time necessary to reach the limiting current; an 80% increase in this value is noted as the rate at which the current rises varies from 0.25 to 20 mA cm^{-2} min^{-1}.

The relative increase in the apparent limiting current is even more important in situations where diffusion alone is involved, for example, where the cathode plate faces downward (Fig. 7). In this case, the observation is that the apparent limiting currents drop steadily to lower values as the current application rate di/dt is decreased. Were it not for edge effects (advection) at the embedded plate, no steady limiting current could be expected.

In the case of free convection, on the other hand, the apparent limiting current will not drop below the steady-state value as di/dt is decreased. For experiments with free convection at horizontal electrodes, Hickman (H3) found that a time

$$t_{ss} = i_L/(di/dt)$$
$$\geq 8 \quad \text{min} \tag{15}$$

is sufficient to guarantee steady-state limiting-current values.

Earlier, Fenech and Tobias (F3) obtained limiting currents for the same free-convection system by increasing linearly the applied potential (cell voltage). Using concentrated solutions, they found that the limiting-current value is essentially independent of the initial voltage applied, provided that the same rate of increase is used. Starting with $\Delta E = 200$ mV, steady reproducible limiting currents were obtained if

$$t_{ss} = 500 \quad \text{mV}/(d\ \Delta E/dt)$$
$$\geq 3 \quad \text{min} \tag{16}$$

This time is considerably shorter than the galvanostatic minimum time (Eq. 15). Since a theoretical analysis of this type of unsteady-state free convection offers severe difficulties, there is no way to confirm this difference except by controlled experiments of the kind undertaken, on a limited scale, by Hickman (H3) and by Fenech (F2).

The experimental evidence discussed so far concerns turbulent free convection at a horizontal surface. Even in the classical example of *laminar* free convection, namely, at a vertical plate, the analysis of unsteady-state transport is very complicated. Only recently have approximate solutions been

TABLE IV

Transition Times (sec) to Steady-State Mass Transfer in Laminar Free and Forced Convection along a Planar Electrode, for a Solution of 0.05 M $CuSO_4$, 1.5 M H_2SO_4 at 25°C[a]

Distance from the leading edge (cm)	0.5	1	5	10	40
Free convection at a vertical plate	42 (2.84)	60 (2.40)	133 (1.61)	189 (1.35)	377 (0.95)
Forced convection between parallel plates (1 cm separation)					
Re = 10	95 (1.31)	151 (1.04)	444 (0.61)	707 (0.48)	1777 (0.30)
100	21 (2.82)	33 (2.24)	96 (1.31)	151 (1.04)	382 (0.65)
1000	4.4 (6.03)	7.1 (4.83)	21 (2.82)	33 (2.24)	83 (1.41)
2000	2.8 (7.63)	4.4 (6.08)	13 (3.56)	21 (2.82)	52 (1.78)

[a] Current densities indicated between parentheses (in mA cm^{-2}).

proposed for the two basic unsteady-state boundary conditions: a concentration (temperature) step, or a flux step, applied at the plate surface. These solutions, insofar as they relate to electrolytic solutions (Sc ≫ 1), have been reviewed and compared with a few transition times reported in work on laminar free-convection mass transfer (S8, S10).

Some transition times calculated for this type of free convection, following a concentration step in 0.05 M $CuSO_4$ solution with excess H_2SO_4, are given in Table IV. It can be seen that the transition times (to a flux 1% in excess of the steady-state flux) vary appreciably along the plate; also in forced convection (which is discussed below) the transition times are generally shorter, except at very low flow rates.

Of course, in free-convection mass transfer the transition time is dependent on the density difference generated at the electrode. The dimensionless time variable of the transient process is

$$\tau_{\text{free conv}} = t(Dg\,\Delta\rho/\mu x)^{1/2} \tag{17}$$

The transition times in very dilute solutions, for example, < 0.01 M $CuSO_4$, become so long that serious difficulty is encountered in obtaining any

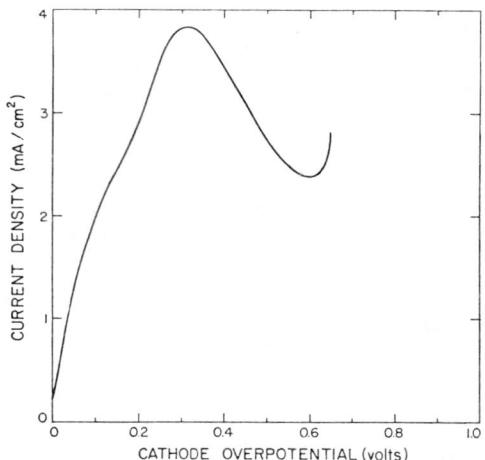

FIG. 8. Limiting-current curve for low concentration of the reacting ion in free convection at a horizontal electrode, recorded by linear increase of applied potential. [From Fenech (F2).]

steady-state value at all, especially when the transport process provides the driving force for secondary convection patterns which may extend over larger distances (or volumes). Examples are high-Schmidt number free convection at a horizontal plate submerged in a deep pool of electrolyte, and laminar-flow free convection in a narrow vertical slit open at the top and bottom to a large reservoir of solution (B9).

An unusual limiting-current curve for a horizontal electrode, obtained by Fenech (F2) in 0.01 M $CuSO_4$, is shown in Fig. 8. Although the current maximum could be decreased by applying a lower $d\ \Delta E/dt$ (cf. Eq. 16), it could not be made to disappear altogether. The maximum is probably caused by the unsteady-state diffusion process close to the surface. The development of free-convection patterns is slow and does not reach steady state by the time the externally controlled rising potential allows hydrogen evolution to begin. Estimation of a limiting current from such a "camelback" curve is unreliable; there is no plateau, nor can the maximum or the minimum be relied upon to reflect a true steady-state transport.

The risk inherent in long transition times, characteristic of free convection with small driving force, is twofold:

1. The bulk solution may become depleted. In free-convection experiments, the cathode and anode compartments are often separated by a diaphragm to prevent interaction of the convection patterns. Under these conditions, replenishment of the catholyte does not take place. Examples of sagging limiting-current plateaus caused by bulk depletion can be found in the work of Boëffard (B8) with ferri–ferrocyanide in free convection at horizontal electrodes.

2. *In copper deposition, surface roughness may lead to an increase in the apparent limiting current.* Measurements at a rotating-disk electrode (see Section IV,F) show that surface roughness is more likely to develop with potentiostatic current control than with approach to limiting current by a series of current steps. This comparison for forced-convection mass transfer is probably true for free convection as well. Although it is convenient in free-convection experiments to apply a negative potential large enough to generate the limiting-current condition, and to let the current rise gradually to its steady-state value, this practice entails the risk of a relatively serious roughness effect.

2. *Forced Convection*

Unsteady-state effects and transition times are also significant in forced convection. Whenever the mass-transfer boundary layer is large anywhere on the working electrode surface, the commonly employed experimental time range of a few minutes may not be adequate to reach the steady-state limiting current.

Table IV includes theoretical transition times (C2, R14, S17c) in laminar flow between parallel plates, following a concentration step at the wall (Lévêque mass transfer). Clearly, in laminar flow (Re $\approx$ 100 or lower), transition times are comparable to those in laminar free convection. Here, however, the dependence on concentration (through the diffusivity) is weak. The dimensionless time variable in unsteady-state mass transfer of the Lévêque type is

$$\tau_{\text{forced conv}} = t(S^2 D/x^2)^{1/3} \qquad (18)$$

where S is the velocity gradient at the wall. For this case the transition time is more strongly dependent on the distance from the leading edge x than it is in free convection at a vertical plate.

Experimental data relative to unsteady-state mass transfer as a result of a concentration step at the electrode surface are not available. However, for a linear increase of the current to parallel-plate electrodes under laminar flow, Hickman (H3) found that steady-state limiting-current readings were obtained only if the time to reach the limiting current at the trailing edge of the plate (see Section IV,E), expressed in the dimensionless form of Eq. (18), is

$$\tau_{ss} = \left(\frac{i_L}{di/dt}\right)\left(\frac{S^2 D}{x^2}\right)^{1/3}$$
$$\geq 27 \qquad (19)$$

This value is an order of magnitude larger than the transition time following a concentration or current step at the plate; approach to the steady-state flux following a concentration step is complete to within 1% at $\tau = 1.25$ (S17c).

Experimental results obtained at a rotating-disk electrode by Selman and Tobias (S10) indicate that this order-of-magnitude difference in the time of approach to the limiting current, between linear current increases, on the one hand, and the concentration-step method, on the other, is a general feature of forced-convection mass transfer. In these experiments the limiting current of ferricyanide reduction was generated by current ramps, as well as by potential scans. The apparent limiting current was taken to be the current value at the inflection point in the current-potential curve.

The minimum time necessary to obtain steady-state limiting currents by a current ramp was found to be

$$t_{ss(galv)} = \frac{i_L}{di/dt} \geq 60 Sc^{1/3} \omega^{-1} \tag{20}$$

where ω is the rotation rate of the disk in rad sec^{-1}. In the case of linearly decreasing potential,

$$t_{ss(pot)} = \frac{350 \text{ mV}}{d\eta/dt} \geq 20 Sc^{1/3} \omega^{-1} \tag{21}$$

which is only a third of that for a current ramp. These results are illustrated in Figs. 9 and 10. Significantly, the transition times following a concentra-

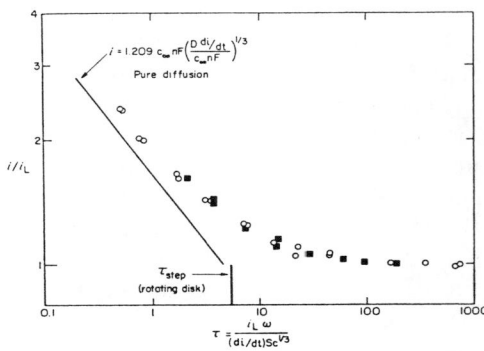

FIG. 9. Logarithmic plot of apparent limiting-current density as a function of current increase rate at a rotating-disk electrode; i = apparent limiting current density; i_L = true steady-state limiting current density; di/dt = current increase rate (A cm^{-2} sec^{-1}); ω = rotation rate (rad sec^{-1}). [From Selman and Tobias (S10).]

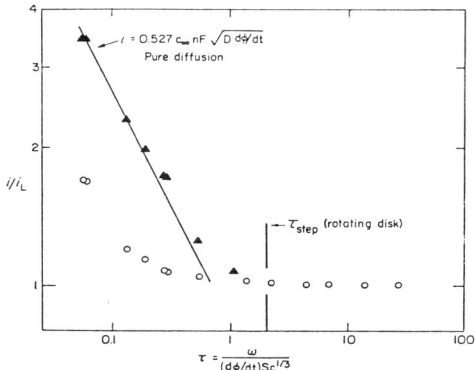

FIG. 10. Logarithmic plot of apparent limiting current density as a function of potential scan rate at a rotating-disk electrode; i = apparent limiting current (or peak current) density; i_L = true steady-state limiting current density; $d\phi/dt$ = potential scan rate expressed in units RT/nF; ω = rotation rate (rad sec^{-1}). [From Selman and Tobias (S10).]

tion step or current step at a rotating disk are both one order of magnitude smaller.

The current peaks observed when fast potential ramps are applied appear similar to the one shown in Fig. 8. The peak currents (triangles in Fig. 10) can be satisfactorily interpreted in terms of a pure-diffusion model (S10); in equimolar ferri–ferrocyanide solution,

$$i_{max} = 0.5274 c_b n F (D\ d\phi/dt)^{1/2} \qquad (22)$$

where $d\phi/dt$ is the potential scan rate in units RT/nF (in this case, $n = 1$).

As shown above, potentiodynamic generation of limiting currents is more rapid and, therefore, preferable in principle to the galvanodynamic technique. However, during a linear decrease of potential to the limiting-current condition, the current density initially rises more rapidly than when a current ramp is used. Therefore, in the case of copper deposition at the cathode, a linear potential ramp tends to yield a rougher deposit and a less well-defined plateau than a linear current ramp (see Section III,F).

E. Interaction with the Potential Distribution at Elongated Electrodes

Analysis of nonstationary convective mass transfer, under well-defined hydrodynamic conditions, may be helpful to understand the way in which the limiting current is established at electrodes of appreciable dimensions. As referred to in the previous section, the transition time to steady-state mass transfer in flow past extended electrodes is distance-dependent (Eq. 17 and 18); this knowledge has been combined with experimental results to yield criteria of practical interest, such as Eq. (19). However, two factors limit the validity of this generalization.

1. Gradual Approach to the Limiting Current

In most electrochemical measurements the limiting current is not imposed abruptly but approached gradually by controlling either the current or the electrode potential (see Section III,E). For example, in the previous section, the experimental linearly controlled approach to the limiting current was reviewed for the relatively simple case of the rotating-disk electrode. It is worthwhile to explore whether a theoretical analysis of the gradual approach to limiting current is possible, since this might simplify experimentation.

In fact, since the unsteady-state transport equation for forced convection is linear, it is possible in principle to derive solutions for time-dependent boundary conditions, starting from the available step response solutions, by applying the superposition (Duhamel) theorem. If the applied current density varies with time as $i(t)$, then the local surface concentration at any time $c_0(x, t)$ is given by

$$c_0(x, t) = \int_0^t \partial i/\partial t(x, t_1) c_0^*(x, t - t_1) \, dt_1 \tag{23}$$

The function $c_0^*(x, t)$ is the response of the local surface concentration to a uniform stepwise change in current density, given by Soliman and Chambré (S17c), as a somewhat involved analytic expression that combines space and time dependence in the dimensionless variable of Eq. (18):

$$\tau = t(S^2 D/x^2)^{1/3}$$

where S is the velocity gradient at the electrode.

Equation (23) implies that the current density is uniformly distributed at all times. In reality, when the entire electrode has reached the limiting condition, the distribution of current is not uniform; this distribution will be determined by the relative thickness of the developing concentration boundary layer along the electrode. To apply the superposition theorem to mass transfer at electrodes with a nonuniform limiting-current distribution, the local current density throughout the approach to the limiting current should be known.

2. Effect of Surface Overpotential on the Current Distribution Below the Limiting Current

In many cases mass transfer is not the sole cause of unsteady-state limiting currents, observed when a fast current ramp is imposed on an elongated electrode. In copper deposition, in particular, as a result of the appreciable surface overpotential (see Section III,C) and the ohmic potential drop between electrodes, the current distribution below the limiting current is very different from that at the true steady-state limiting current.

The development of sectional limiting-current curves at a horizontal-plate electrode in laminar forced convection, as recorded by Hickman (H3), is illustrated by Fig. 11. Here the total current is increased linearly with time, and the potential at selected electrode segments is recorded. The limiting-current plateau is formed first at the trailing edge of the plate, then in the center, and finally at the leading edge. The time difference is on the order of 4 min, that is, approximately one-third of the time required to reach i_L everywhere. It is clear that if sectional limiting currents had been read with respect to the *central* overpotential, inaccurate local current values would have resulted for the edge segments. If the total current had been recorded against the central overpotential, the correct total i_L might have been read (due to compensating errors), but the plateau would not be well defined. Figure 11, therefore, emphasizes the practical importance of sectional measurements on extended electrodes, which were first applied systematically by Fenech (F2).

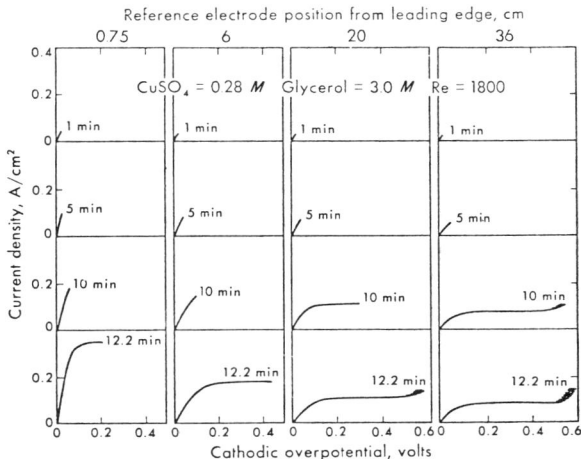

FIG. 11. The development of local cathodic overpotentials at a segmented horizontal plate electrode in forced laminar convection. [From Hickman (H3).]

From the viewpoint of unsteady-state mass transfer the following points should be noted. An optimum rate of increase exists, as found by Hickman (H3) in a model series of experiments, for approaching the limiting current (Eq. 19); any higher rate of increase causes the apparent limiting current to reflect unsteady-state effects. On this basis one would expect correct measurements to be quasi-static, that is, deviations from the steady state should be insignificant during the passage to the limiting current. Nevertheless, the limiting current in different segments is not reached simultaneously. An apparent limiting current due to depletion at the electrode is reached earliest where the time to reach the steady-state concentration profile would be longest, in accordance with Eq. (18), that is, at the last segment (trailing edge) of the electrode.

To explain this, one has to take into account that the applied current is neither distributed uniformly at all times, nor similar to the limiting-current distribution. If steady state is essentially maintained during the transition, the trailing edge will have a fairly high current density throughout (see

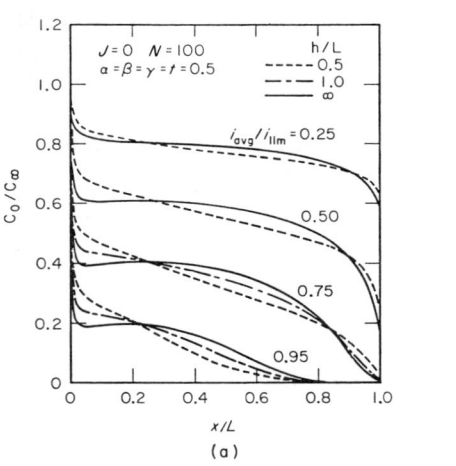 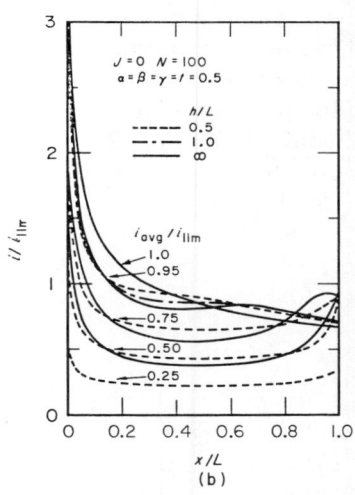

(a) (b)

FIG. 12. (a) Concentration of the reacting ion at the surface of a plane cathode of length L with parallel counterelectrode at distance h for various average current levels, at steady state, below the limiting current. The value, $J = 0$, of the dimensionless exchange current corresponds to Tafel polarization. The dimensionless average current density $N = 100$ pertains, for example, to copper deposition from a 0.15 M $CuSO_4$, 0.5 M H_2SO_4 solution flowing at a speed of 10 cm sec^{-1} past electrodes of 5-cm length, 2.5-cm separation ($i_{lim} = 50$ mA cm^{-2}). The corresponding current distributions are shown in Part b. (b) Current distribution at a plane cathode of length L with parallel counterelectrode at distance h. See Part a for the meaning of symbols. [From Parrish and Newman (P1). Reprinted by permission of the publisher, The Electrochemical Society, Inc.]

Fig. 12). This will lower the minimum time required to establish a steady-state limiting current, compared with the values suggested by Table IV. The distribution of this minimum time along the plate will probably be more uniform than that following a $x^{2/3}$ distribution, as required by Eq. (18). Nevertheless, it is likely that in the final stage of the passage to limiting current, where mass transfer dominates the potential distribution, each small increase of the current (now distributed approximately as $x^{-1/3}$) will require a local transition time distributed as $x^{+2/3}$. At the same time, as Fig. 12 illustrates, the steady-state reactant concentration in certain cases can reach a virtual zero level near the trailing edge of the plate before the limiting current is reached. Therefore, even if the passage to the limiting current is quasistatic for the plate as a whole until very close to the limiting current (e.g., 95%), the attainment of the limiting-current condition, that is, $c_R \to 0$, is nonstationary and therefore nonsimultaneous.

Sectional limiting currents near the trailing edge of the plate will tend to be somewhat higher than expected, unless the passage to the limiting current is made very slowly near the limiting current. The minimum time allowable to reach steady-state limiting current by applying a current ramp should be equal to, or greater than, that required for the *trailing* section.[4] Note that no consideration is given to the formation of surface roughness or to hydrogen evolution, conditions that may enhance the tendency to instability, that is, create an oscillating unsteady state. Recently White and Newman (W3b) analyzed simultaneous reactions at a rotating-disk electrode in the steady state and showed that hydrogen gas bubbles may be formed near the edge of the disk while the copper deposition at the center is still below its limiting rate.

F. Effect of Surface Roughness

Limiting currents measured for a deposition reaction may be excessively high due to surface roughness formation near the limiting current. Rough deposits in the case of copper deposition have been mentioned several times in previous sections, since this reaction is one commonly used in limiting-current measurements. However, many other metals form dendritic or powdery deposits under limiting-current conditions, for example, zinc (N1b) and silver. Processes of electrolytic metal powder formation have been reviewed by Ibl (I2).

The mechanism of surface-roughness formation at the limiting current has

---

[4] The time in which the limiting current is reached at the trailing end of the electrode is an order of magnitude larger than the transition time calculated from Eq. (18), using $\tau = 1.25$, and stated in Table IV. See Section IV,E , 2.

not been fully explained, but all available evidence indicates that it is a direct result of depletion of the diffusion layer. Ibl (I4) has given the following interpretation. If an electrode has protrusions smaller than the diffusion layer thickness, the current distribution between peaks and recesses will be determined approximately by the balance between surface overpotential and ohmic potential drop in the cell. However, as the current increases to the limiting value and concentration overpotential becomes dominant, the local diffusion-layer thickness determines the current distribution. The peaks receive a greater current density than the recesses; therefore, the initial profile will be amplified. In laminar flow the protrusions eventually disturb the velocity profile at the electrode surface and disrupt the laminar diffusion layer. The microscale turbulence thus generated increases the current toward the recesses in the wake of the peaks, as well as to the peaks themselves. A steep increase of the total current results, which is clearly noticeable when the electrode potential is held constant (Fig. 13).

In turbulent flow, the effect of a growing surface roughness on the limiting current during the growth process depends on the degree to which turbulent eddies are able to penetrate the recesses between peaks. This picture, developed by Levich (L8), has been confirmed experimentally (M1). Results obtained with artificial roughness profiles of fixed pitch and variable height (R17) suggest that the limiting current in turbulent flow may reach a maximum when the height/pitch ratio is 0.5.

The origin of the initial microroughness and the events leading up to its final amplification by mass-transfer limited deposition, have not been clarified definitively (P4a). It has been shown (I9a) that preliminary electropolishing, to assure a smooth surface, does not prevent surface roughness at

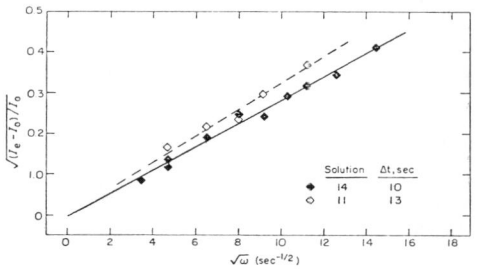

FIG. 13. Effect of rotation rate on the increase in current, between −500-mV potential and the inflection point, for a slow potential scan (20 mV sec^{-1}) during the time interval Δt indicated. I_0 = initial current at −500-mV potential, I_e = current after time Δt. Solution 14: 0.295 M CuSO$_4$, 1.492 M H$_2$SO$_4$. Solution 11: 0.05 M CuSO$_4$, 0.492 M H$_2$SO$_4$. [From Selman (S8).]

the limiting current. Microscopic heterogeneities, such as crystallographic defects or oxide and sulfide inclusions, appear to provide sufficient nuclei for eventual preferential deposition. In deposition on single crystals, the nature of the crystal face perpendicular to the current direction has a direct effect on the shape and distribution of "seeds" on the cathode surface at the limiting current (K14).

Since certain additives in plating baths have a leveling effect on the deposits (see also Section III,B), the question arises as to whether the growth of initial microroughness into the roughness observed near the limiting current may be suppressed by suitable additives. Ibl and co-workers (I9a, I9b, J5a) investigated the effect of thiourea, which is codeposited at transport-controlled rates from acidified $CuSO_4$ solutions. In dilute solutions of $CuSO_4$, for example, 0.05 M $CuSO_4$ or less, in excess H_2SO_4 the addition of sufficient thiourea (200 mg liter^{-1}) prevents the formation of rough deposits even at currents 30% in excess of the limiting current (I9a).

In subsequent experiments (I9a), however, it was found that below the limiting current the deposit in the presence of thiourea is unexpectedly rough, much more so than an equally thick deposit without thiourea. At current levels approaching the limiting current, the surface roughness is less; it becomes virtually independent of the current as the limiting current is exceeded. Without thiourea, the deposit roughness increases steeply with the level of the current, in the neighborhood of the limiting current and above. This complicated behavior appears to be related to low current efficiency and to the formation of fine nonadherent copper powder, which was also observed below the limiting current in dilute $CuSO_4$ solutions containing thiourea.

In concentrated solutions, codeposition of the additive cannot compete with the rate of copper deposition. Surface roughness does develop, although apparently at a lower rate than without the additive (I9a, I9b).

It therefore appears that surface roughness cannot be avoided, at least in concentrated $CuSO_4$ solutions. Moreover, its effect is most prominent in concentrated solutions, which can sustain high limiting currents. In experiments at a rotating-disk electrode where the limiting current was generated by applying a potential step, Selman (S8) observed that the subsequent current increase, which reflects the increase in effective surface area, was proportional to the initial limiting current (see Fig. 13).

Two questions need to be answered if one wants to develop a method of generating a limiting current gradually while minimizing the effect of surface roughness:

1. At which point in the approach to the limiting current (a certain

potential, or a certain fraction of the limiting current) does the increase of effective surface area begin?

2. How rapid is the increase in effective surface area?

Unfortunately, these questions are not susceptible to simple answers. Ibl and Schadegg (I7) studying roughness formation at a rotating-disk electrode, characterized the growth of effective surface quantitatively by impedance measurements. When applying the limiting current in a single current step, they found an induction period of approximately 30 sec before the formation of surface roughness. The electrode potential attained a steady value after 4 min. Apparently, the limiting-current condition no longer exists; the increased effective area is able to sustain a larger current than the current applied, and consequently no new rough deposit is formed.

When they applied the limiting-current condition by a single potential step, they observed a similar induction period before surface-roughness formation; but its duration was generally longer, particularly in dilute solutions (up to 4 min in 0.05 M $CuSO_4$). The induction period also depended on the applied electrode potential. Ibl and Schadegg (I7) observed that the formation of roughness in concentrated $CuSO_4$ solutions starts at more positive potentials than in dilute solutions. and likewise occurs earlier at high rotation rates than at low ones. During the induction period, in concentrated solutions, a surface rearrangement appears to take place that causes the current to decrease initially. Thus, a current minimum is found at the end of the induction period, at least for rotation rates and potentials low enough to suppress surface roughening.

If the limiting current is not applied instantaneously (as in Ibl and Schadegg's experiments), but is generated gradually, the effect of surface roughness clearly depends on the method and the rate of approach to the limiting current. Several investigators (E5, I7) have observed that galvanodynamic generation leads to less surface roughness than the potentiodynamic method, within the same time span. This can be understood from the different condition at the electrode surface in the two cases. Linear decrease of the potential generates rapidly a nonstationary limiting-current condition at the electrode, thereby inducing rough deposition, and sustains limiting-current conditions until the steady-state current is established. Under current control, the electrode potential falls more gradually, and the limiting-current condition is established only at the end of the transition period.

Measurements of rotating-disk effective diffusivities (H7, S8) for Cu^{2+} ion (see Section IV,C) indicate that a greater increase in surface area takes place when the limiting current is generated by linear potential decrease than when the current is stepped up to the limiting value. In these measurements, the time in which the limiting current is generated should be long enough to avoid excessive (apparent) limiting currents due to transient mass transfer, but short enough to avoid erroneously high limiting currents due to roughness formation. In most cases of forced convection, where transition to steady-state mass transfer is relatively fast (see Table IV), a certain range of times is permissible, at least for dilute solutions ($\leq 0.1\ M\ CuSO_4$). Conversely, a fairly reproducible value of the effective diffusivity is found in such solutions. The mobility product $\mu D/T$, corrected for migration (see Section IV,B), is not significantly dependent on the $CuSO_4$ concentration or on the ionic strength.

In more concentrated solutions, the time limit imposed by the roughness effect will approach the lower time limit set by the diffusional unsteady state. In some cases a true steady state may not be realized. This is reflected in effective diffusivities, which depend on the $CuSO_4$ concentration, and to some degree on the method of generating the limiting currents from which they were calculated (Fig. 14). To eliminate migration and dissociation effects, the effective diffusivity corrected for migration (see Fig. 4) has been plotted against true ionic strength; the latter has been calculated using the stoichiometric dissociation constant of HSO_4^- ion given by Hsueh and Newman (H10).

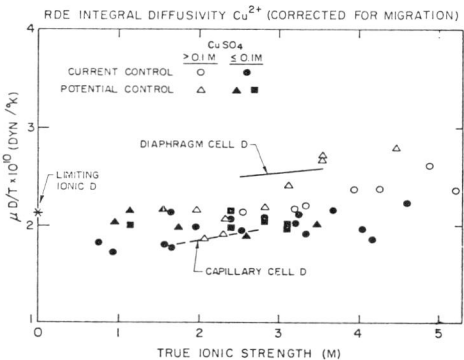

FIG. 14. Integral diffusivities of Cu^{2+}, corrected for migration effect, as a function of true ionic strength, I_t. Circles and triangles indicate values reported by Selman (S8) and Hsueh (H7), squares indicate results of Arvía et al. (A5). Diaphragm cell diffusivities according to Fenech (F3) and capillary cell diffusivities according to Hsueh (H7) are also shown. [From Selman (S8).]

V. Conditions for Valid Measurement and Interpretation of Limiting Currents

In order to obtain meaningful limiting-current measurements, it is desirable that the current plateau be well defined. From the foregoing, it is clear that this cannot always be achieved in systems where the electrode surface undergoes a change, for example, the copper-deposition reaction (in contrast to the ferri–ferrocyanide reaction). The following conditions work against achieving a clearly defined limiting-current plateau:

1. The reaction may be characterized by slow surface kinetics, leading to shortening of the plateau. Compare, for example, ferricyanide reduction and copper deposition at a rotating disk (shown in Fig. 3a and b).

2. Progress of the electrodeposition reaction may alter the surface morphology of the test area. This phenomenon causes two adverse effects: on the one hand, a loss of definition of the plateau because of progressive increase of the actual surface area during deposition; on the other hand, a higher level of limiting current than that corresponding to the projected (superficial) surface area. If one works with relatively concentrated solutions of the metal ion, especially at high current densities, the tendency to roughening of the surface can only be decreased by special addition of leveling agents to the electrolyte. Deposition from low concentrations (less than $0.1\ M$) and at low current levels yields relatively smooth surfaces, even in the absence of leveling agents.

3. Use of low concentrations of the reacting ionic species in free-convection studies leads to significant depletion of the bulk concentration during limiting-current experiments if, as is usually the case, separate cathode and anode compartments are used. This problem is aggravated by the relatively more protracted unsteady-state period associated with the smaller driving force generated in dilute solutions.

Certain conditions that adversely affect the precision of limiting current measurements can be avoided rather easily These include:

1. Unsuitable position of the reference electrode resulting in inclusion of a high ohmic potential drop between reference and working electrode. Moreover, when extended surfaces are used over which the mass transfer boundary layer thickness depends on position, a suitable number of independent reference electrodes should be used to measure local overpotentials on electrically isolated segments of the working electrode.

2. Changes in bulk reactant concentration during the limiting current measurement, due, for example, to variations in gas pressure (oxygen reduction) or to the presence of other species susceptible to reduction at the

potential of the limiting current (oxygen, in I_3^- reduction). For an accurate estimate of the bulk concentration during the measurement, it is recommended that bulk concentrations be determined before and after each limiting-current run.

3. Unsteady-state mass transfer caused by excessively fast current or potential ramps. This is especially likely to occur in measurements involving laminar flow past elongated surfaces and in free-convection studies, in which the establishment of secondary flow patterns may require long times. A compromise between the time sufficient to reach steady-state transport and the time necessary to avoid bulk depletion and surface roughening (in metal deposition) is required, and is found most reliably by preliminary experimentation.

4. Poorly defined plateaus resulting from limiting-current measurements on large elongated electrodes in forced convection. Use of segmented electrodes and current or potential ramps slow enough to allow steady-state transport to be reached at the trailing section of the electrode leads to well-defined, correct, segmental limiting-current plateaus and thus to accurate local mass-transfer coefficients.

Finally, when calculating mass-transfer coefficients from limiting-current values, the following factors should be considered:

1. The contribution of transport under the influence of the electric field (migration), which, if appreciable, should be subtracted from the total mass flux. The use of excess inert (supporting) electrolyte is recommended to suppress migration effects. However, it should be remembered that this changes the composition of the electrolyte solution at the electrode surface. This is particularly critical in the interpretation of free-convection results, where the interfacial concentration of the inert as well as the reacting ions determines the driving force for fluid motion.

2. The use of excess inert electrolyte so as to reduce differences in transport properties of the solution at the electrode surface and in the bulk. In such a solution, the ionic diffusivity of the reacting ion, for example, Cu^{2+} or $Fe(CN)_6^{3-}$, should be employed in the interpretation of results, and not the molecular diffusivities of the compounds, for example, $CuSO_4$ or $K_3Fe(CN)_6$.

VI. Review of Applications

For convenience, experimental results of mass transfer by the limiting-current method are summarized in eight subsections. Table VII in Section VI,I presents a collection of important correlations and their references in an easily accessible form.

Not included in this survey are limiting currents at porous electrodes, because they usually are not controlled exclusively by convective diffusion [for exceptions, see (H11b)], limiting currents due to limited gas solubility at an electrode (N8b), or limiting currents recorded by electrochemiluminescence (H6c, C12b).

A. SIMPLE LAMINAR FLOWS

For a number of flow situations, the mass-transfer rate can be derived directly from the equation of convective diffusion (see Table VII, Part A). The velocity profile near the electrode is known, and the equation is reduced to a simpler form by appropriate similarity transformations (N6). These well-defined flows, therefore, are being exploited increasingly by electrochemists as tools for the kinetic characterization of electrode reactions. Current distributions at, or below, the limiting current, transient mass transfer, and other aspects of these flows are amenable to analysis. Especially noteworthy are the systematic investigations conducted by Newman (review until 1973 in N7; also N9b, N9c, H6b and references in Table VII), by Daguenet and other French workers (references in Table VII), and by Matsuda (M4a–d). Here we only want to comment on the nature of the velocity profile near the electrode, and on the agreement between theory and mass-transfer experiment.

In the case of the rotating disk, there is a uniform axial velocity toward the disk that depends only on the axial distance from the disk, and consistent with this the mass-transfer rate is also uniform. If the center of the disk is nonreactive, the electrode no longer is uniformly accessible, the average flux to the active ring is higher than otherwise, and the rate is a maximum at the inner edge of the active ring, where the mass-transfer boundary layer begins.

For flow parallel to an electrode, a maximum in the value of the mass-transfer rate occurs at the leading edge of the electrode. This is not only the case in flow over a flat plate, but also in pipes, annuli, and channels. In all these cases, the parallel velocity component in the mass-transfer boundary layer is practically a linear function of the distance to the electrode. Even though the parallel velocity profile over the hydrodynamic boundary layer (of thickness δ_h) or over the duct diameter (with equivalent diameter d_e) is parabolic or more complicated, a linear profile within the diffusion layer (of thickness δ_d) may be assumed. This is justified by the extreme thinness of the diffusion layer in liquids of high Schmidt number:

$$\delta_d/\delta_h \sim \mathcal{O}(\mathrm{Sc}^{-1/3}). \qquad (24)$$

Similarity transformations lead then to two types of correlations: the

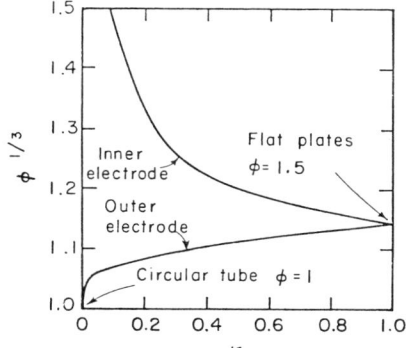

FIG. 15. The constant ϕ in Eq. (27) for an annulus as a function of the radial ratio $\kappa = r_i/r_0$.

boundary layer type,

$$\text{Sh} = C\,\text{Re}^{1/2}\,\text{Sc}^{1/3} \tag{25}$$

valid for boundary layers which have transverse velocity components, and the Lévêque type,

$$\text{Sh} = C(\text{Re}\,\text{Sc}\,d_e/x)^{1/3} \tag{26}$$

valid for parallel flows. The rotating-disk correlation, Eq. (1a), in Table VII, is an example of the first. Equations (5) and (6) in Table VII belong to the Lévêque type. All are characterized by the one-third power dependence on the Schmidt number, corresponding to the ratio of boundary layer thicknesses (hydrodynamic/mass transfer) given by Eq. (24).

The general expression for the average mass-transfer rate over length L in annular channels is given by Eq. (5a) in Table VII

$$\text{Sh} = 1.615(\phi\,\text{Re}\,\text{Sc}\,d_e/L)^{1/3} \tag{27}$$

where ϕ is a function of the geometric parameter κ, the ratio between inner and outer radius r_i/r_0. In Fig. 15 $\phi^{1/3}$ is shown as a function of κ; Re and Sh are based on the equivalent diameter d_e, that is, the difference between the inner and outer diameter.

Equation (5b), in Table VII, established by Lin et al. (L9) for coaxial flow in annuli with $\kappa = 0.5$, was originally taken by them to be a striking confirmation of the theoretically derived correlation, Eq. (27), with $\phi = 1$. The latter condition, however, corresponds only to the limiting case of Eq. (27), at $\kappa \to 0$, that is, mass transfer to the wall of an outer cylinder, without an inner cylinder present. On these grounds, and because of other experimental conditions, the correlation of Lin et al. (L9) was criticized by Friend and Metzner (F9), who calculated that the constant in Eq. (5b) of

Table VII should be 1.939 for the ratio $\kappa = 0.5$. Part of the 17% discrepancy between the results of Lin et al. (L9) and Eq. (27) may be ascribed to the use of incorrect diffusivities. An estimate of the errors is possible for part of their experiments. The value of the product $\mu D/T$ of $K_3Fe(CN)_6$ based on the electric mobility at infinite dilution as used by Lin et al. is 11% too high, according to more recent measurements of the effective ionic diffusivity of $Fe(CN)_6^{3-}$ by Gordon et al. (G5). Similarly, the mobility product of $K_4Fe(CN)_6$ is 16% too high, and that of O_2 no less than 26% too high, compared with data of Davis et al. (D7) (see Table III). According to Eq. (27) the value of D would have to be 27% too high to account fully for a coefficient that is 17% too high; consequently, the discrepancy cannot be attributed entirely to incorrect diffusivities.

A recent measurement of mass transfer in coaxial annular flow of radial ratio $\kappa = 0.5$ by Ross and Wragg (R18, W13a) confirmed the coefficient 1.94 derived from Eq. (27). This experiment, employing copper deposition from acidified $CuSO_4$ solution, again brought out the importance of using correct diffusivities. The original correlation of their data lay about 10% below Eq. (27). In a later review they ascribed this discrepancy to the use of diaphragm-cell diffusivities of $CuSO_4$, measured by Cole and Gordon (C12a) (see also Section IV,C). Recorrelating their results using the rotating-disk integral diffusivities measured by Arvía et al. (A5), they obtained satisfactory agreement with Eq. (27).

Flow between parallel plates corresponds to the other limiting case of Eq. (27) for $\kappa \to 1$ ($r_i = r_0$, $d_e = 2h$, where h = channel height):

$$Sh = 1.849(ReScd_e/L)^{1/3} \quad (28a)$$

This agrees well with results of Tobias and Hickman (T2) in an experimental channel with an aspect ratio γ of 10–15. For channels with a smaller aspect ratio, the coefficient in Eq. (28a) should be multiplied by a factor ϕ_2:

$$Sh = 1.849\phi_2(Re\,Sc\,d_e/L)^{1/3} \quad (28b)$$

Values of ϕ_2 calculated from the velocity profile are given in Table V for various aspect ratios and also for cases in which the electrodes occupy only part of the channel width. Rousar et al. (R21a) have obtained good experimental agreement with Eq. (28b).

Besides the recent work of Acosta et al. (A1a) and Landau and Tobias (L1b), referred to in Table VII, theoretical and experimental work by Meyer et al. (M6b) on the Ag-AgCl electrode in flow channels should be mentioned here.

The effectiveness of the limiting-current methods for mass-transfer measurement was demonstrated quite early by Tobias and co-workers (W4, W5)

TABLE V

VALUES OF ϕ_2 IN EQ. (28b) AS A FUNCTION OF ASPECT RATIO
AND ELECTRODE-TO-CHANNEL WIDTH RATIO[a]

Electrode-to-channel width ratio	Aspect ratio (height/width)					
	2	1	0.5	0.2	0.1	0.05
0	0.8956	0.9285	0.9668	0.9840	0.9899	0.9944
0.1	0.8949	0.9279	0.9664	0.9840	0.9899	0.9944
0.2	0.8929	0.9259	0.9652	0.9893	0.9899	0.9944
0.3	0.8894	0.9226	0.9630	0.9838	0.9899	0.9944
0.4	0.8844	0.9178	0.9598	0.9835	0.9899	0.9944
0.5	0.8777	0.9113	0.9552	0.9829	0.9899	0.9944
0.6	0.8689	0.9027	0.9489	0.9818	0.9899	0.9944
0.7	0.8575	0.8915	0.9403	0.9796	0.9897	0.9944
0.8	0.8426	0.8767	0.9281	0.9752	0.9890	0.9944
0.9	0.8223	0.8563	0.9099	0.9657	0.9859	0.9940
0.95	0.8085	0.8422	0.8968	0.9567	0.9811	0.9925
1.0	0.7883	0.8214	0.8760	0.9388	0.9671	0.9829

[a] From Rousar et al. (R21a).

in free convection at vertical electrodes, Eq. (8a) in Table VII. Their results showed excellent agreement with the theoretically predicted correlation:

$$Sh_L = 0.670(Gr_L Sc)^{1/4} \qquad (29a)$$

These experiments were carried out with the copper deposition reaction, which creates a large density difference at high concentrations. The diffusivity data used were the diaphragm cell data of Cole and Gordon (C12a). These data, as correlated by Fenech (F3), were also used in later work on horizontal electrodes. The question whether molecular diffusivity data are adequate for the work in question, has been discussed in Section IV,C.[5]

Later measurements by Fouad and Gouda (F5), who used the ferri–ferrocyanide redox reaction, yielded a much lower coefficient in the free-convection correlation, Eq. (29a). However, uncertainty about the interfacial composition in free convection (S9a) may be responsible for the discrepancy observed in the results obtained with the redox system. Taylor and Hanratty (T1b) showed that the data of Fouad and Gouda could be

---

[5] In this connection it should be noted that the most careful studies (N9a, D2) undertaken to confirm the rotating-disk correlation, Eq. (1a) in Table VII, used the iodine reduction reaction with very small reactant concentrations, and employed diaphragm cell diffusivities of KI_3 in KI.

made to agree better, though not completely, with the theoretically expected Eq. (29a).

The application to free-convection mass transfer is a particularly severe test of the limiting-current method. The driving-density difference term $\Delta\rho$ of the Grashof number ($\mathrm{Gr} = \Delta\rho g L^3/\nu^2\rho$) in the correlation for free convection, Eq. (8) in Table VII, is influenced by the accumulation or depletion of supporting electrolyte at the electrode. Thus, there is interaction not only between diffusion and convection (expressed in the coupling of the equation of motion and the equation of convective diffusion), but also between electromigration and convective diffusion. This is true for well-supported electrolytes, as well as for high-concentration of $CuSO_4$ as used frequently in free-convection studies.

In free-convection mass transfer at electrodes, as well as in forced convection, the concentration (diffusion) boundary layer δ_d extends only over a very small part of the hydrodynamic boundary layer δ_h. In laminar free convection, the ratio of the thicknesses is

$$\delta_d/\delta_h \sim \mathrm{Sc}^{-1/4} \tag{29b}$$

The hydrodynamic boundary layer has an inner part where the vertical velocity increases to a maximum determined by a balance of viscous and buoyancy forces. In fluids of high Schmidt number, the concentration diffusion layer thickness is of the same order of magnitude as this inner part of the hydrodynamic boundary layer. In the outer part of the hydrodynamic boundary layer, where the vertical velocity decays, the buoyancy force is unimportant. The profile of the vertical velocity component near the electrode can be shown to be parabolic.

Different again, with respect to the velocity profile in the region of varying concentration, is the case of mass transfer at a moving boundary, for example, at a moving continuous electrode, Eq. (9) in Table VII. Next to the uniformly moving continuous flat plate, the flow generated by a moving, continuous (e.g., extruded) cylinder is the simplest of a class of laminar boundary layer flows first investigated by Sakiadis (S1). In high Schmidt number liquids, the diffusion layer is contained in a region near the moving electrode, where the liquid velocity is practically equal to that of the cylinder. The mass-transfer rate should then follow a penetration-type correlation, involving the Peclet number, $\mathrm{Pe} = vL/D$, where v is the electrode velocity, L the transfer length, and D the diffusion coefficient:

$$\mathrm{Nu} = (2/\sqrt{\pi})\mathrm{Pe}^{1/2} \tag{30}$$

However, experiments in which the electrode is circulated in and out of the solution are difficult to carry out without electrolyte and current leaks.

Rotte et al. (R20), using a nickel wire circulated through ferri–ferrocyanide solution, found only approximate agreement with Eq. (30); experimental mass-transfer rates were mostly low compared with Eq. (30), and showed considerable scatter. Rao and Sharma (R4) accounted for turbulence in the hydrodynamic boundary layer over part of the cell length, obtaining good agreement between the predicted values and their own data, as well as those of Rotte et al.

Recently, Chin and co-workers (C5a, C6b) and Nadebaum and Fahidy (N1a) have used a rotating cylinder with one or more wiper blades to simulate a moving continuous flat surface; they obtained good agreement with Eq. (30) in the laminar range, which was shown to extend very far, in spite of the turbulence in the bulk of the liquid.

B. Laminar Flows with Complications

Under this heading, a number of correlations for mass transfer to rotating electrodes is given in Part B of Table VII. These are comparable to Eqs. (1)–(3) in Table VII but the flow geometries are not as well characterized hydrodynamically, for example, because of secondary flows in a restricted cell volume. There is a considerable discrepancy between the rigorously tested rotating disk correlation Eq. (1a) of Table VII and that given in Eq. (10b), or between rotating cone correlation Eq. (1e) of Table VII and that given in Eq. (11a). This may be due to inaccurate execution of the measurements. Recent work by Despic et al. (D11) has shown that the limiting current at a rotating disk electrode is only affected by secondary convection if the interelectrode gap is less than the radius of the rotating disk; in that case it is less than the theoretical current. The difference decreases with increasing rotation rate. If a central fluid intake is provided for, such as in the interesting parallel disk cells studied by Jansson and Ashworth (J4b), the rotating disk behaves essentially as a free disk even at very small interelectrode distances, Eq. (10c) of Table VII. Mohr and Newman (M13g) have analyzed the effect of eccentricity on the mass-transfer rate to an off-center rotating disk electrode. The effect is felt only when the relative eccentricity exceeds 0.6377 times the radius of the mass-transfer surface. Experimental data were in good agreement with the numerical solution, approximated by Eq. (10f) of Table VII. Of particular interest is Eq. (10e) in Table VII, indicating the existence and magnitude of an intercept in the i versus $\omega^{1/2}$ plot for a rotating electrode that has inactive sites.

It is remarkable that boundary layer flow along a flat plate, much studied and well understood, has not been used in electrochemical mass-transfer studies. The average Sherwood number for an electrode of length L in that case is

$$\text{Sh}_L = 0.677 \text{Re}_L^{1/2} \text{Sc}^{1/3} \qquad (\text{Re}_L < 10^5) \qquad (31)$$

This expression has been used to correlate results obtained in a rectangular channel, Eq. (14) in Table VII, the hydrodynamic entrance length of the channel ($L_e = 0.0575$ d Re_d) being too short to assure a fully developed flow. The results were still 24% high, compared with Eq. (31) modified by a relaxation assumption:

$$\text{Sh}_{x_0+L} = 0.677 \text{Re}_{x_0+L}^{1/2} \text{Sc}^{1/3} \left[\left(1 + \frac{x_0}{L}\right)^{3/4} - \left(\frac{x_0}{L}\right)^{3/4} \right]^{2/3} \left(1 + \frac{x_0}{L}\right)^{1/2} \qquad (32)$$

where x_0 is the inactive plate length preceding the mass-transfer section of length L. The merging of the boundary layers apparently made the velocity profile at the channel wall steeper than it would be at a flat plate.

Equation (31) has also been used to correlate results obtained in annular channels with insufficient hydrodynamic entrance length (B2, B3). Some of the data fit Eq. (31) with a slightly smaller coefficient. A similar expression, including terms for the hydrodynamic entrance length, was used by Pickett and Ong (P3a).

Coeuret and co-workers (C9, C10) measured the limiting current at a short cylindrical electrode in the center of a large tube; their results were of the general form of Eq. (31). Carbin and Gabe (C1c) in a similar case found better agreement with Eq. (26). Wragg (W8b) has critized this interpretation of Carbin and Gabe's data, and suggested that Eq. (31) should be more representative of the developing-flow regime in their experiment. Birkett and Kuhn (B4d) found, at low flow velocities in short rectangular channels, reasonable agreement with a flat-plate boundary-layer correlation similar to Eq. (31).

Unfortunately, considerable experimental effort is sometimes expended on flow cells that are neither well characterized hydrodynamically nor of evident practical interest (D12, D13).

In view of its fundamental importance, Eq. (31) should be tested in a flow channel of sufficient height, provided with a suitable flow-straightening section. The results should be relevant to many corrosion processes occurring in maritime structures. The closest approach to such a test has been the work of Coeuret and Vergnes (C8) on a stationary wire in parallel flow.

The Lévêque-type correlation, Eq. (26), has been used for mass-transfer from a liquid film falling under gravity (I10, W13b), where it holds for Reynolds numbers in the laminar range, in spite of the presence of surface waves on the film. The latter caused small local oscillations of the mass-transfer rates, of the order of 2% in the laminar range (but higher in the turbulent range; see below). In these measurements, Iribarne et al. (I10) found rates only slightly below those predicted by the Lévêque relation.

Of considerable interest is the use of small isolated electrodes, in the form of strips or disks embedded in the wall, to measure local mass-transfer rates or rate fluctuations. Mass-transfer to spot electrodes on a rotating disk is represented by Eqs. (10g–i) of Table VII. Analytical solutions in this case have to take account of curved streamlines. Despic et al. (D11d) have proposed twin spot electrodes as a tool for kinetic studies, similar to the ring-disk electrode; applications of disk and ring-disk electrodes for kinetic studies are discussed in several monographs (A3b, P4b). In fully developed channel or pipe flow, mass transfer to such electrodes is given by the following equation based on the Lévêque model:

$$k_{avg} = 0.8075(D^2 S/L)^{1/3} \tag{33}$$

where S is the velocity gradient at an electrode of length L. Equations (5a) in Table VII and Eq. (27) are special forms of this general relation. The actual rates found by Iribarne et al. (I11a) at an array of electrodes (1.5 mm in flow direction, 1 cm wide, 0.5-cm spacing in flow direction) were considerably below those predicted by Eq. (33); probably the electrode spacing was not large enough for complete decay of the concentration profile between electrodes.[6]

Hanratty and co-workers (M7, R12, R13, V1, V2) made particular use of strip and spot electrodes to record mass-transfer fluctuations in turbulent flows, where the viscous sublayer governs the behavior. They used 0.38–1.63-mm diameter disks (effectively equivalent to rectangles of length L, equal to 0.82 times the disk diameter), and also rectangular shapes 0.51–1.57 mm wide, and 0.076–0.53 mm long in the flow direction. Knowledge of the mass-transfer intensity (i.e., root-mean-square fluctuation) provides information about the turbulent velocity intensity near the wall. In turbulent flow, the diffusion layer is an order of magnitude smaller than the viscous sublayer, at least for electrolytic solutions. The velocity fluctuations, as well as the average velocity transverse to the wall, are therefore negligible compared with the longitudinal quantities; the Lévêque model applies here again, but includes a time dependence.

---

[6] Interestingly enough, the use of extremely short electrodes in high Schmidt systems may complicate the Leveque mass-transfer model, which neglects longitudinal diffusion. At the leading edge of the electrodes this neglect is not correct and the flux is actually increased over a distance of the order of $d_e/\sqrt{Re\ Sc}$. Newman (N7) has treated this problem for the case of pipe flow.

For small frequencies $\tilde{f}$, Reiss and Hanratty (R12, R13) deduced a simple relationship between fluctuating mass transfer and velocity gradient. This relationship makes use of the dimensionless frequency $\tilde{n}$:

$$\tilde{n} = 2\tilde{f}L^{2/3}/D^{1/3}S^{2/3} \tag{34}$$

Thus they were able to calculate the velocity intensity from the mass-transfer intensity and the spectral distribution function of mass-transfer fluctuations. By measuring and correlating mass-transfer fluctuations at strip electrodes in longitudinal and circumferential arrays, information was obtained about the structure of turbulent flow very close to the wall, where hot wire anemometer techniques become unreliable. A concise review of this work has been given by Hanratty (H2).

When electrically insulated strip or spot electrodes are embedded in a large electrode, and turbulent flow is fully developed, the steady mass-transfer rate gives information about the eddy diffusivity in the viscous sublayer very close to the electrode (see Section VI,C below). The fluctuating rate does not give information about velocity variations, and is markedly affected by the size of the electrode. The longitudinal, circumferential, and time scales of the mass-transfer fluctuations led Hanratty (H2) to postulate a surface renewal model with fixed time intervals based on the median energy frequency.

The basic Lévêque mass-transfer relation, Eq. (33), can also be used to deduce surface velocity gradients from local mass-transfer measurements. Hanratty and co-workers (S18, D15) studied forced flow, transverse to a cylinder, in the ranges $60 < \text{Re} < 360$ (where vortex shedding first occurs) and $5000 < \text{Re} < 10^5$ (region of constant drag). Strip electrodes were embedded in the surface of a nonconducting cylinder and local velocity gradients determined from the time-averaged current density, at various angles to the approach velocity. The shear–stress distribution thus obtained allowed calculation of the distribution of mass-transfer rates over the circumference of the cylinder. Local mass-transfer rates then were measured at insulated-strip and spot electrodes embedded in a conducting cylinder surface, with the entire cylinder surface kept at the same cathodic potential. The recorded rate distribution was in good agreement with that predicted from the shear–stress distribution on the inert cylinder. Only at very low Reynolds numbers did free convection interfere with the velocity field, in particular at the forward and rear stagnation points.

When velocity gradients are small, for example, near the boundary layer separation point and at the rear of a cylinder in separated flow, Eq. (33) is inaccurate. The separation point was determined with an accuracy of 1 degree by using twin strip electrodes of 125 μm length, separated by

50–70-μm insulation. As long as one electrode is downstream from the boundary layer of the other, it gives a diminished signal when the other electrode is activated. With a proper choice of electrode configuration, this procedure can be employed to measure quantitatively the direction of surface velocity gradients in three-dimensional boundary layers (K2a, P6). The extreme care with which embedded electrodes for such hydrodynamic measurements have to be prepared is discussed in a paper by Son and Hanratty (S18).

Altogether, Hanratty's investigations represent a remarkable example of the versatility of the electrochemical mass-transfer technique. The method is considered so reliable that generalized correlations, for example, Eq. (33), can be used to measure convection profiles at a reacting surface.

A survey of the work of French investigators using microelectrodes, in particular on rotating disks, is given by Aïmeur et al. (A3a).

Overall mass-transfer rates at cylinders and spheres are important for chemical engineering applications. Overall mass transfer at a cylinder was measured first by Dobry and Finn (D16a). Ranz (R3) intended to use the limiting current at a wire as a measure of the flow velocity in analogy to hot-wire anemometer principles. However, a simple velocity correlation could not be established in these early investigations. Subsequent work by Grassmann et al. (G7) and by Vogtländer and Bakker (V6) established the importance of the cylinder diameter-to-duct diameter ratio and of free convection at low forced-flow rates, represented in Eq. (20-4) in Table VII by a term in the Schmidt number only. Mass-transfer rates measured by Vogtländer and Bakker at gauzes, Eqs. (21-2) and (21-1) in Table VII, are 10–20% higher than at a single wire, based on the approach velocity.

Overall mass-transfer rates at a sphere in forced flow, and mass-transfer rate distribution over a sphere as a function of the polar angle have been measured by Gibert, Angelino, and co-workers (G2, G4a) for a wide range of Reynolds numbers. The overall rate dependence on Re exhibited two distinct regimes with a sharp transition at Re = 1250. Local mass-transfer rates were deduced from measurements in which the sphere was progressively coated by an insulator, starting from the rear.

C. Free Convection and Mixed Convection

Free convection flow around horizontal cylinders and spheres is laminar for moderate values of GrSc (see Table VII, Part C); mass-transfer rates obey correlations of the same type as that for a vertical plate electrode, Eq. (29a):

$$\text{Nu} \sim \mathcal{O}(\text{GrSc})^{1/4}. \tag{35}$$

Schütz (S6) observed local transition to turbulence at cylinders above $GrSc = 3 \times 10^8$, by the use of an insulated strip electrode. Taylor and Hanratty (T1b) observed two types of instabilities at large electrolyte concentrations: one involved unsteady flow at the rear of the cylinder, the other at both front and rear. The average mass-transfer rate, as well as the local mass-transfer profile, were in good agreement with laminar boundary layer theory. Since an unsupported electrolyte was used, the upper GrSc limit was not as high as in the experiments of Schütz (S6) and Smith and Wragg (S15a). These workers found a significantly higher average mass-transfer rate than Taylor and Hanratty, possibly because of local transition to turbulence at the highest GrSc number experiments. Weder (W3a), who also found higher rates than expected at high GrSc numbers, accounted for this by adopting the exponent 0.3 on GrSc.

Measurements on arrays of horizontal cylinders were reported by Smith and Wragg (S15a).

Schütz's correlation for free convection at a sphere, Eq. (25) in Table VII, takes pure diffusion into account by means of the constant term $Sh = 2$. According to his measurements using local spot electrodes, the flow here is not laminar but already in transition to turbulence.

Free convection at vertical cylinders differs from that along vertical plates because of the influence of curvature, but the effect is important only if the diffusion layer thickness is appreciable compared with the diameter. This can be expressed as

$$x/R \sim \mathcal{O}(Gr_x Sc)^{1/4} \tag{36}$$

where x is the vertical coordinate and R the cylinder radius. Ravoo et al. (R10) developed an approximate solution for this case and, using wires of 0.1–2.64-mm diameter and length 1–374 cm, found fairly good agreement with theory (Fig. 16).

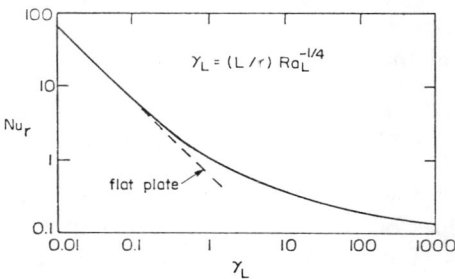

FIG. 16. Relation between Nusselt number and free convective curvature parameter for a vertical cylinder. [After Ravoo et al. (R10).] See also Eq. (27) in Table VII.

Another case of free convection with some complications, but amenable to solution, is that due to combined temperature and concentration differences. De Leeuw den Bouter et al. (D10) experimented with such combined free convective transfer, assuming complete analogy of heat and mass transfer if the Grashof number employed is of the form

$$\text{Gr}^{**} = [\alpha \, \Delta c + \zeta \, \Delta T (\text{Le})^{1/2}] g L^3 / \nu^2 \qquad (37)$$

where α is the densification coefficient due to concentration and ζ that due to temperature.

Their results for cooperating body forces come very close to a complete confirmation of this analogy, since Eqs. (28b) and (28c) in Table VII are also valid for the heat-transfer rate, which was measured simultaneously:

$$\text{Nu} = 0.64 (\text{Gr}_h^{**} \text{Pr})^{0.252} \qquad (38)$$

where

$$\text{Gr}_h^{**} = [\zeta \, \Delta T + \alpha \, \Delta c (\text{Le})^{-1/2}] g L^3 / \nu^2 \qquad (39)$$

Marchiano and Arvía (M3) also measured mass transfer by thermal and diffusional free convection at a vertical plate. They derived on theoretical grounds a combined Grashof number as follows:

$$\text{Gr}^* = [(\alpha \, \Delta c)^{3/4} + (\zeta \, \Delta T)^{3/4} (\text{Le})^{1/4}]^{4/3} g L^3 / \nu^2 \qquad (40)$$

They found fair agreement with experimental data, except for an unstable flow region when the temperature densification was of opposite sign but not large enough to cancel the concentration effect. For stronger cooling of the cathode, the flow was again stable and mass-transfer rates satisfied the theory.

The Grashof number given by Eq. (40) appears to have a weaker theoretical basis than that given by Eq. (37), since it is based on an analysis that approximates the profile of the vertical velocity component in free convection, for example, by a quadratic function of the distance to the electrode. The choice of an appropriate Grashof number, as well as the experimental conditions in the work of de Leeuw den Bouter et al. (D10) and Marchiano and Arvía (M3), has been reviewed critically by Wragg and Nasiruddin (W10). They measured mass transfer by combined thermal and diffusional, turbulent, free convection at a horizontal plate [see Eq. (31) in Table VII], and correlated their results satisfactorily with the Grashof number of Eq. (37).

Combined thermal and diffusional free convection at a horizontal cylinder was investigated by Weder (W3a). He used a Grashof number of the type of Eq. (37), but corrected for the higher than 0.25 exponent of Gr found in isothermal conditions [see Eq. (24b) in Table VII], by defining a combined Rayleigh number

$$Ra^* = (GrSc)^{1.07} + 1.49\, Gr_h Sc(Le)^{0.5} \qquad (41)$$

Weder's experiments were carried out with opposing body forces, and large current oscillations were found as long as the negative thermal densification was smaller than the diffusional densification. [Note that the Grashof numbers in Eq. (41) are based on absolute magnitudes of the density differences.] Local mass-transfer rates oscillated by 50%, and total currents by 4%. When the thermal densification dominated, the stagnation point moved to the other side of the cylinder, while the boundary layer, which separates in purely diffusional free convection, remained attached.

Turbulent free convection attracted early attention in limiting-current studies, because the laminar, free convection boundary layer at a vertical plate becomes turbulent at a height characterized approximately by the Rayleigh number

$$Ra_L = Gr_L Sc \approx 10^{11}\text{--}10^{13} \qquad (42)$$

Fouad and Ibl (F6) have suggested that the transfer rate in free-convective turbulent flow should follow a $Ra^{1/3}$ dependence. Experimentally they found a slightly lower dependence, probably due to the partly laminar regime on such plates. Because the exponent is not determined by theoretical considerations, there is considerable variation in the coefficients of turbulent, free convection correlations. Among other factors, inaccuracies in the calculation of the density-driving force may be responsible for discrepancies. This is particularly likely for the case of the ferricyanide reduction reaction.

It is interesting to note that free convection on a rough surface [see Eq. (35) in Table VII] follows a higher GrSc dependence than that on a smooth plate for the same Ra range. This suggests that surface roughness eases the transition to turbulence.

Fouad and Ahmed (F4) investigated free convection at inclined plates, and expressed their results in the form of a dependence on $(CGrSc)^{0.28}$ in Eq. (34) of Table VII. The constant C corresponds roughly to $\cos\theta$ (θ = angle of inclination from the vertical), as long as the densification is stable, that is, in the direction of gravity. Hence, for a vertical plate, $\theta = 0$ and $C = 1$. For unstable densification, C increases sharply with θ, and least-squares analysis gives a GrSc dependence which approaches one-third power as the plate becomes increasingly horizontal.

Free convection on a horizontal plate has been studied extensively. Fenech and Tobias (F3) first established the $(GrSc)^{1/3}$ dependence for this kind of turbulent transport. The rate per unit area is independent of plate dimension, provided edge effects are unimportant. Another condition investigated by these authors was the effect of a second flat plate (diaphragm) placed parallel above the first one. This plate did not influence the mass transfer unless its distance from the other plate was less than approximately 0.5 cm. An alternative correlation was given, Eq. (30a4) in Table VII, which takes advection into account in the case of a horizontal cathode embedded in an inert surface.

At low Rayleigh numbers, Wragg (W6) found a smaller Ra dependence, resembling more the dependence in laminar free convection. In this range of Ra numbers, a cellular flow pattern is believed to exist, analogous to that of thermal and surface tension-driven cellular convection (Bénard cells; F3). In the range where the convection is turbulent, the $Ra^{1/3}$ dependence has been confirmed over seven powers of Ra by Ravoo (R9), who used a centrifuge to vary the body force at constant bulk composition.

As a general observation on free-convection limiting currents, it should be remembered that in solutions where the driving force is small, several experimenters (B9, F3) have found it difficult to obtain satisfactory limiting-current plateaus. This question has been discussed in Section IV,D. Caution is justified in interpreting results at low Ra values.

Free convection mass transfer in restricted spaces has been investigated by Böhm et al. (B9) and earlier by Schmidt (S5). The latter author restricted the distance between parallel-plate electrodes in halogen redox reactions (without supporting electrolyte) to such small values that convection could hardly develop. Similar experiments, also using closed cells but with larger electrode distance, were reported by O'Brien and Mukherjee (O1). Böhm et al. (B9) restricted the liquid movement by varying the distance between the vertical cathode and the vertical diaphragm from 6 to 0.15 mm; the bulk electrolyte could flow into and out of this gap. Experiments with impeded gap accessibility were also reported. A critical distance between vertical electrode and diaphragm was found, below which the current is less than that for a single vertical plate with a distant counterelectrode; this distance is

$$a_{cr} = 5.48(GrSc)^{-0.3} \tag{43}$$

The mass-transfer correlation obtained by Böhm et al. (B9), Eq. (33), in Table VII, is conspicuous for its remarkably high exponent (0.85) on the GrSc product. Since the current is almost independent of diffusivity, this must mean that the reacting ion is depleted at the downstream end of the narrow slit between the cathode and diaphragm. The total current then is determined largely by the convective transport of reactant into the slit, which, in turn, depends on the density difference but not on diffusivity.

Combinations of forced and free convection mass transfer (mixed convection) are amenable to theoretical analysis, if the free-convection flow is in the same direction as the forced "aiding flow," for example, free convection with upward forced flow in vertical channels. For this situation Wragg and Ross (W12) found experimentally that a smooth transition occurred from the forced-convection Leveque correlation, Eq. (27), to the typical $(GrSc)^{1/4}$ dependence for laminar free convection. In opposing flows, a more complex behavior was found (R15, W8a), the mass-transfer rate having a minimum at a certain critical combination of length, Reynolds number, and Grashof number. It is very likely that, at this minimum, the leading edge of the diffusion layer moves from one end of the transfer section to the other. Marked current oscillations and poor reproducibility indicate the occurrence of flow instability.

Tobias and Hickman (T2), the only investigators to date to study combined free and forced convection in horizontal channel flow, found a remarkably sharp separation between forced- and free-convection dominated mass transfer. In forced convection, the critical Grashof number, based on the diffusion layer thickness, is

$$\text{Gr}_\delta = \frac{g\alpha(\Delta c)\,dx}{v\bar{v}} = 920. \tag{44}$$

For $\text{Gr}_\delta < 920$, mass transfer could be represented by the forced-convection correlation; and for $\text{Gr}_\delta > 920$, by the free-convection correlation of Fenech and Tobias (F3). Tobias and Hickman (T2) also inferred the existence of cellular vortex flow near the electrode from deposition patterns, the induction length for this behavior agreeing with Eq. (44).

As discussed in Section IV,E, it is necessary, as well as convenient, in studies such as these to divide the electrode into several insulated sections in order to make accurate limiting-current measurements; otherwise the limiting current may be obscured. A limiting-current curve is then obtained for each insulated section. This technique, first introduced by Fenech and Tobias (F3), requires that the potential be kept equal on the various sections as limiting current is approached. Sectional currents are therefore measured by means of potential drop through low-ohmic precision resistors.

D. Turbulent Forced Convection

Turbulent mass-transfer relations concerning forced convection are of interest for two main reasons (see Table VII, Part D): (a) because of their practical importance, since turbulence promotes increase of transfer rates; and (b) they afford an indirect means of gaining insight into the mechanism of turbulence very close to the wall, a region in which the hydrodynamic probes usually employed (e.g., hot wire anemometers) disturb the measured object itself (the viscous sublayer).

Numerous turbulent mass-transfer relationships are given in Eqs. (39)–(50), Table VII. Although the most important ones in practical applications are those for channels and tubes, several other configurations also have been investigated because of their hydrodynamic interest. Generally, it is not possible to predict mass-transfer rates quantitatively by recourse to turbulent flow theory. An exception to this is for the region of developing mass transfer, where a Lévêque-type correlation between the mass-transfer coefficient and friction coefficient f can be established:

$$k/\bar{v} = 0.8075(D^2 f/2\nu L)^{1/3} \tag{45}$$

or, in dimensionless terms,

$$k^+ = k/v^* = 0.8075(L^+)^{-1/3} \text{Sc}^{-2/3} \tag{46}$$

where the plus superscript denotes quantities made nondimensional with respect to wall-shear stress τ_0 by use of the "friction velocity" $v^* = (\tau_0/\rho)^{1/2} = \bar{v}(f/2)^{1/2}$.

By substituting the well-known Blasius relation for the friction factor, Eq. (45) in Table VII results. Van Shaw et al. (V2) tested this relation by limiting-current measurements on short pipe sections, and found that the Re and (L/d) dependences were in accord with theory. The mass-transfer rates obtained averaged 7% lower than predicted, but in a later publication this was traced to incorrect flow rate calibration. Iribarne et al. (I10) showed that the Leveque relation is also valid for turbulent mass transfer in falling films, as long as the developing mass-transfer condition is fulfilled (generally expressed as $L^+ < 10^3$) while $\text{Re} > 10^3$. The fundamental importance of the Lévêque equation for the interpretation of microelectrode measurements is discussed at an earlier point.

When the concentration profile is fully developed, the mass-transfer rate becomes independent of the transfer length. Spalding (S20a) has given a theory of turbulent convective transfer based on the hypothesis that profiles of velocity, total (molecular plus eddy) viscosity, and total diffusivity possess a universal character. In that case the transfer rate k^+ can be written in terms of a single universal function of the transfer length L and fluid properties (expressed as a molecular and a turbulent Schmidt number):

$$k^+ = k^+(L^+, \text{Sc}, \text{Sc}_t). \tag{47}$$

For small x^+, Eq. (46) is the asymptotic form of Eq. (47). For large x^+, the specific form of Eq. (47) can only be derived with use of a hypothesis about the distribution of eddy diffusivity in the turbulent boundary layer next to the wall, particularly in the inner part of the viscous sublayer. Since the mass-transfer rate attains a constant value for very large x^+ ($> 10^4$), the dependence of this asymptotic k^+ on the Schmidt number has been the objective of numerous experimental investigations employing the limiting-current technique.

An equally important objective of these turbulent mass-transfer measurements was the establishment of the exact dependence of k on the Reynolds number. The well-known empirical Chilton–Colburn equation for turbulent transfer (C3) predicts that

$$\text{Nu} = (f/2)\text{Re}(\text{Sc})^{1/3} \tag{48a}$$

which implies an explicit dependence of k_+ on f, that is, on the Reynolds number. Semiempirical modifications of the Chilton–Colburn equation have been proposed (V5), which also imply this additional dependence. Some expressions believed valid in the limit of large L^+ and high Sc are listed in Table VI; a more detailed list has been given by Lengas (L5a).

Son and Hanratty (S19) reviewed the experimental evidence from electrochemical and other model experiments. They concluded that eddy diffusivity varies with the fourth power of the distance from the wall, assuming that the friction factor takes care of the Reynolds number dependence. Shaw and Hanratty (S11a) recently corroborated this conclusion by further experiments that led to the equation (47b, (5)) in Table VII, which is equivalent to

$$k^+ = 0.0889\ \text{Sc}^{-0.704} \tag{48b}$$

Sirkar and Hanratty (S13) showed, by means of refined measurements using strip electrodes at different orientations with respect to the mean flow, that transverse velocity fluctuations play a significant part in the turbulent transport very close to the wall, and that the eddy diffusivity may well be dependent on the cube of the distance y^+, leading to a $\text{Sc}^{1/3}$ dependence of mass-transfer correlations, which is often found experimentally.

Hubbard and Lightfoot (H11a) earlier reported a $\text{Sc}^{1/3}$ dependence on the basis of measurements in which the Schmidt number was varied over a very large range. The data did not exclude a lower Reynolds number exponent than 0.88, and reaffirmed the value of the classical Chilton–Colburn equation for practical purposes. Recent measurements on smooth transfer surfaces in turbulent channel flow by Dawson and Trass (D8) also firmly suggest a $\text{Sc}^{1/3}$ dependence and no explicit dependence of k^+ on the friction coefficient, with Sh thus depending on $\text{Re}^{0.875}$. The extensive data of Landau

TABLE VI
Expressions for Fully Developed Turbulent Mass Transfer

Ref.	Expression for Sh	Expression for k^+	Basis D_t	Type
Chilton and Colburn (C3)	$= (f/2)\text{Re}(\text{Sc})^{1/3}$	$= (f/2)^{1/2}\text{Sc}^{-2/3}$	$\sim (y^+)^3$	Empirical
Deissler (D9)	$= 0.0789 f^{1/2}\text{Re}(\text{Sc})^{1/4}$	$= 0.112\text{Sc}^{-3/4}$	$\sim (y^+)^4$	Semianalytical
Lin et al. (L10a)	$= 0.057(f/2)^{1/2}\text{Re}(\text{Sc})^{1/3}$	$= 0.081\text{Sc}^{-2/3}$	$\sim (y^+)^3$	Semianalytical
Levich (L8)	$\sim f^{1/2}\text{Re}(\text{Sc})^{1/4}$	$\sim \text{Sc}^{-3/4}$	$\sim (y^+)^4$	Semiempirical
	$\sim f^{1/2}\text{Re}(\text{Sc})^{1/3}$	$\sim \text{Sc}^{-2/3}$	$\sim (y^+)^3$	Semiempirical
Vieth et al. (V5)	$= 4.586(v_{\max}/v_{\text{avg}})(f/2)\text{Re}(\text{Sc})^{1/3}$	$= 4.586(v_{\max}/v_{\text{avg}})(f/2)^{1/2}\text{Sc}^{-2/3}$	$\sim (y^+)^3$	Semiempirical
Spalding (S20a)	$= 0.053 f\text{Re}(\text{Sc})^{1/4}$	$= 0.0746\text{Sc}^{-3/4}$	$\sim (y^+)^4$	Semiempirical

and Tobias (L1b), which are in agreement with the $Sc^{1/3}$ dependence, indicate an even higher exponent, 0.92, for the Re dependence of Sh in fully developed turbulent flow.

Turbulent mass-transfer studies with the rotating disk and the rotating cylinder have also yielded valuable, though not completely definitive, information about turbulent transport. The laminar boundary layer on a rotating disk becomes turbulent at $Re_r \approx 2 \times 10^5$. The range of instability has been investigated by Chin and Litt (C6a), using microelectrodes as indicators. At rotating disks in turbulent flow, the mass-transfer rate depends on the 0.9 power of the rotation rate.

The dependence on Sc is less clear. Ellison and Cornet (E4) found a dependence on $Sc^{1/4}$, in accordance with Deissler's model (see Table VI). However, Daguenet (D2), who used a higher Schmidt number range, and Hamann et al. (H1), using a rotating ring, observed a dependence on $Sc^{1/3}$.

At rotating rings of decreasing width and large radius, the mass-transfer dependence on rotation rate ω was shown to decrease from $\omega^{0.9}$ to $\omega^{0.6}$ as radial convection becomes more important; in the limit of very thin rings, a Lévêque-type expression, Eq. (40-3) in Table VII, is obtained [compare also Eq. (42a1) in Table VII]. Mass transfer at a rough disk likewise exhibits a lower dependence on rotation rate than $\omega^{0.9}$ as a result of significant radial transport.

Flow and mass transport between concentric rotating cylinders has been reviewed by Gabe (G1a). The most reliable correlations for turbulent mass transfer, established by Eisenberg et al. (E1), and confirmed by Arvía and Carrozza (A4) and Robinson and Gabe (R16), suggest a $Sc^{1/3}$ rather than $Sc^{1/4}$ dependence. The overall results for the turbulent regime are in good agreement with those obtained by dissolving or corroding cylinders of salt or metal (K3, K4a).

The influence of surface roughness on turbulent mass transfer was first investigated by dissolution of rotating cylinders (M2a). More recent limiting-current measurements, Eq. (44) in Table VII, are in good agreement with the earlier results. The extensive investigation of turbulent mass transfer at rough surfaces in channel flow by Dawson and Trass (D8) has shown that the effect of roughness is complicated, and only can be accounted for by a simple correlation if the roughness is much larger than the thickness of the viscous sublayer.

To summarize, a comprehensive understanding of turbulent transport is not yet achieved, and information will be needed from optical as well as from further mass-transfer measurements. The latter will have to be made at high Reynolds numbers ($> 50,000$ in channel flow) and at very high Schmidt numbers ($> 10,000$) to yield critical information about the transfer process.

E. Oscillating Flows

The comparative ease with which mass transfer at electrodes in vibration can be studied experimentally has prompted several investigations on this subject (see Table VII, Part E).

Mass transfer in pulsating flow, perpendicular to a stationary plate, was correlated (R1) by (1) a laminar-type mass-transfer expression at low frequencies, and (2) by a turbulent one at higher frequencies. The pulsating flow is generated by a vibrating diaphragm, which forms the bottom of a cylindrical cell with an open upper end. Mass transfer to the horizontal disk electrode appears to depend neither on the distance to the diaphragm (provided this distance is not very small) nor on free liquid height.

Rao et al. (R5) and Raju et al. (R2) also investigated mass transfer at vibrating electrodes for low vibration frequencies (higher frequencies would cause cavitation). Mass transfer follows a laminar-type correlation both for a transverse vibration of a vertical cylinder and for a vertical plate vibrating parallel to the face. In the case of the plate, the Reynolds number is based on width, indicating the predominance of form drag. When vibrations take place perpendicular to the thickness, skin friction predominates and the Reynolds number is then preferably based on the equivalent diameter (total surface area divided by transverse perimeter).

Noordsij and Rotte (N10) studied mass transfer at a vibrating sphere. The oscillating motion was achieved by means of an attached rod, held by a spring against an eccentric wheel. The results, showing considerable scatter, are represented approximately by a $Pe^{1/2}$ dependence, with an additive constant term accounting for pure diffusion.

Of particular interest are measurements on a vibrating sphere in forced flow by Gibert and Angelino (G3) because of the careful experimental execution and wide range of frequencies and Re numbers investigated. Below a certain critical frequency, the mass-transfer rate to the sphere is not affected by vibration.

Goren and Mani (G6) investigated the enhancement of oxygen mass transfer through very thin liquid layers in standing-wave motion. Contrary to theoretical expectations, they found the increase in the rate to be linearly proportional to the amplitude of the imposed standing wave. The stationary (zero-frequency) rate of oxygen transfer was found to be four to five times larger than would correspond to pure diffusion, possibly due to capillary motion in the electrolyte or simultaneous CO_2 transfer.

Recently significant advances have been made in the analytical solution of mass transfer to a sinusoidally modulated rotating disk electrode. The resulting expressions, confirmed by refined experimental techniques, allow determination of the diffusivity of the reacting ion as well as reaction kinetics (T4a, D11c).

F. Stirred Cells

Under this heading are grouped a number of investigations in which a finite volume of electrolyte is stirred either by rotation of the electrode, or where this stirring (or circulation) is accomplished independently (see Table VII, Part F). In principle, the boundary separating these types of flows from laminar flows discussed earlier is not sharp.

Complications arise, mainly because of secondary flows in cells of restricted volume. Such steady secondary flows occur also by the action of vibrating electrodes, leading to the mass transfer rates correlated in the previous subsection.

In work by Okada et al. (O3) on a rotating-disk flow, Eqs. (10a) and (10b) in Table VII, the electrolyte was completely enclosed between the rotating disk and the counterelectrode. Mass transfer was measured at the rotating as well as at the stationary disk, and the distance between disks was varied. At low rotation rates, the flux at the rotating disk was higher than predicted by the Levich equation, Eq. (1a) in Table VII. The flux at the stationary disk followed a relation of the Levich type, but with a constant roughly two-thirds that in the rotating-disk equation.

Homsy and Newman (H6a) have analyzed mass transfer at a plane below a rotating disk. They indicated a zone of decreasing mass transfer radially inward from the edge of the disk. The coefficient that they predict for the Levich-type average mass-transfer correlation, Eq. (2) in Table VII, is much higher than that suggested by the results of Okada et al. (O3), due obviously to the experimental geometry; the stationary disk and the rotating disk each occupy almost completely the circular cross section of the cell. The local current distribution predicted by Homsy and Newman (H6a) would be expected when the stationary disk has a much smaller diameter than the rotating disk.

The opposite situation occurs if the rotating disk is much smaller than the cell diameter. Krishna and co-workers (K8, S2) measured mass transfer at a cylindrical rotor, as well as at ring electrodes inserted concentrically at the bottom of the cell. An impeller attached to the free end of the rotor did not change the mass-transfer rate. They found that mass transfer to the rings was greatest for rings of the same radius as the rotor (or impeller). Beyond this radius, mass-transfer rates decreased. A correlation for this region of decreasing mass-transfer rates is given by Eq. (57b) in Table VII.

Another factor that complicates the application of laminar convection theory to stirred vessels is the possibility that boundary layers on the stirring electrodes may become detached. Local turbulence can be generated near electrodes or inert stirrers. Of interest is the mass transfer to the stirrer, as well as to selected points of the cell. The first has been investigated to a greater extent, but it is clear that valuable information about convection patterns in stirred tanks can be obtained by the refinement of such measurements as discussed above at ring electrodes in the bottom of a stirred cell.

The most extensive and refined application of microelectrodes, both in the wall of a cell and as a moveable probe, was made by LeLan and Angelino (L2, L3, L4), who charted convection patterns and mass-transfer rate distributions in cylindrical cells with and without baffles.

Flow in stirred vessels was also investigated by Holmes *et al.* (H5), who simulated mass transfer in a diaphragm diffusion cell stirred by magnetic stirrer bars. This is a good example of a simple model study with a direct practical purpose. A minimum stirring speed in such cells is necessary to avoid appreciable errors in the cell constant. The experiment permits this stirring speed to be related to the solution properties.

An interesting modification of coaxial flow in annular cells was studied by Regner and Rousar (R11), who created swirling flow by means of a tangential inlet. Their mass-transfer correlation combines features of the boundary layer correlations (dependence on $L^{-1/2}$) with those of turbulent flow between rotating cylinders (dependence on $Re^{0.78}$), although the Reynolds numbers do not reach the range at which flow is turbulent. Tangential inflow into a cell is also a feature of some of the wire-plating cells studied by Tvarusko (T4b), Eq. (17) in Table VII. Related is the flow in a hydrocyclone investigated by Ravoo (R8).

G. Convection by Gas Evolution

Electrolytic gas evolution has been used as a means of generating convection in heat-transfer experiments (M8), where the primary purpose is to gain an insight into the mechanism of subcooled boiling (see Table VII, Part G). However, it has only been realized more recently that electrochemical mass transfer itself may serve as a convenient method for the study of heat transfer in the neighborhood of an "active" bubble-generating point. Weder (W2, W3a) measured limiting currents of ferricyanide reduction at concentric horizontal ring electrodes surrounding a single orifice through which gas was bubbled. Dependence on $Q^{0.31}$ (where Q is the gas volume flow rate) was observed, in excellent agreement with earlier heat-transfer results.

Earlier, Venczel, Ibl, and co-workers (V3, I8) measured limiting currents in the reduction of ferric and ceric ions at a vertically oriented cathode where evolution of hydrogen also occurred. The bulk solution was titrated after the experiment to establish the partial current due to hydrogen ion reduction. In this more complex situation, two effects can be expected to occur at the same time: convection of the solution due to the rising bubbles and turbulent mixing near the cathode subsequent to the disturbance set up by the detachment of a single bubble. The experimental dependence of mass-transfer rate on $(Q/A)^{0.56}$ (where A is the electrode area) shows that the second effect is predominant. Ibl (I3) proposed a modified "surface renewal" theory to explain this dependence, and for hydrogen evolution at copper electrodes at least fair agreement was found with experiment.

Janssen and Hoogland (J3, J4a) made an extensive study of mass transfer during gas evolution at vertical and horizontal electrodes. Hydrogen, oxygen, and chlorine evolution were visually recorded and mass-transfer rates measured. The mass-transfer rate and its dependence on the current density, that is, the gas evolution rate, were found to depend strongly on the nature of the gas evolved and the pH of the electrolytic solution, and only slightly on the position of the electrode. It was concluded that the rate of flow of solution in a thin layer near the electrode, much smaller than the bubble diameter, determines the mass-transfer rate. This flow is affected in turn by the incidence and frequency of bubble formation and detachment. However, in this study the mass-transfer rates could not be correlated with the square root of the free-bubble diameter as in the surface renewal theory proposed by Ibl (I8).

H. Particulate Electrodes

The limiting-current method has been used widely for studies in packed and fluidized beds (see Table VII, Part H). Limiting current measurements in these systems overlap in part with the design and analysis of packed-bed and fluidized-bed electrochemical reactors; in particular the potential distribution in, and the effectiveness of, such reactors (for example, for metal removal from waste streams) is an extensive area of research, which cannot be covered in this review. For a complete discussion of porous flow-through electrodes the reader is referred to Newman and Tiedemann (N8d).

In Table VII, Section H, some limiting current measurements on or in packed beds are represented. These measurements were initiated by Sioda (S12a–d) and others (V6, W13b) in cells with stacked grids or spheres.

Rao et al. (R7) first made local mass-transfer measurements, by ring electrodes embedded in the perforated plates between which a packed bed was contained. They measured the local mass-transfer rate at ring electrodes embedded in the perforated plates in between which a packed bed was contained. They employed the conventional dimensionless quantities for a fluidized bed system:

$$j_D = k/\bar{v} Sc^{2/3} \tag{49a}$$

and

$$Re_p = \bar{v} d_p/(1-\varepsilon)v \tag{49b}$$

with the result that

$$j_D \sim Re_p^{-0.38} \tag{49c}$$

Very refined measurements at various positions of the packed bed were made by Jolls and Hanratty (J6), who used an active sphere (electrode) in a packed bed consisting of 1-inch inert spheres. The overall mass-transfer data for the turbulent flow regime suggest a dependence of

$$j_D \sim Re_d^{-0.42} \tag{50a}$$

or

$$j_D \sim Re_d^{-0.44} \tag{50b}$$

outside the low Reynolds number flow region ($Re_p < 35$), where the flow-rate dependence is uncertain. Karabelas et al. (K2b) later investigated this low Reynolds number region and found a dependence of j_D on $Re^{-2/3}$. A transition to turbulence occurred between $Re_d = 110$ and 150(J6); however, no conspicuous accompanying changes in the mass transfer ensued, because of the large flow fluctuations which occurred over the whole range.

The rate expressions, Eq. (70a1 and 70a2) in Table VII, are valid only for the case of a single, active sphere in a bed of inert spheres. In the entire bed of active spheres, these are bathed in the detached diffusion layers of other spheres, and the rates are therefore lower. This, of course, constitutes a general objection to the application of this method in the study of single activated elements of a particulate system. Nevertheless, such a study can yield valuable information about flow patterns, and variations throughout the system. Appel and Newman (A3f) investigated mass transfer to an entire packed bed at very low flow rates ($Pe < 100$) and found the rates to be below the prediction based on the well-known correlation of Wilson and Geankoplis; see Eq. (70b(2)) in Table VII.

Hicks and Mandersloot (H4) investigated flow systems with turbulence promoters, where the orientation of the promoters (or packed irregular particles) is an important parameter. In such systems the Reynolds number,

$d_p \bar{v}/(1 - \varepsilon)v$, used for packed beds of spheroid particles is not effective. Instead, Hicks and Mandersloot correlated their data with the "viscous part" of the pressure drop, based on the consideration that the average transfer rate in turbulent flow over a flat plate is proportional to the square root of the shear stress [see Eqs. (2)–(4) in Table VI].

The modified Reynolds number therefore is based on the velocity in the void fraction $\bar{v}/\varepsilon$, the kinematic viscosity v, and an equivalent diameter s/a, where s is total area per unit volume and a is the dimensional coefficient derived from a correlation of pressure drop data:

$$\Delta P/L = a\mu\bar{v} + b\rho\bar{v}^2 \tag{51}$$

Using these parameters, the modified Reynolds number is

$$\text{Re}_m = s\bar{v}/a\varepsilon v \tag{52}$$

and the mass-transfer correlation then takes the general form:

$$j_D = \text{const. Re}_m^{-1/2} \tag{53}$$

This expression has been satisfactorily confirmed by limiting current measurements.

In 1962 Jottrand and Grunchard (J7) reported on mass transfer to a small rectangular nickel plate immersed in a liquid fluidized bed of sand particles. Mass-transfer rates were five to ten times higher than those measured in an open pipe flow; a maximum rate was measured at a bed porosity of 0.58. Le Goff *et al.* (L1c) later showed that this maximum is directly related to a maximum in the average kinetic energy of the fluidized particles per unit bed volume.

Measurements of the limiting current at small electrodes immersed in fluidized beds were also made by Coeuret *et al.* (C9, C10) and by Carbin and Gabe (C1a). Their results are in good agreement but deviate from those of Jottrand and Grunchard; the discrepancy is discussed by Carbin and Gabe (C1d).

Measurements at the wall of fluidized beds were made by Krishna *et al.* (K7, K11), who used an annular cell with varying radial ratios to test the influence of inner tubes on fluidization. Only at low r_i/r_0 values is the inner wall coefficient appreciably higher than the outer wall coefficient (which is itself somewhat lower than in a pipe bed of the same radius).

In these wall-transfer measurements, stagnant liquid regions may play an important part. It is interesting to note the differences obtained by measurements at different positions of the bed.

Hutin and Coeuret (H11b) have investigated the potential distribution in a fluidized bed and indicated the existence of dissolution zones. Carbin and Gabe (C1f) studied the crystallographic characteristics of copper deposited from a fluidized bed.

The concept of a fluidized bed consisting of electrically conducting particles as a statistically continuous electrode was first discussed by Le Goff et al. (L1c). Interesting similarities with heat-transfer studies in fluidized beds may be exploited to advantage by use of the limiting current method.

These references and others given in Table VII are by no means exhaustive and serve only to suggest that the fluidized-bed electrode is an active area of research in which the limiting-current method has made important initial contributions.

VII. Concluding Remarks

In this chapter the theory and practice of limiting-current technique for the measurement of mass-transport coefficients have been described. The selective discussion and tabular compilation of results of investigations that used limiting-current measurements should be indicative of the widespread use of this relatively novel method.

The authors have aimed at developing the necessary background to understand the inherent advantages, as well as the limitations and characteristic difficulties, of this technique. Detailed critical evaluation of results of past investigations has been provided for the most important flow situations and geometries.

Following careful consideration of the first four sections, which provide the necessary fundamental background, the reader should be well equipped to exercise his own judgment in weighing the reliability of results published in the literature.

By providing a better understanding of the method, this chapter is intended to inspire more precise and more widespread use of the limiting-current technique in studies of transport phenomena.

Acknowledgment

This work was supported by the Division of Material Sciences, Office of Basic Energy Sciences, United States Department of Energy.

Dr. T. E. Lengas, while a guest scientist at the Lawrence Berkeley Laboratory, contributed suggestions regarding the organization of this chapter.

TABLE VII

CORRELATIONS ESTABLISHED BY LIMITING-CURRENT MEASUREMENTS

System	Parameters	Correlation	Range	Reactants[a]	Ref.
A. SIMPLE LAMINAR FLOWS					
1a. Laminar flow at a rotating disk	ω = rotation rate r = disk radius r_i = inner radius of ring r_o = outer radius of ring $Re = r^2\omega/\nu$ $Sh = rk/D$	$Sh = 0.621 Re^{1/2} Sc^{1/3}$	$Re_r < 2.7 \times 10^5$	0 1 3 5 2 5 5	L6, N3 S14 O3 N9a H9 D2 E6
1b. Laminar flow at a rotating ring		$Sh = Sh_{disk, r_o}$ $\times (1 - r_1^3/r_o^3)^{2/3}$	$Re_{r_o} < 2.7 \times 10^5$	5 3	D2 D11a
1c. Laminar flow at a rotating hemisphere	r = radius of spherical electrode	$Sh = 0.474 Re^{1/2} Sc^{1/3}$	$Re < 15,000$	3	C4
1d. Laminar flow at a rotating spherical segment ($\theta < 40°$)	θ = angle of segment with axis of rotation	See System 1a ($\pm 5\%$)		3, 4, 2	C7
1e. Laminar flow at a rotating cone	r = base radius of cone θ = cone angle (half opening angle)	$Sh = 0.621(\sin\theta)^{1/2}$ $Re^{1/2} Sc^{1/3}$	$Re < 25,000$ $25° < \theta < 60°$	6	K4b
2. Laminar rotational flow at a stationary disk	r = radius of stationary disk	(1) $Sh = 0.761 Re^{1/2} Sc^{1/3}$ (2) $Sh = 0.761 Re^{1/2} Sc^{1/3}$	$Re < 8000$	0 3	M4b, H6a B11b, M4b

3a. Laminar stagnation flow at a stationary disk	r = diameter disk v = velocity main flow $Re = vr/\nu$ $Sh = kr/D$	$Sh = 0.532 Re^{1/2} Sc^{1/3}$	$40 < Re < 1200$ $Sc = 1550$ $r_o = 0.5$–1.5 mm	11	M4c
3b. Laminar stagnation flow at a wedge	θ = wedge angle $\beta = \theta/180$ L = length transfer section L_1 = length wedge surface V = velocity main flow $Re = vL/\nu$ $Sh = kL/D$	$Sh = K_\beta Re^{1/2} Sc^{1/3}$ $\times (L/L_1)^{\beta/(2(2-\beta))}$ $\theta = 0° \ K_\beta = 0.677$ $\theta = 90° \ K_\beta = 0.714$ $\theta = 180° \ K_\beta = 0.586$	$200 < Re < 700°$ $Sc = 1550$	11	M4d
4a. Laminar Couette flow between concentric cylinders; mass transfer to rotating inner cylinder	d = inner cylinder diameter $Re = \omega d^2/\nu$ $Sh = kd/D$	(a1) Sh independent of Re (Sh = 37 at Sc = 460) (a2) Sh = 3.5	$5 < Re < 150$ $Re < 10$	1 3	C13 C11a
4b. Laminar Couette flow between concentric cylinders; mass transfer to section of rotating inner cylinder	ω = rotation rate $\kappa = r_i/r_o$ = ratio of inner to outer radius L = length transfer section $Re = L^2\omega/(1-\kappa^2)\nu$ $Sh = kL/D$	(b1) $Sh = 1.0174(Re Sc)^{1/3}$ $+ 0.20002 L/R_i$ (b2) $Sh \sim \omega^{1/3} L^{2/3}$ ^{b}	— $0.2 \ sec^{-1} < \omega < 3 \ sec^{-1}$ $0.03 \ cm < L < 0.2 \ cm$ $R_i = 0.47$ or 0.7 cm $R_o = 2.5$ cm	0 3, 4	M13f K2c

^{a} Reactants: 0, theoretical value; 1. O_2; 2. Cu^{2+}; 3. $Fe(CN)_6^{3-}$; 4. $Fe(CN)_6^{4-}$; 5. I_3^-; 6. quinone; 7. Hg^{2+}; 8. Ce^{3+}; 9. Ce^{4+}; 10. Ag^+; 11. Fe^{3+}; 12. Cd^{2+}.
^{b} See also Gabe and Robinson (G1b). Correct expression given by Mohr and Newman (M13f).

(continued)

TABLE VII (continued)

System	Parameters	Correlation	Range	Reactants	Ref.
5. Laminar parallel flow in annular duct	$d_e = r_o - r_i$ r_i = inner tube radius r_o = outer tube radius $Re = \bar{v}d_e/\nu$ $Sh = kd_e/D$	(1) $Sh = 1.615(\phi ReScd_e/L)^{1/3}$ ϕ, see Fig. 15		1	N6
		(2) $Sh = 1.62(ReScd_e/L)^{1/3\,(b)}$ for $r_i/r_o = 0.5$	$1.6 \times 10^4 < ReScd_e/L$ $< 1.1 \times 10^6$ $300 < Sc < 3000$	1, 3, 4, 6	L9
		(3) $Sh = 1.94(ReScd_e/L)^{1/3}$ for $r_i/r_o = 0.5$	$1.7 \times 10^4 < ReScd_e/L$ $< 9 \times 10^7$	2 2	R18 W13a
6. Laminar parallel flow in rectangular duct (electrodes in 2 opposing walls)	h = electrode separation w = electrode width d_e = equivalent diameter $Re = \bar{v}d_e/\nu$ $Sh = kd_e/D$ $\gamma = h/w$	(1) $Sh = 1.85(ReScd_e/L)^{1/3}$ for $d_e/h > 1.85$	$75 < Re < 7000$ $600 < Sc < 12{,}000$ $0.05 < d_e/L < 20$	2	T2
		(2) $Sh = 1.85\phi_2(ReScd_e/L)^{1/3}$ ϕ_2, see Table V for electrode/duct width < 1, $\gamma = 1$	$80 < Re < 380$ $Sc = 1271$	7, 10	R21a
		(3) $Sh = 2.54Re^{1/3}Sc^{0.29}$ $\times (d_e/L)^{1/3}$ for electrode/duct width ≈ 0.9, $\gamma = 0.16$	$200 < Re < 2000$ $1000 < Sc < 3500$ $0.025 < d_e/L < 1$	2	P3b
		(4) $Sh = 1.847(ReScd_e/L)^{1/3}$ for $\gamma = 0.02$–0.05	$1300 < Re < 2800$ $Sc = 3150$	3	A1a
		(5) $Sh = 1.864(ReScd_e/L)^{0.331}$ for $\gamma = 0.07$	$400 < Re < 2500$ $1200 < Sc < 25{,}100$	2	L1b
7. Falling liquid film (laminar flow)	δ = film thickness $Re = (g\delta^3 \sin\theta)/3\nu$ $Sh = k\delta/D$	$Sh = 1.16(Re_\delta Sc\delta/L)^{1/3}$	$Re < 30$ $0 < \theta < 89°$	5	C1a

282

8. Laminar free convection at a vertical plate (or large-diameter cylinder)	L = height $Gr = g\,\Delta\rho L^3/\rho\nu^2$	(1) $Sh = 0.67(GrSc)^{1/4}$ (2) $Sh = 0.45(GrSc)^{1/4\,(c)}$	$5 \times 10^6 < GrSc < 5 \times 10^{12}$ $500 < Sc < 80{,}000$ $9 \times 10^9 < GrSc < 4.6 \times 10^{11}$ $2350 < Sc < 3650$	2 2 3	W1 W4, W5 F5
9a. Laminar flow at a moving continuous cylinder (reentrant wire)	v = wire velocity L = transfer length $Pe = vL/D$ $Sh = kL/D$	$Sh = 1.13Pe^{1/2}$	$4000 < Pe^{1/2} < 4 \times 10^4$ $Sc = 700, 1800$	0, 3	R20
9b. Laminar flow at a moving continuous flat surface (reentrant cylindrical sheet)	v = sheet velocity = ωr_i y_i = radius cylinder L = transfer length $Pe = vL/D$ $Sh = kL/D$	(b1) $Sh = 1.128Pe^{1/2}$ (b2) $Sh = 1.128Pe^{1/2}$	$1500 < Re < 70{,}000$ $1800 < Sc < 29{,}300$ $Pe < 1.5 \times 10^8$ $r_i = 2.54$ cm $v < 12.5$ cm/sec $Re < 5000$ $Sc = 392, 1237$ $r_i = 2.5$ cm	3 1, 3	C5b, C6b N1a

B. FORCED LAMINAR FLOWS WITH COMPLICATIONS

10a. Rotating and stationary disk in restricted volume	h = distance of disks r = radius of disks ω = rotation rate	None $i_{stat} < i_{rot}$	$30 < \omega < 3000$ (sec^{-1}) $r = 6$ cm $h = 0.19$–2 cm	2	O3
10b. Rotating disk in restricted volume	$Re = \omega r^2/\nu$ $Sh = kr/\nu$	$Sh_r = 0.345Re_r^{0.5}Sc^{0.4}$ $(\pm 10\%)$	$r = 0.5$–2.5 cm $h = $ small $5 \times 10^3 < Re_r < 2 \times 10^5$	3	S21

[b'] Coefficient doubtful; expected value 1.939 (see System 5, Correlation 3).

[c] Coefficient doubtful. Expected value 0.67 (see System 8, Correlation 1). Recalculation of correlation by Taylor and Hanratty (T1b) yielded coefficient 0.62.

(*continued*)

TABLE VII (continued)

System	Parameters	Correlation	Range	Reactants	Ref.
10c. Rotating and stationary disk with throughflow (pump cell)	r_i = inner radius (stator) r_o = outer radius (both disks) $R\phi = \omega r_o^2/\nu$ $Re = Q/h\nu$ Q = rate of volume throughflow	$Sh \sim [Re R\phi h^3/r_o(r_o^2 - r_i^2)]^{1/2}$	$4 \times 10^6 < Re R\phi h^3/r_o(r_o^2 - r_i^2)$ $< 30 \times 10^6$	3	J4b
10d. Rotating disk with grooves	W = groove width G = distance between grooves r = disk radius	$Sh \sim [(W + G)/r]^{-0.2}$	$0.08 < (W + G)/r < 0.3$	3	K13
10e. Rotating disk with partially inactive surface	$Sh = kr/\nu$ r = disk radius r_1 = radius of active site r_2 = radius (active site + inactive ring) δ = diffusion layer thickness	$Sh^{-1} = Sh_{Levich}^{-1}$ $+ [(r_2)/r]a_n \cdot tgh$ $(x_n \delta/r_2)$ $Sh_{Levich} = Eq.$ (1a) of this table a_n = tabulated function of (r_1/r_2) x_n = zeros of $J_1(x)$	$0.02 < \omega < 16$ (sec^{-1}) $r = 0.5$ cm 5×10^{-2} cm $< r_1 < 6 \times 10^{-4}$ cm $0.25 < r_1/r_2 < 0.75$	3	S4a, S4a, S4b
10f. Rotating disk with off-center electrode	ε = eccentricity/radius disk	$Sh = Sh(1.027\varepsilon^{1/3}$ $+ 0.044\varepsilon^{-5/3})$ $(\pm 1\%)$	$\varepsilon > 0.8$	2	M13g

10g.	Rotating disk: mass transfer to a circular spot electrode on the disk	r_o = radial distance to center of electrode d = electrode diameter $Sh_d = kd/D$ $Re_{r_o} = \omega r_o^2/\nu$	$Sh_d = 0.83 Re_{r_o}^{0.5} Sc^{1.3}$ $\times (d/r_o)^{2/3(a)}$ $(d \ll r_o)$	$8 \times 10^4 < Re_{r_o} < 4 \times 10^5$ $d = 0.1$ cm, $r_o = 4$ cm $Sc = 773$	5 3	D5 M13i
10h.	Rotating disk: mass transfer to an annular sector	r_i = inner radius ring r_o = outer radius ring ϕ = sector angle $Re = \omega r_o^2/\nu$	$Sh = Sh_{ring} f(\phi, r_i, r_o)$ Sh_{ring}: see Eq. (1b), this table $f(\phi, r_i, r_o)$: see reference	$0.8 < r_i < 1.5$ cm $1.1 < r_o < 1.7$ cm $0.29 < \phi < 3.14$ rad	3	D14b
10i.	Rotating disk: mass transfer to rectangular spot electrodes oriented along streamlines	See Eq. (1b) of this table	$Sh = Sh_{ring}$ See Eq. (1b) of this table	$Re < 5 \times 10^4$ $r_i = 0.72$ cm	3	
11a.	Rotating cone	l = cone slant height α = cone apex half-angle ω = rotation rate $Re_l = \omega l^2 (\sin \alpha)/\nu$	$Sh = 0.345 Re_l^{0.5} Sc^{0.4}$ $(\pm 10\%)$	$10^3 < Re_l < 10^5$ $\alpha > 40°$ $0.6 < l < 2.2$ cm	3	S21
11b.	Rotating cone in non-Newtonian solution		See Smith and Greif (S15b)		1	P2a, S15b
11c.	Rotating sphere	d = sphere diameter ω = rotation rate	(c1) $Sh = 0.25 Re_d^{0.5} Sc^{0.4}$ $(\pm 10\%)$ (c2) $Sh = 10 + 0.43 Re^{0.5} Sc^{1.3}$	$6 \times 10^3 < Re_d < 2 \times 10^5$ $1.3 < d < 2.5$ cm $800 < Re < 27{,}000$ $500 < Sc < 2000$ $1.2 < \omega < 31$ (sec^{-1}) $d = 2.53$ cm	3 2, 3	S21 N10

(continued)

[a] Calculated from Fig. 2 of Daguenet et al. (D5).

TABLE VII (continued)

	System	Parameters	Correlation	Range	Reactants	Ref.
12.	Rotating cylinders (Couette flow in transition regime) or rotating wire	d = inner cylinder diameter ω = rotation rate $\mathrm{Re} = \omega d^2/\nu$	$\mathrm{Sh} \sim (\mathrm{ReSc})^{1/3}$	$150 < \mathrm{Re}_d < 1500$ (rotating inner cylinder)	3	C11a
13.	Developed laminar flow in open channel	d_e = equivalent diameter $= 2w/(1+\gamma)$ w = channel width = electrode separation h = liquid height $\gamma = w/h$ $\mathrm{Sh} = kd_e/D$ $\mathrm{Re} = vd_e/\nu$ L = electrode length	$\mathrm{Sh} = 1.85(\mathrm{ReSc}d_e/L)^{1/3}f(\gamma)$ $f(\gamma) = \left[\dfrac{1 - 0.271\gamma}{(1 - 0.314\gamma)(1+0.5\gamma)}\right]^{1/3}$	$\mathrm{Re} < 2100$	0	M13d
14.	Developing flow in rectangular channel ($L_e/d_e < 0.01\mathrm{Re}$)	L_e = entrance length d_e = equivalent diameter L = transfer length h = height $\mathrm{Re} = vd_e/\nu$ $\mathrm{Re}_L = vL/\nu$ $\mathrm{Sh}_L = kL/D$	(1) No general correlation $k \sim v^{1/2}$ (2) $\mathrm{Sh}_L = 0.66\,\mathrm{Re}_L^{0.5}\mathrm{Sc}^{1/3}$	$825 < \mathrm{Re} < 3000$ $d_e/L = 0.343$ $d_e/h = 1.14$ See reference	3 8	W14 B4c
15.	Developing flow in annular channel ($L_e/d_e < 0.01\mathrm{Re}$)	$\mathrm{Re}_L = vL/\nu$ $\mathrm{Re}_d = vd_e/\nu$ d_i = diameter inner electrode $\mathrm{Re}_{d_i} = vd_i/\nu$	(1) $\mathrm{Sh} = 0.647\mathrm{Re}_L^{1/2}\mathrm{Sc}^{1/3\,(d)}$ (2) $\mathrm{Sh} = 0.52\mathrm{Re}_L^{1/2}\mathrm{Sc}^{1/3}$ $\times (L/d)^{3.4(e)}$ (3) $\mathrm{Sh} = 0.45\mathrm{Re}_{d_i}^{0.53}\mathrm{Sc}^{1/3}$ (4) $\mathrm{Sh} = 3.93\mathrm{Re}_d^{0.32}$ $\times \mathrm{Sc}^{1/3}(d_e/L)^{0.35}$	 $d_i = 0.5$ cm $d(\text{tube}) \gg d_i$ $25 < \mathrm{Re}_{d_i} < 225$ $\mathrm{Sc} = 1230$ $6 \times 10^4 < \mathrm{ReSc} < 10^6$ $\mathrm{Sc} = 750, 1680$	2 3 3 2	B2 B3 C9, C10 C1c

16. Flow parallel to a semi-infinite wire	x_0 = start of transfer length r = wire radius L = transfer length $Sh = k(x_0 + L)/D$ $Re = v(x_0 + L)/\nu$	$Sh \sim Re^{1/2} Sc^{1/3} \{[1 + x_0/L)^{3/4} - (x_0/L)^{3/4}]^{2/3} \times (1 + x_0/L)^{1/2}$	$r = 1.5$ cm $4 < x_0 + L < 13$ cm $0.3 < x_0/L < 55$ $Sc = 1040$	3	C8
17. Flow parallel to a wire in cell with axial, radial, or tangential inflow	d_e = equivalent diameter L = transfer length	$Sh \sim Re^{0.5}(d_e/L)^{0.5}$	$3 \times 10^4 < ReScd_e/L < 10^6$ $Sc = 1730$	3	T4b
18. Developed channel flow: mass transfer to strip electrodes in the wall	d_e = equivalent diameter L = transfer length	$Sh = 1.43(ReScd_e/L)^{1/3}$	$d_e/L = 8.1$ or 9.2 $L = 0.15$ cm $4 \times 10^5 < ReScd_e/L < 1.6 \times 10^7$	2	111a
19. Falling liquid film	δ = film thickness g = acceleration of gravity $Sh = k\delta/D$ $Re = g\delta^3/3\nu^2$	(1) $Sh = 1.21(ReSc\delta/L)^{1/3}$ (2) $Sh = 1.11(ReSc\delta/L)^{1/3}$	$70 < Re < 700$ $1400 < Sc < 18{,}500$ $15 < Re < 550$ $Sc = 3030, 3550$	0, 3 3 3	110 W13c R19
20. Forced flow around a cylinder	d = cylinder diameter $Re = vd/\nu$	(1) $Sh \sim Re^{0.385}$ (2) $Sh = 1.08Re^{0.44}Sc^{1/3}$ for $d/d_{channel} = 0.2$ (3) $Sh = 1.20Re^{0.42}Sc^{1/3}$ for $d/d_{channel} = 0.1$	$0.02 < Re < 10$ $Sc = 1200$ $160 < Re < 1000$ $160 < Re < 1000$	3, 4 3 3	D16a G7 G7

a Validity doubtful. See Section VI.B and Eq. (32) in text.

(continued)

TABLE VII (continued)

System	Parameters	Correlation	Range	Reactants	Ref.
		(4) $Sh = 0.38 Sc^{-0.2}$ $+ (0.56 Re^{0.5}$ $+ 0.001 Re) Sc^{0.33}$ for $d/d_{channel} < 0.05$	$5 < Re < 100$ $1300 < Sc < 2000$	3	V6
		(5) $Sh = 0.508 Re_d^{0.5} Sc^{1/3}$ for $d/d_{channel}$ $= 0.05–0.15$	$150 < Re < 1500$	3	R6
		(6) $Sh \sim Re^{0.4}$ (See also entry 17)	$0.112 < Re < 4.40$ $1422 < Sc < 3433$	2, 3	T4c
21. Forced flow through a gauze	d = diameter of gauze wire v = approach velocity $Re = vd/\nu$	(1) $Sh = 0.64 Sc^{0.2} + 0.52$ $\times Re^{0.5} Sc^{1/3}$, open area 66%	$1 < Re < 50$ $1300 < Sc < 2000$	3	V6
		(2) $Sh = 1.03 Sc^{0.2} + 0.61$ $\times Re^{0.5} Sc^{1/3}$, open area 56%		3	V6
22. Forced flow at a sphere: overall mass transfer	d = sphere diameter	(1) $Sh = 0.882 Re^{0.452} Sc^{1/3}$	$400 < Re < 1250$ $0.053 < d/d_{channel} < 0.26$	3	G4a
		(2) $Sh = 0.447 Re^{0.538} Sc^{1/3}$	$1250 < Re < 12,500$ $0.053 < d/d_{channel} < 0.26$		
23. Forced flow at a sphere: local mass transfer		$Sh = (A_1 + A_2 \log Re) Re^n Sc^{1/3}$, see Gibert and Angelino (G2)	$2000 < Re < 12,000$	3	G2

C. FREE CONVECTION AND MIXED CONVECTION

24a. Free convection at a horizontal cylinder	d = cylinder diameter $Sh = kd/D$ $Gr = g\,\Delta\rho d^3/\nu^2\rho$	(a1) $Sh = 0.53(GrSc)^{0.25}$	$1.2 \times 10^7 < GrSc < 10^9$ transition at $GrSc = 3 \times 10^8$	2	S6
		(a2) $Sh = 0.23(GrSc)^{0.3}$	$2.3 \times 10^6 < GrSc < 10^9$	3	W3a
		(a3) $Sh = 0.42(GrSc)^{0.25}$	$1.7 \times 10^6 < GrSc < 7.1 \times 10^8$ $Sc = 925$	3	T1b
		(a4) $Sh = 0.56(GrSc)^{0.25}$	$3.4 \times 10^6 < GrSc < 10^9$ $2470 < Sc < 2860$	2	S15a
24b. Combined forced and free convection at a horizontal cylinder		$(ShRe^{1/2})(Re^2/Gr)^{1/2}$ $= f(Gr/Re^2Sc^{1/3})$	$5 \times 10^{-5} < Gr/Re^2Sc^{1/3} < 50$	1, 3	T4b
25. Free convection at a sphere	d = sphere diameter	$Sh = 2 + 0.59(GrSc)^{0.25}$	$2.3 \times 10^8 < GrSc < 1.5 \times 10^{10}$	2	S6
26. Free convection at a vertical gauze	d = diameter of gauze wire	$Sh = 0.45(GrSc)^{0.25}$	$2 \times 10^5 < GrSc < 5 \times 10^{10}$	3	W7
27. Free convection at a vertical wire	L = height R = radius $\gamma_L = (L/R)(GrSc)^{-1/4}$	$Sh = f(\gamma_L)$ see Fig. 16	$0.11 < \gamma_L < 6.8$ $2 \times 10^7 < GrSc < 6 \times 10^{12}$	3	R10
28. Combined heat and mass transfer at a vertical plate	L = height Gr^*, see Eq. (40) $Sh = kL/D$ Gr^{**}, see Eq. (37)	(1) $Sh = 0.636(Gr^*Sc)^{0.25}$, opposing and cooperating body forces	$4 \times 10^7 < Gr^*Sc < 2 \times 10^8$ $Sc = 2000$ $\Delta T = 0\text{--}13°C$	2	M3
		(2) $Sh = 0.79(Gr^{**}Sc)^{0.243}$, opposing and cooperating body forces	$6 \times 10^9 < Gr^{**}Sc < 2 \times 10^{11}$ $2 \times 10^6 < Gr^{**}Pr < 5 \times 10^7$	2	D10

(continued)

TABLE VII (continued)

System	Parameters	Correlation	Range	Reactant	Ref.
29. Combined heat and mass transfer at a horizontal cylinder	d = cylinder diameter Ra^*, see Eq. (41)	(3) $Sh = 0.64(Gr^{**}Sc)^{0.25}$, cooperating body forces only	As in Correlation (2)	2	D10
		$Sh = 0.23(Ra^*)^{0.28}$, opposing body forces	$2.3 \times 10^6 < GrSc < 1.5 \times 10^{10}$ $5.5 \times 10^3 < Gr_h Pr < 8.7 \times 10^6$ $2 \times 10^7 < Ra^* < 4 \times 10^{10}$	3	W3a
30a. Free convection at a horizontal plate; (1) laminar regime; (2) transitional regime; (3–7) turbulent regime	$Gr = g\alpha \Delta c L^{*3}/\nu^2$ $L^* = A/p$ = surface/perimeter L = width of plate or diameter disk	(a1) $Sh = 0.54(GrSc)^{1/4}$ (a2) $Sh = 0.72(GrSc)^{0.25}$ (disk)	$2.2 \times 10^4 < GrSc < 8 \times 10^6$ $3 \times 10^4 < GrSc < 3 \times 10^7$	2 2	L10b W6, W9, P2b
		(a3) $Sh = 0.19(GrSc)^{1/3}$	$10^8 < GrSc < 1.4 \times 10^{12}$ $2100 < Sc < 52{,}000$	2 3	F3 K8
		(a4) $Sh = 0.122(GrSc^{9/8})^{1/3}d_n$, d_*, see Fenech (F2)	GrSc, Sc same as Corr. (3). $d_n = 1.0$ if plate surrounded by vertical walls; $d_n < 1.0$ if plate embedded	2	F2
		(a5) $Sh = 0.18(GrSc)^{1/3}$	$3 \times 10^7 < GrSc < 10^{12}$ $2200 < Sc < 2300$	2	W6, W9, P2b
	$Gr = (aL^3 \Delta\rho)/\rho\nu^2$ a = centrifugal acceleration	(a6) $Sh = 0.152(GrSc)^{1/3}$ (disk)	$6 \times 10^6 < GrSc < 5 \times 10^{12}$ $2000 < Sc < 40{,}000$ $1 < a/g < 420$	2, 3, 4	R9
	$Gr = g\alpha \Delta c L^{*3}/\nu^2$ $L^* = A/p$ = surface/perimeter	(a7) $Sh = 0.15(GrSc)^{1/3}$	$8 \times 10^6 < GrSc < 1.6 \times 10^9$	2	L10b
30b. Free convection at a horizontal screen	$Gr = g \Delta\rho L^3/\nu^2 \rho$ L = screen width	$Sh = 0.375(GrSc)^{0.305}$	$22 \times 10^8 < GrSc < 26 \times 10^{10}$	1	S11b

	System	Equation	Variables	Range	Notes	Ref.
31.	Combined heat and mass transfer to a horizontal plate	(1) $Sh = 1.75(Gr^*Sc)^{0.227}$ (2) $Sh = 0.163(Gr^*Sc)^{0.33}$	L = width of plate	$10^7 < Gr^*Sc < 5 \times 10^9$ $5 \times 10^9 < Gr^*Sc < 10^{11}$	2 2	W10 W11
32.	Turbulent free convection at a vertical cylinder with strong curvature	$Sh_R = 0.57(Gr_R Sc)^{0.11}$	R = radius of cylinder $Gr_R = g\,\Delta\rho R^3/\nu^2\rho$	$2.9 < Gr_R Sc < 400$ $2500 < Sc < 3800$ $3700 < L/R < 78{,}000$	3	R10
33.	Free convection at a vertical plate in a cell of restricted volume (free convection pump cell)	$Sh_L = 0.0225(a/L)^2$ $\times (Gr_L Sc)^{0.85}$	L = height of plate a = distance of electrode to diaphragm	$2 \times 10^6 < GrSc < 2 \times 10^8$ $0.005 < a/L < 5.48(GrSc)^{-0.3}$	2, 3	B9
34.	Free convection at an inclined plate	$Sh_L = 0.3(CGr_L Sc)^{0.23}$ $C = f(\text{angle of inclination})$ see Fouad and Ahmed (F4)	L = height of plate	$6 \times 10^{10} < GrSc < 7 \times 10^{13}$ $1750 < Sc < 3000$	2	F4
35.	Free convection at a vertical plate with surface roughness	$Sh_L = A(Gr_L Sc)^{0.32}$ $A = f(h/\delta)$	L = height of plate h = height of roughness δ = diffusion layer thickness	$2.2 \times 10^{10} < GrSc < 2.5 \times 10^{13}$ $2400 < Sc < 3400$	3	F7c
36.	Turbulent free convection at a vertical plate	(1) $Sh = 0.31(GrSc)^{0.28}$ (2) $Sh = 0.15(GrSc)^{0.29}$	L = height of plate	$4 \times 10^{13} < GrSc < 10^{15}$ $4.57 \times 10^{11} < GrSc < 10^{14}$ $2400 < Sc < 3700$	2 3	F6 F5

(continued)

TABLE VII (continued)

System	Parameters	Correlation	Range	Reactants	Ref.
37. Combined free and forced convection in a vertical annulus	L = transfer length r_i = inner radius r_o = outer radius $d_e = 2(r_o - r_i)$	(1) $Sh = 1.96[Re_e Sc d_e/L$ $+ 0.04(Gr_L Sc d_e/L)^{0.75}]^{1/3}$ for aiding flow (2) No correlation for opposing flow	$1.3 \times 10^7 < Gr Sc d_e/L$ $< 1.1 \times 10^9$ $7 \times 10^3 < Re Sc d_e/L$ $< 2.5 \times 10^5$	2 2 2	W12 W8a R15
38a. Combined free and forced convection at a horizontal wall in channel flow	d_e = equivalent diameter L = transfer length v = flow velocity $Gr_L = g \Delta\rho L^3/\rho v^2$ $Gr_\delta = g \Delta\rho L d_e/\mu v$ $Re = v d_e/\nu$	(a1) $Sh = 0.85(ReScd_e/L)^{1/3}$ for $Gr_\delta < 920$ (a2) $Sh = 0.19(Gr_L Sc)^{1/3}$ for $Gr_\delta > 920$	$75 < Re < 3000$ $600 < Sc < 12{,}000$ $0.05 < d/L < 20$	2	T2
38b. Combined forced and free convection at a vertical wall in horizontal channel flow	i_L = limiting current density $i_{LNC} = i_L$ in natural convection model $i_{LFC} = i_L$ in forced convection model	$i_L/i_{LNC} = [1 + (i_{LFC}/i_{LNC})^{1/n}]^n$ $0.67 < n < 2$	$i_{LFC}/i_{LNC} = 0$ to 0.4	2	M13d

D. TURBULENT FORCED CONVECTION

System	Parameters	Correlation	Range	Reactants	Ref.
39a. Turbulent flow at a smooth rotating disk (transition regime)	ω = rotation rate r = radius $Re = \omega r^2/\nu$ $Sh = kr/D$	$Sh = 0.89 \times 10^5 Re^{-1/2} Sc^{1/3}$ $+ 9.7 \times 10^{-15} Re^3 Sc^{1/3}$	$2 \times 10^5 < Re < 3 \times 10^5$	1, 2	M13h

#	System	Definitions	Equation	Range	N	Ref
39h.	Turbulent flow at a smooth rotating disk (fully developed turbulence)	ω = rotation rate r = radius $Re = \omega r^2/\nu$ $Sh = kr/D$	(b1) $Sh = 0.00707 Re^{0.9} Sc^{1/3}$ ^{f}	$2.7 \times 10^5 < Re < 1.5 \times 10^6$ $345 < Sc < 6450$	5	D2
			(b2) $Sh = 0.0117 Re^{0.896} Sc^{0.249}$	$8.9 \times 10^5 < Re < 1.18 \times 10^7$ $34 < Sc < 1400$	1	E4
			(b3) $Sh = 0.89 \times 10^5 Re^{-1/2} Sc^{1/3}$ $+ 9.7 \times 10^{15} Re_o^3 Sc^{1/3}$	$2.0 \times 10^5 < Re_o < 3.0 \times 10^5$	2	M13h
40.	Turbulent flow at a smooth rotating ring	ω = rotation rate r_i = inner radius r_o = outer radius r_m = mean radius $Re = \omega r^2/\nu$ $Sh = kr/D$	(1) $Sh_{r_o} = Sh_{disk, r_o}$ $\times [1 - (r_i/r_o)^{2.5}]^{0.8}$	$2.7 \times 10^5 < Re < 1.5 \times 10^6$ $345 < Sc < 6450$	5	D2
			(2) same as (1)	$4 \times 10^5 < Re_{r_o} < 1.5 \times 10^6$ $0.1 < r_i/r_o < 0.9$ $Sc = 3000$	3	D11a
			(3) $Sh_{r_o} \sim Re^{0.6} Sc^{1/3}$ $\times [r_i/(r_o - r_i)]^{1/3}$	$2.7 \times 10^5 < Re_{r_o} < 3.5 \times 10^5$ $r_i/r_o \to 1$ $Sc = 3000$	3	D11a
			(4) $Sh_{r_m} = 0.0177 Re_{r_m}^{0.878} Sc^{0.34}$	$2.7 \times 10^5 < Re_{r_m} < 10^6$ $r_i/r_o = 0.96$ $500 < Sc < 2000$	1	H1
			(5) Radial drag component from k (see reference)	$3 \times 10^4 < Re_o < 10^6$	3	D11b
41.	Turbulent flow at a rotating hemisphere	ω = rotation rate r = radius $Re = \omega r^2/\nu$ $Sh = kr/D$	(1) $Sh = 0.019 Re^{0.8} Sc^{1/3}$, $r_{support\ rod} = r_{hemisphere}$	$40{,}000 < Re < 100{,}000$ $910 < Sc < 6300$	5	G5a
			(2) $Sh = 0.092 Re^{0.67} Sc^{1/3}$, $r_{support\ rod} > r_{hemisphere}$	$15{,}000 < Re < 40{,}000$	3	C4

^{f} Correlation developed from experimental data on ring electrodes. System 40. for the hypothetical case of a disk without laminar flow region.

(*continued*)

TABLE VII (*continued*)

System	Parameters	Correlation	Range	Reactants	Ref.
42a. Turbulent flow at a smooth rotating disk: mass transfer to a spot electrode on the disk	r_o = radial distance to center of electrode d = diameter of electrode $Sh_d = kd/D$ $Re_{r_o} = \omega r_o^2/\nu$	(a1) $Sh_d = 0.27 Re_{r_o}^{0.6} Sc^{1/3}$ $\times (d/r_o)^{2/3 g}$	$Re_{r_o} > 8 \times 10^5$ $d = 0.1$ cm $r_o = 4$ cm	5	D5
42b. Turbulent flow at a rough rotating disk: mass transfer to a spot electrode in the valleys of macroroughness ($h \gg \delta_d$)	h = height of roughness (pyramidal)	(b1) $Sh_d = 0.10 Re_{r_o}^{2/3} Sc^{1/3}$ $\times (h/r_o)^{2\beta/3 g'}$	$h = 0.2$–0.3 cm $d = 0.01$ cm $r_o = 3$–4 cm $h \gg \delta_d$	5	D6a
42c. Turbulent flow at a rough rotating disk: mass transfer to a spot electrode on the peaks of macroroughness ($h \gg \delta_d$)	h = roughness height (pyramidal)	(c1) $Sh_d = 1.2 Re_{r_o}^{4/9} Sc^{1/3}$ $\times (h/r_o)^{2\beta/3 g\,i}$	$h = 0.3$ cm $d = 0.04$ cm $r_o = 3$–4 cm $h \gg \delta_d$	5	M5
42d. Turbulent flow at a rough rotating disk: mass transfer to entire surface	h = roughness height (pyramidal) r = radius disk $Sh = kr/D$ $Re = \omega r^2/\nu$	(d1) $Sh \sim Re^{2/3} Sc^{1/3}(h/r)^{2\beta/3\,i}$	$h = 0.2$ cm $r = 4.5$ cm $h \gg \delta_d$	5	M6a
43a. Turbulent Couette flow between a rotating inner cylinder and stationary outer cylinder	ω = rotation rate d_i = diameter inner cylinder d_o = diameter outer cylinder $Sh = kd_i/D$ $Re = \omega d_i^2/\nu$	(a1) $Sh = 0.0791 Re^{0.7} Sc^{0.356}$ (rotor)	$10^3 < Re < 10^5$ $835 < Sc < 11{,}490$ $0.093 < d_i/d_o < 0.83$	3, 4	E1
		(a2) $Sh = 0.079 Re^{0.69} Sc^{0.41}$ (rotor)	$29{,}000 < Re < 388{,}000$ $830 < Sc < 2210$	2	G1a, R16
		(a3) $Sh = 0.033 Re^{0.875} Sc^{0.013\,j}$ (rotor)	$32 \times 10^4 < Re < 5.2 \times 10^6$ $2000 < Sc < 50{,}000$	3	K5a

	System	Correlation	Range of variables	β^h (%)	Ref.
43b.	Turbulent Couette flow between a rotating outer cylinder and a stationary inner cylinder	$Sh = kd_i/D$ $Re = \omega d_i^2/\nu$	$Sh = 0.0791 Re^{0.7} Sc^{0.356}$ $\times (d_i/d_o)^{0.7}$ (stator)	$1000 < Re < 5.3 \times 10^4$ $Sc = 2450$ $0.22 < d_i/d_o < 0.68$	2	A4
43c.	Turbulent radial flow in a capillary gap between disks	$Re = Q/h\nu$ $Q =$ volumetric flow rate $h =$ interelectrode gap $r_o =$ outer radius $r_i =$ inner radius $Sh = Ih/nFcD$ (based on entire disk current I)	$Sh \sim [Re\,h^2/(r_o^2 - r_i^2)]^{0.8}$	$Re > 4 \times 10^5$ $r_o/r_i = 1.75$–5.33 $h = 0.05$–0.1 cm $r_o = 8.4$–23 cm	3	A6b
44.	Turbulent Couette flow at a rotating rough cylinder	$Sh = kd_i/D$ $Re = \omega d_i^2/\nu$ $h =$ roughness height	$Sh = \dfrac{Re(Sc)^{0.356}}{[1.25 + 5.76 \log(d_i/2h)]^2}$ if $Re > Re_{crit}$ $= (11.8 d_i/h)^{1.18}$	$600 < Re < 2.5 \times 10^5$ $d_i/h > 87$ $Sc = 400$	1	K1c
45.	Mass-transfer entrance regime in turbulent channel flow ($L/d < 5000 Re^{-7/8}$)	$v =$ velocity $d =$ diameter $L =$ transfer length $Re = vd/\nu$ $Sh = kd/D$	(1) $Sh = 0.276 Re^{0.58} (ScL/d)^{1/3}$	$5000 < Re < 75{,}000$ $0.018 < L/d < 4.31$ $Sc = 2400$	3 2 3	V2 S7 S22
		$Sh = kd/D$ $Re = vd/\nu$	(2) $Sh \sim Re^{0.58} Sc^{0.33}$	$2000 < Re < 10{,}000$ $600 < Sc < 3100$	11	D16b

[a] Calculated from Fig. 2 of Daguenet *et al.* (D5).
[g] β undetermined, < 0.01.
[i] β undetermined, < 0.01.
[j] Derived from $Nu = 0.0589 \sqrt{f/2} Sc^{1/4}$ in Kishnevskii (K5a), with $f = 0.099 Re^{-0.25}$ for $Re > 20{,}000$. The power on Sc was fixed *a priori* by theoretical considerations.

(*continued*)

TABLE VII (continued)

System	Parameters	Correlation	Range	Reactants	Ref.
	$Sh = kd/D$ $Re = vd/v$	(3) $Sh = 0.276 Re^{0.58}(Sc)^{1/3}(d/L)^{1/3}$	$5900 < Re < 135{,}000$ $1200 < Sc < 4600$ $0 < L/d < 0.5$ $Re < 10{,}000$ $0 < L/d < 0.1$ $Re > 100{,}000$	3	B4b
	$Sh_x = kx/D$ $Re = vd/v$	(4) $Sh = 0.067 Re^{0.75} Sc^{1/3}(d_h/x)^{0.3}$	$0 < x/d_h < 0.73$	2	L1b
46. Mass-transfer entrance regime in turbulent falling film	δ = film thickness $Re = g\delta^3/3v^2$	$Sh = 1.080 Re Sc(\delta/L)^{1/3}$	$Re > 1000$ $L/\delta \leqslant 900 v/(g\delta^3)^{1/2}$	3	I10
47a. Developed mass transfer in turbulent flow (annulus)	v = velocity $d = d_o - d_i$ d_o = outer diameter d_i = inner diameter $Sh = kd/D$ $Re = vd/v$	$Sh = 0.023 Re^{0.8} Sc^{1/3}$	$2100 < Re < 30{,}000$	1, 3, 4, 6	L9
				12	L10a
47b. Developed mass transfer in turbulent flow (pipe)	f = friction factor d = diameter	(b1) $Sh = 0.115(f/2)^{0.5} Re Sc^{0.25}$ (Deissler)^k	$5000 < Re < 50{,}000$ $Sc = 2170$	2	S7
	$Sh = kd/D$ $Re = vd/v$	(b2) $k = 3.52 \times 10^{-4}(f/2)^{0.5}v$	$8000 < Re < 50{,}000$ $Sc = 2400$	3	V1, S19
	$Sh = kd/D$ $Sh^+ = Sh Re^{7/8} Sc^{-1/3}$	(b3) $Sh \sim Re^{0.74} Sc^{0.33}$ or $Sh^+ = 0.14$	$2 \times 10^4 < Re < 2 \times 10^5$ $600 < Sc < 3100$ $L/d > 0.285$	3	D16b
	$Sh = kd/D$ $Re = vd/v$ $\xi_L = 0.021 Sc^{-1/4} Re \sqrt{f/2}(L/d)$	(b4) $Sh = 0.115 Re_v \sqrt{f/2} Sc^{1/4}$ $\times [\text{cth}(1.17\xi_L^{1/2}/d)]^{2/3}$	$4500 < Re < 25{,}000$ $1300 < Sc < 28{,}000$	3	K5b

		(b5) $k = 0.0889 v_* \sqrt{f/2} \text{Sc}^{0.296}$	$10{,}000 < \text{Re} < 150{,}000$ $1730 < \text{Sc} < 37{,}000$	3.5	S11a
		(b6) $\text{Sh} = 0.0165 \text{Re}^{0.86} \text{Sc}^{0.33}$	$8000 < \text{Re} < 200{,}000$ $1000 < \text{Sc} < 6000$	3	B4b
47c. Developed mass transfer in turbulent flow (channel)	d = equivalent diameter $\text{Sh} = kd/D$ $\text{Re} = vd/\nu$	(c1) $\text{Sh} = (f/2)\text{Re}(\text{Sc})^{1/3}$ (Chilton–Colburn)	$7000 < \text{Re} < 60{,}000$ $1700 < \text{Sc} < 30{,}000$	3	H11a
		(c2) $\text{Sh} \sim \text{Re}^{0.9} \text{Sc}^{1/3}$	$3000 < \text{Re} < 80{,}000$ $800 < \text{Sc} < 15{,}000$	3	M12
		(c3) $\text{Sh} = 0.0153 \text{Re}^{0.88} \text{Sc}^{0.32}$	$3000 < \text{Re} < 1.2 \times 10^5$ $400 < \text{Sc} < 4600$	3	D8
		(c4) $\text{Sh} = 0.0278 \text{Re}^{0.875} \text{Sc}^{0.21}$	$4000 < \text{Re} < 16{,}000$ $1000 < \text{Sc} < 4000$	3	P3b
		(c5) $\text{Sh} = 0.0113 \text{Re}^{0.87} \text{Sc}^{0.35}$	$2800 < \text{Re} < 12{,}000$ $1200 < \text{Sc} < 25{,}100$	2	L1b
		(c6) $\text{Sh} = 0.0100 \text{Re}^{0.92} \text{Sc}^{0.336}$	$12{,}000 < \text{Re} < 125{,}000$ $1200 < \text{Sc} < 25{,}100$	2	L1b
47d. Developed mass transfer in turbulent flow (falling film)	δ = film thickness $\text{Sh} = k\delta/D$ $\text{Re} = g\delta^3/3\nu^2$	$\text{Sh} = 0.1155(f/2)^{0.5} \text{Re} \text{Sc}^{0.25}$ (Spalding–VanDriest)	$\text{Re} > 1000$ $1000 < L/\delta(g\delta^3/\nu^2)^{1/2} < 5000$ $1400 < \text{Sc} < 18{,}500$	3	I10
48a. Developed mass transfer at a rough surface in turbulent flow (channel)	h = roughness height $\text{Sh} = kd/D$ $h^+ = hv(f/2)^{0.5}/\nu$	(a1) $\text{Sh}_{\text{rough}}/\text{Sh}_{\text{smooth}}$ $= 1.94 \text{Sc}^{0.09}/(h^+)^{0.10}$ $\text{Sh}_{\text{smooth}}$ see Eq. (47c, 2) of this table	$3000 < \text{Re} < 1.2 \times 10^5$ $400 < \text{Sc} < 4600$ $h^+ > 25$	3	D8
		(a2) No universal correlation	$h^+ < 25$	3	D8
		(a3) $\text{Sh} = 0.0039 \text{Re}^{1.06} \text{Sc}^{0.5}(h/d)^{0.21}$	See reference	2	L1b
48b. Flow in channel with turbulence promoters		See references	—	3	S20c, L1d

[k] Power of Sc fixed *a priori* on theoretical grounds.

(continued)

TABLE VII (continued)

System	Parameters	Correlation	Range	Reactants	Ref.
48c. External turbulent flows at cylinders and spheres	—	See reference	—	1	G1d
49a. Mass transfer in hydrodynamic entrance region of turbulent pipe flow	—	Complex correlation of k_x with $(\tau_x)_{wall}$ (measured by spot electrodes)	—	3	M11, M13a
49b. Mass transfer in hydrodynamic entrance region of turbulent channel flow	$Re = vd/v$ d = equivalent diameter f = friction factor	$Sh = f/2 Re Sc^{1/3}$ (Chilton–Colburn)	$10 < Re f/2 < 75$ $Re > 10{,}000$ $2000 < Sc < 3200$	3	A1a
50a. Turbulent flow at a moving continuous cylinder	—	Comparison with approximate numerical solution	—	3	R4
50b. Turbulent flow at a moving continuous sheet	—	See reference	—	1, 3	N1a

E. OSCILLATING FLOWS

System	Parameters	Correlation	Range	Reactants	Ref.
51a. Mass transfer at a vertical plate in vibration perpendicular to the width of the plate	w = width t = thickness l = height $d = l + \omega t/(\omega + t)$ a = amplitude $\bar{f}$ = frequency $Re_w = \bar{f} a w / v$	$Sh_w = 0.476 Re_w^{0.56} Sc^{1/3}$ $\times (1 + w/l)^{0.7}$	$w > 1$ cm, $l < 4$ cm 0.13 cm $< a < 2.0$ cm 1.6 sec$^{-1} < \bar{f} < 23.3$ sec^{-1} $165 < Re_w < 8850$	3	R2

	System	Correlation	Range of variables	Electrodes	Ref.
51b.	Mass transfer at a vertical plate in vibration perpendicular to the thickness	$Re_d = \tilde{f}ad/\nu$	$Sh_d = 0.437 Re_d^{0.74} Sc^{1/3}(w/t)^{0.5}$	$145 < Re_d < 4160$	3	R2
51c.	Mass transfer at a vertical plate in longitudinal vibration	$E = i_{\text{lim, vibr}}/i_{\text{lim, convection}}$ a = amplitude $\tilde{f}$ = frequency L = plate height	$E = 0.0174 a^{0.42} \tilde{f}^{1.09}$	$0 < \tilde{f} < 48 \text{ sec}^{-1}$ $0 < a < 8 \text{ mm}$ $1 < L < 5 \text{ cm}$	2	A3d, 111b, A3c
52.	Mass transfer at a vertical cylinder in vibration perpendicular to the length	d = diameter $Re = \tilde{f}ad/\nu$ $Sh = kd/D$	$Sh = 0.41 Re^{0.62} Sc^{1/3}$	$0.18 \text{ cm} < d < 1.02 \text{ cm}$ $0.24 \text{ cm} < a < 3.7 \text{ cm}$ $1.6 \text{ sec}^{-1} < \tilde{f} < 25 \text{ sec}^{-1}$ $60 < Re < 10^4$ $800 < Sc < 2240$	4	R5
53a.	Mass transfer at a horizontal circular plate in transverse vibration	d = diameter $Re = \tilde{f}ad/\nu$ $Sh = kd/D$	(a1) $Sh = 0.24 Re^{0.62} Sc^{1/3}$ if $Re < 150$ (a2) $Sh = 0.066 Re^{0.87} Sc^{1/3}$ if $150 < Re < 960$	$0.05 \text{ cm} < a < 0.77 \text{ cm}$ $3.3 \text{ sec}^{-1} < \tilde{f} < 42 \text{ sec}^{-1}$ $18 < Re < 960$ $850 < Sc < 1100$	3	R1
53b.	Same	$Re_{\text{rms}} = 2\pi \tilde{f} a/2^{1/2}$	$Sh = 0.07 Re_{\text{rms}}^{0.75} Sc^{0.5}$	$8 < Re_{\text{rms}} < 800$ $850 < Sc < 19{,}000$	3, 4	P5
53c.	Mass transfer at a horizontal circular plate (disk) in sinusoidal oscillating rotation superimposed on steady rotation	$Z = (i_{\text{lim}} - i_{\text{lim, s.s.}})/\Delta \omega$ ω = rotation rate $\Delta \omega$ = amplitude of sinusoidal oscillation of ω ω_0 = steady rotation rate $\tilde{f}$ = frequency of oscillation $Z_0 = i_{\text{lim, s.s.}}/\omega_0$	$Z = f(Z_0, Sc, \omega_0, \tilde{f})$ (see reference)	$0 < \tilde{f} < 20 \text{ sec}^{-1}$	3	D11c

(*continued*)

TABLE VII (continued)

System	Parameters	Correlation	Range	Reactants	Ref.
54. Mass transfer at a vibrating sphere	d = diameter a = amplitude f = frequency $\mathrm{Re} = \tilde{f}ad/\nu$ $\mathrm{Sh} = kd/D$	$\mathrm{Sh} = 2 + 0.42(\mathrm{ReSc})^{0.5}$	d = 2.53 cm 0.075 cm < a < 0.157 cm 3 sec^{-1} < $\tilde{f}$ < 35 sec^{-1} 100 < Re < 1600 500 < Sc < 2000	2, 3	N10
55. Mass transfer to a vibrating sphere in forced flow	d = diameter v = velocity forced flow a = amplitude $\tilde{f}$ = frequency $\widetilde{\mathrm{Re}} = 4\tilde{f}ad/\nu$ $\overline{\mathrm{Re}} = vd/\nu$ $\mathrm{Sh} = kd/D$ $V = (\widetilde{\mathrm{Re}}/\overline{\mathrm{Re}})(d/2a)^{0.45}$	(a) $V < 0.06$ $\mathrm{Sh} = \overline{\mathrm{Sh}}$(forced flow only), see Eq. (22) of this table (b) $V > 0.06$ $\mathrm{Sh} = \overline{\mathrm{Sh}} + \mathrm{Sh}$ $= 0.477\overline{\mathrm{Re}}^{0.538}$ $\mathrm{Sc}^{1/3}[1 + 1.05$ $\times (V - 0.06)^{1.26}]$	$0.06 < V < 0.60$ $1250 < \overline{\mathrm{Re}} < 12{,}000$ $0 < \widetilde{\mathrm{Re}}/\overline{\mathrm{Re}} < 2/\pi$ d = 1.75, 2.45 cm Sc not given^f	3	G3
56. Mass transfer through a horizontal liquid film in standing wave motion	h = film thickness a = peak-to-peak amplitude of wave generator L = width of film in wave direction $\tilde{f}$ = frequency $\mathrm{Sh} = kh/D$	(a) $\tilde{f} < 60$ sec^{-1} $\widetilde{\mathrm{Sh}} = \mathrm{Sh} - \overline{\mathrm{Sh}}$ $= K a\tilde{f}^{0.75} h^{0.5} L^{-0.5}$ $K = 35$ sec$^{0.75}$ cm^{-1} (b) 60 sec^{-1} < $\tilde{f}$ < 90 sec^{-1} $\widetilde{\mathrm{Sh}} = \mathrm{Sh} - \overline{\mathrm{Sh}}$ $= K_1 a h^{0.5} L^{-0.5}$ $\times \{1 - \exp(-K_2 \tilde{f}^{0.75})\}$ $K_1 = 430$ cm^{-1} $k_2 = 0.081$ sec$^{0.75}$	3 sec^{-1} < $\tilde{f}$ < 90 sec^{-1} 0.1 cm < h < 0.4 cm 0.03 cm < a < 0.14 cm 3.5 cm < L < 15 cm	1	G6

F. STIRRED CELLS

57a. Flow at a rotating cylinder in a large cylindrical cell with baffles	ω = rotation d = rotor diameter d_o = cell diameter $h = (d_o/d)/2$ $\mathrm{Re} = \omega dh_r/2\nu$ $\mathrm{Sh} = kd_o/D$	$\mathrm{Sh} = 1.83\,\mathrm{Re}^{0.58}\mathrm{Sc}^{1/3}$	$2000 < \mathrm{Re} < 40{,}000$ Sc not varied	2, 4	K8
57b. Flow at ring electrodes in bottom of cell as in System 57a (region of decreasing mass-transfer rate)	d_r = ring diameter $h_r = (d_o - d_r)/2$ l = rotor to bottom distance $\mathrm{Re} = \omega dh_r/2\nu$ $\mathrm{Sh} = kd_o/D$	$\mathrm{Sh} = 0.511\,\mathrm{Re}^{0.62}\mathrm{Sc}^{1/3}$ $\times (d_o/l)^{0.32}$	$6 \times 10^3 < \mathrm{Re} < 4 \times 10^5$ d_o = 15.3, 21.6 cm d = 5, 7, 10 cm l = 5, 7.6 cm	3	S2
58a. Flow at the vertical wall of a cylindrical cell with turbine impeller and without baffles	ω = rotation rate d = rotor diameter d_o = cell diameter b = vertical dimension of impeller blade H = liquid height $\mathrm{Re} = \omega d^2/\nu$ $\mathrm{Sh}_d = kd/D$	$\mathrm{Sh}_d = 0.228\,\mathrm{Re}^{2/3}\mathrm{Sc}^{1/3}$	$10^2 < \mathrm{Re} < 3 \times 10^4$ $0.2 < b/H < 0.7$ d = 6.6 cm $H = d_o$ = 10 cm	2	M10
58b. Same	l = rotor-to-bottom distance h = center of electrode to bottom distance $\mathrm{Sh}_{d_o} = kd_o/D$	$\mathrm{Sh}_{d_o} = A\,\mathrm{Re}^m\mathrm{Sc}^{1/3}$ m independent of l/h but slightly dependent on d/d_o ($m = 0.59 \to 0.65$ as $d/d_o = 0.22 \to 0.475$); broad maximum of A at $l/H = 0.5$	$3 \times 10^3 < \mathrm{Re} < 8 \times 10^4$ $0.22 < d/d_o < 0.475$ $0.2 < l/d_o < 0.8$	3	L3

[1] Apparently Sc was not varied.

(continued)

TABLE VII (continued)

System	Parameters	Correlation	Range	Reactants	Ref.
58c. Same as (a) and (b), but with baffles		$Sh_{d_o} = ARe^m Sc^{1/3}$ $m = 0.65$ if $d/d_o > 0.3$; sharp maximum of A at $l/H = 0.5$	$3 \times 10^3 < Re < 8 \times 10^4$ $0.22 < d/d_o < 0.475$ $0.2 < l/d_o < 0.8$	3	L3
59a. Flow at wall of a diaphragm cell[m] with magnetic bar stirrer	d = stirrer bar length ω = rotation rate $Re = \omega d^2/\nu$ $Sh_d = kd/D$	$Sh_d = 0.050 Re^{0.79} Sc^{0.38}$	$200 < Re < 4000$ $2500 < Sc < 62,000$ $d = 2.5$ cm cell volume, 125 cm^3	2	H5
59b. Same in free convection (vertical diaphragm)	r = diaphragm radius $Sh_r = kr/D$ $Gr_r = g\,\Delta\rho r^3/\rho\nu^2$	$Sh_r = 0.57(GrSc)^{0.25}$	$5.4 \times 10^9 < GrSc < 10^{10}$	2	H5
60a. Mass transfer to a spherical probe in a stirred cylindrical cell without baffles	d_s = sphere diameter d_o = cell diameter d = impeller diameter ω = rotation rate $Sh_d = kd/D$ $Re = \omega d d_s/2\nu$	$Sh_d = ARe^m Sc^{1/3}$, $0.49 < m < 0.56$ if m chosen = 0.5, A is independent of ω by $\pm 10\%$; charted for cells of various d_o	$7 \times 10^3 < Re < 10^5$ 6.2 cm $< d < 13.4$ cm $d_s = 0.42$ cm 18.6 cm $< d_o < 40$ cm	3	L4
60b. Same with baffles				3	L2
61a. Annular cell with tangential inflow	r_i = inner radius r_o = outer radius L = length Q = volume flow rate $A = \pi(r_o^2 - r_i^2)$ A_i = flow entrance area $d_e = 2(r_o - r_i)$ $Re = Q d_e/\nu A$	$Sh_d = 0.24 Re^{0.78} Sc^{1/3}$ $\times (d_e/L)^{0.5}(A/A_i)^{1/3}$	$48 < Re < 500$ $1.3 < L/d_e < 12$	4	R11

61b. Mass transfer at the wall of a hydrocyclone[n]	L = length d = diameter d_i = inlet diameter h = electrode-to-inlet distance θ = half-angle distance of apex Q = volume flow rate $Re = Q/vd$	$Sh_d = 14.1\ \mathbf{ReRe}^{2/3}Sc^{1/3}G$ If $\theta > 13°$ $G = (d - 2h\tan\theta/d)^{2/3}$ $\quad \times (d/d_1)^{1/6}(\tan\theta)^{4/3}$; If $\theta° < \theta \leq 13°$ $G = 0.14(d - 0.46h/d)^{2/3}$ $\quad (d/d_1)^{1/6}$	$Re > 20$	3	R8
62. Mass transfer at a vertical plates continuously wiped by nets	L = length d = interthread distance v = wiping velocity $Re_L = vL/v$ $Sh_L = kL/D$	(1) $Sh_L = 29.45(Re_L L/d)^{0.369}$ $\quad$ (Upward motion) (2) $Sh_L = 25.07(Re_L L/d)^{0.377}$ $\quad$ (Downward motion)	$10 < Re_L(L/d) < 3 \times 10^7$ $d = 0.05$ and 0.25 cm	3	S3

G. CONVECTION BY GAS EVOLUTION

63a. Mass transfer at a vertical electrode[n] evolving hydrogen	Q = volume rate of gas generation A = electrode surface area h = electrode height d = bubble diameter	$k \sim (Q/A)^p$ $p = 0.50 \to p = 0.53$ $\quad$ as $i < 5$ mA cm^{-2} $\quad \to i > 30$ mA cm^{-2} k independent of h and d		3, 9	V3, 18
63b. Same[o]		$k \sim (Q/A)^p$ $p = 0.50 \to p > 0.75$ $\quad$ as $i < 40$ mA cm^{-2} $\quad \to i > 200$ mA cm^{-2} k decreases with h and with d		8, 9	J3

(continued)

---

[m] As used for diffusivity measurement.
[n] Various metals.
[o] Platinum.

TABLE VII (continued)

System	Parameters	Correlation	Range	Reactants	Ref.
63c. Same evolving oxygen[p]		$k \sim (Q/A)^p I^{-0.13}$ $p = 0.25$ for 7 mA cm^{-2} $< i < 100$ mA cm^{-2}	—	8, 9	J3
63d. Mass transfer at a vertical electrode-evolving hydrogen	i_G = gas-evolution current density i_R = reaction current density	$i_R \sim i_G^{2/3}$	$10^{-3} < i_G < 10^{-2}$ A/cm^2	7	R21b
63e. Mass transfer at one wall of a rectangular cell with bubble-driven convection generated at the opposing wall	L = electrode height $Sh = kL/D$ Sh_B = Sh without bubble-driven convection Q = volume rate of gas generation A = cell cross section W = cell width d_e = cell equivalent diameter $Re_g = Qd_e/Av$	$Sh = Sh_B + \alpha Re_g^\beta$ $Sc^{1/4}(L/W)^\gamma$	$0.2 < Re_g < 5$	2	M13f
63f. Mass transfer with gas evolution at a rotating ring electrode	—	(1) Same as Eq. (1b) of this table	$Q < 250$ cm^3/cm^{2}hr $\omega = 2950$ rpm	13	K5a
63g. Mass transfer with gas evolution at a rotating double-ring electrode	—	See reference	$Q < 250$ cm^3/cm^{2}hr $\omega = 2950$ rpm	3, 13	K5b

304

63h. Mass transfer at a cylindrical gas-evolving electrode[a]	d_c = cylinder diameter (horizontal)	$k \sim (Q^{0.11}/d_c^{0.08})^{2.17}$	—	3	F7a
63i. Mass transfer at a rough vertical electrode[q] evolving hydrogen or oxygen		no correlation; k increases faster than true area	—	3, 4	F8
64a. Mass transfer at a horizontal electrode[p] evolving hydrogen	Q = volume rate of gas generation A = electrode surface area	$k \sim (Q/A)^p$ $p = 0.62$ for $i > 30$ mA cm^{-2} in acid $p = 0.36$ for $i > 30$ mA cm^{-2} in alkali	—	3, 4, 8, 9	J4a
64b. Same evolving oxygen[p]		$k \sim (Q/A)^p$ $p = 0.57$ for $i > 30$ mA cm^{-2} in acid $p = 0.87$ for $i > 30$ mA cm^{-2} in alkali $p = 0.33$ for $i < 30$ mA cm^{-2} in alkali	—	3, 4, 8, 9	J4a
64c. Mass transfer at horizontal gas-evolving screen electrodes	Q = volume rate of gas generation	(c1) $k \sim Q^{0.17}$ (hydrogen) (c2) $k \sim Q^{0.29}$ (oxygen)	—	3 2	F7b F7b
65. Mass transfer in two-phase channel flow	—	See reference	—	2	J5b

[p] Platinum.
[q] Nickel.

(continued)

TABLE VII (continued)

	System	Parameters	Correlation	Range	Reactants	Ref.
66.	Mass transfer at a horizontal plate having a single orifice at which gas sparging takes place	Q = volume rate of gas sparging	$k \sim Q^{0.31}$	0.04 cm^3 sec$^{-1} < Q$ < 10 cm^3 sec^{-1} 0.04-cm orifice diameter	3, 4	W2
67.	Mass transfer at a horizontal plate having a single active site at which nucleate boiling takes place	—	No correlation	—	3	B6, B7
68.	Mass transfer and simultaneous nucleate boiling at a horizontal plate	—	No correlation	—	2	W10

H. PARTICULATE ELECTRODES

	System	Parameters	Correlation	Range	Reactants	Ref.
69a.	Mass transfer to the wall in a packed bed	ϵ = porosity d_p = particle diameter $\bar{v}$ = superficial velocity $\text{Re}_p = \bar{v}d_p/\nu$ $j_D = (k/\bar{v})\text{Sc}^{2/3}$	$j_D = 0.822\text{Re}_p^{-0.38}$	—	4	K6, K12
69b.	Mass transfer to a screen inserted perpendicular to the flow direction in a packed bed	—	$j_D = 0.822\text{Re}_p^{-0.38}$	$30 < \text{Re}_p < 2400$ Sc not varied	4 3, 4 3	K6 K10 R7

306

System	Variables	Correlation	Range		Ref.
70a. Mass transfer to a single sphere in a packed bed of spheres	d = sphere diameter $\bar{v}$ = superficial velocity $Re = \bar{v}d/\nu$ $Sh = kd/D$	(a1) $Sh = 1.44 Re^{0.58} Sc^{1/3}$ for $35 < Re < 140$ (a2) $Sh = 4.58(ReSc)^{1/3}$ ($Re < 10$)	d = 1 cm $6 < Re < 1500$ $1580 < Sc < 1890$ $Re < 10$	3 3	J6 K2b
70b. Mass transfer to a packed bed of spheres	$j_D = (k/\bar{v})Sc^{2/3}$ $Sh = kd/D$ $Re = \bar{v}d/\nu$ d = sphere diameter	(b1) $j_D = 0.83 Re^{-0.44}$ (b2) $\epsilon Sh = 1.09(ReSc)^{1/3}$ ($ReSc > 100$) See reference for $ReSc < 100$	$23 < Re < 520$ $2830 < Sc < 3040$ $3 < ReSc < 70$	2 3	P3c A3f
70c. Mass transfer to a packed bed of Raschig rings	$Sh = kd_p/D$ $Re = \bar{v}L/\nu$ $Gr = g\alpha\,\Delta c d_p^3/\nu^2$	(c1) $Sh = 0.658(Re/Gr^{1/2})^{0.0036} \times (GrSc)^{0.25}$ (Aiding flow) (c2) $Sh = 0.638(Re/Gr^{1/2})^{0.0053} \times (GrSc)^{0.25}$ (Opposed flow)	$2 \times 10^{-4} < Re/Gr^{1/2} < 2.5 \times 10^{-2}$ $2 \times 10^{-4} < Re/Gr^{1/2} < 5.5 \times 10^{-2}$	3 3	M2b M2b
71a. Mass transfer in a packed bed, or a channel with turbulence promoters	ϵ = porosity $\bar{v}$ = superficial velocity s = area/unit volume a = see Eq. (51) of text $Re_m = \bar{v}s/a\nu\epsilon$ $j_D = (k/\bar{v})Sc^{2/3}$	$j_D = 0.121 Re_m^{-0.49}$	$60 < Re_m < 500$ $Sc = 2500$	3	H4
71b. Mass transfer to a packed bed of screens in a flow channel	$j_D = (k/\bar{v})Sc^{2/3}$ $Re_d = \bar{v}d/\nu\epsilon$ a = area/unit volume d = wire diameter ϵ = porosity	$j_D = 0.904 Re_d^{-0.64}$	$0.02 < Re_d < 0.3$ $Sc = 1350$	3	S12d

(continued)

307

TABLE VII (continued)

System	Parameters	Correlation	Range	Reactants	Ref.
71c. Mass transfer to a packed bed of screens in a flow channel	$j_D = (k/\bar{v})Sc^{2/3}$ $Re_l = \bar{v}l/\nu\epsilon$ ϵ = porosity l = interwire distance d = wire diameter	$\epsilon j_D = 1.08 Re_l^{-0.656}(l/d)^{1/3}$	$1 < Re_l < 100$ $0.81 < \epsilon < 0.87$ $2.3 < l/d < 4.7$	3	C1b
72. Mass transfer to the wall of a fluidized bed (annulus)	ϵ = porosity d_p = particle diameter d_e = equivalent diameter annular bed $\bar{v}$ = superficial velocity $Re_p = \bar{v}d_p/\nu(1-\epsilon)$ $j_D = (k/\bar{v})Sc^{2/3}$	(a) $\epsilon j_D = 0.43 Re_p^{-0.38}$ (b1) $j_D = 0.0113$ $\quad \times [(d_e/d_p)$ $\quad \times (r_o/r_{max})(1-\epsilon)]^{0.44}$ $\quad$ for $\epsilon > 0.87$ (b2) $j_D = 0.029$ $\quad \times [(d_e/d_p)$ $\quad \times (r_o/r_{max})(1-\epsilon)^2]^{0.44}$ $\quad$ for $\epsilon < 0.87$	$200 < Re_p < 23{,}000$ $2.67 < r_o/r_i < 4$ $1.6 < r_o/r_i < 53$ $200 < Re_p < 23{,}000$ Sc not varied	4 3 3 3	J1 K9 K11 K7
73a. Mass transfer to a small plate immersed in a fluidized bed	ϵ = porosity d_p = particle diameter $\bar{v}$ = superficial velocity $Re_p = \bar{v}d_p/\nu(1-\epsilon)$ $j_D = (k/\bar{v})Sc^{2/3}$	(a1) $j_D\epsilon = 0.027 Re_p^{-0.375}Sc^{0.292}$ $\quad \times (1-\epsilon)^{0.15}\epsilon^{-1.25}$ (a2) $j_D\epsilon = 1.21 Re_p^{-0.52}$	$200 < Re_p < 2800$ $10 < Re_p < 200$ $Sc = 1230$	3 3	J7 C9, C10
73b. Mass transfer to a small cylinder immersed in fluidized bed	d_e = annular equivalent diameter L = electrode length	(b1) $j_D = 1.24 Re_p^{-0.57}$ (b2) $j_D = 1.24 Re_p^{-0.57}$ $\quad \times (d_e/L)^{0.15}$	$0.1 < Re_p < 70$ $Sc = 760, 1700$ $1 < d_e/L < 5$	2 2	C1c C1e
74. Mass transfer to a rotating cylinder in a suspension	$i_{lim,s}$ = limiting current in suspension $i_{lim,o}$ = limiting current without suspension N = particle density	$i_{lim,s}/i_{lim,o} = 1 + aN^{0.09}$ $(0.10 < a < 0.23)$		5	P3d

Nomenclature[7]

a	Thermal diffusivity (cm^2 sec^{-1})	t_i	Transference number of ionic species i
a	Dimensional coefficient (Eq. 51) (cm^{-2})	u_i	Mobility of ionic species i (cm^2 mole J^{-1} sec^{-1})
a_{cr}	Critical distance electrode diaphragm (Eq. 43) (cm)	x	Distance along electrode in flow direction (cm)
b	Dimensional coefficient (Eq. 51) (cm^{-1})	y	Distance from the electrode, normal to the electrode surface (cm)
c	Concentration (moles cm^{-3})	v	Velocity (cm sec^{-1})
d	Diameter or equivalent diameter (cm)	w	Width of electrode or film (cm)
e^-	Electron	z_i	Charge number of ionic species i
f	Friction coefficient	A	Electrode area (cm^2)
$\tilde{f}$	Frequency (sec^{-1})	C	Numerical constant (Eqs. 25 and 26)
g	Acceleration of gravity (cm sec^{-2})		
h	Heat-transfer coefficient (J cm^{-2} sec^{-1} °K^{-1})	D_i	Diffusion coefficient of species i (cm^2 sec^{-1})
h	Distance between electrodes in Fig. 12a and b (cm)	E	Electrode potential (V)
		$E°$	Standard electrode potential versus nhe (V)
i	Current density (A cm^{-2})	ΔE	Applied cell potential (V)
i_{lim}	Limiting current density (A cm^{-2})	F	Faraday's constant (96, 501C)
i_0	Exchange current density (A cm^{-2})	Gr_L	Grashof number $= gL^3 \, \Delta\rho/\rho v^2$
i_D	Limiting diffusion current density (Fig. 4) (A cm^{-2})	Gr^*	Grashof number defined by Eq. (40)
i_L	Limiting current density (A cm^{-2})	Gr^{**}	Grashof number defined by Eq. (37)
j_D	Dimensionless mass-transfer parameter (Eq. 49c)	Gr_h^{**}	Grashof number defined by Eq. (39)
		I	Ionic strength $= \frac{1}{2}\sum_i c_i z_i^2$ (moles liter^{-1})
k	Mass-transfer coefficient (cm sec^{-1})	I_0, I_e	Currents in Fig. 13 (A)
k_q	Thermal conductivity (J cm^{-1} sec^{-1} °K^{-1})	J	Dimensionless exchange current density in Fig. 12a, b
n	Number of electons in charge-transfer reaction (Eq. 1b)	L	Length of mass or heat transfer section (cm)
$\tilde{n}$	Dimensionless frequency (Eq. 34)	Le	Lewis number, $k_q/c_p D\rho$
r	Radius (r_i, inner; r_o, outer) (cm)	N	Dimensionless average current density in Fig. 12a, b
r	Supporting electrolyte fraction (Figs. 4 and 5)	N_i	Molar flux of species i (moles cm^{-2} sec^{-1})
s	Surface area per unit volume of packed bed (Eqs. 51 and 52) (cm^{-1})	Nu_L	Nusselt number $= hL/k_q$
		Pe_L	Péclet number $= vL/D$
s_i	Number of ions or molecules of species i participating in charge-transfer reaction (Eq. 1b)	Q	Volume flow rate of gas at an electrode (cm^3 sec^{-1})
t	Time (sec)	R	Gas constant (8.3143 CV mole^{-1} °K^{-1})
t	Transference number of reactant species in Fig. 12a, b	Ra_L	Rayleigh number $= Gr_L Sc$

[7] Symbols used in Table VII are defined for each equation in the second column of the table.

Ra*	Rayleigh number defined by Eq. (41)	Δ	Difference of an electrolyte property (density, concentration, temperature) in the bulk and close to the electrode
Re_L	Reynolds number $= vL/v$		
S	Local velocity gradient normal to the surface, at the mass-transfer surface (sec^{-1})	ε	Porosity
		ζ	Densification coefficient of solution due to temperature increase (°K^{-1})
Sc	Schmidt number $= v/D$		
Sh_L	Sherwood number $= kL/D$		
T	Temperature (°K)	η	Overpotential (V)
		η_c	Concentration overpotential (V)
Greek Letters		η_s	Surface overpotential (V)
α	Anodic charge-transfer coefficient (Fig. 12a, b)	κ	Ratio of inner and outer radius in annular flow channel or ring disk
α_a	Anodic charge-transfer coefficient	λ_i	Mobility of ionic species i (cm^2 Ω^{-1} equiv^{-1})
α_c	Cathodic charge-transfer coefficient		
α_i	Densification coefficient of solution due to concentration increase of species i (cm^3 mole^{-1})	μ	Dynamic viscosity (gm cm^{-1} sec^{-1})
		v	Kinematic viscosity (cm^2 sec^{-1})
β	Cathodic charge-transfer coefficient (Fig. 12a, b)	$\mathcal{O}(A)$	Quantity of the order of magnitude of A
γ	Exponent of concentration dependence of exchange-current density	ρ	Density (gm cm^{-3})
		τ	Dimensionless time
		τ_0	Shear stress at the wall (gm cm^{-1} sec^{-2})
γ_L	Free-convection curvature parameter (Fig. 16)	ϕ	Dimensionless potential $= \Phi nF/RT$
δ_d	Mass-transfer boundary-layer thickness (cm)	ϕ	Coefficient in Eq. (27) and Fig. (15)
		ϕ_2	Coefficient in Eq. (28b) and Table V
δ_h	Hydrodynamic boundary-layer thickness (cm)	Φ	Potential (V)
		$\Delta\Phi_\Omega$	Ohmic contribution to overpotential (V)
δ_N	Nernst diffusion-layer thickness (cm)	ω	Rotation rate (rad sec^{-1})

References

A1a. Acosta, R. E., Muller, R. H., and Tobias, C. W., "Mass Transfer under Condition of High-Rate Electrolysis," Ext. Abstr. No. 265, Electrochemical Society Meeting, Washington, D.C., May 1976, 660.

A1b. Agar, J. N., *Discuss. Far. Soc.* **1**, 26 (1947).

A2. Agar, J. N., and Bowden, F. P., *Proc. R. Soc. London* A**169**, 206 (1938).

A3a. Aïmeur, F., Daguenet, M., Kermiche, F., and Meklati, M., *Electrochim. Acta* **18**, 87 (1973).

A3b. Albery, W. J., and Hitchman, M. L., "Ring-disc Electrodes," Clarendon, Oxford, 1971.

A3c. Al-Taweel, A. M., and Ismail, M. I., *J. Appl. Electrochem.* **6**, 559 (1976).

A3d. Al-Taweel, A. M., Ismail, M. I., and El-Abd, M. 2., *Ch-Ing. Techn.* **46**, 861 (1974).

A3e. Angell, D. H., and Dickinson, T., *J. Electroanal. Chem.* **35**, 55 (1972).

A3f. Appel, P. W., and Newman, J., *AIChE J.* **22**, 979 (1976).

A4. Arvia, A. J., and Carrozza, J. S. W., *Electrochim. Acta* **7**, 121 (1962).

A5. Arvía, A. J., Bazán, J. C., and Carrozza, J. S. W., *Electrochim. Acta* **11**, 881 (1966).
A6a. Arvía, A. J., Marchiano, S. L., and Podestá, J. J., *Electrochim. Acta* **12**, 259 (1967).
A6b. Ashworth, G. A., and Jansson, R. E. W., *Electrochim. Acta* **22**, 1295 (1977).
B1. Barnartt, S., *J. Electrochem. Soc.* **99**, 549 (1952).
B2. Bazán, J. C., and Arvía, A. J., *Electrochim. Acta* **9**, 17 (1964).
B3. Bazán, J. C., and Arvía, A. J., *Electrochim. Acta* **9**, 667 (1964).
B4a. Beach, K. W., and Muller, R. H., *J. Electrochem. Soc.* **122**, 59 (1975).
B4b. Berger, F. P., and Hau, K.-F. F.-L., *Int. J. Heat Mass Transfer* **20**, 1185 (1977).
B4c. Bird, R. B., Stewart, W. E., and Lightfoot, E. N., "Transport Phenomena." Wiley, New York, 1960.
B4d. Birkett, M. D., and Kuhn, A. T., *Electrochim. Acta* **22**, 1427 (1977).
B5. Blaedel, W. J., and Schieffer, G. W., *J. Electroanal. Chem.* **80**, 259 (1977).
B6. Bode, H., *Chem. Ing. Tech.* **43**, 293 (1971).
B7. Bode, H., *Waerme Stoffuebertrag.* **5**, 134 (1972).
B8. Boëffard, A. J. L. P. M., UCRL-16624; M.Sc. thesis, University of California, Berkeley, January 1966.
B9. Böhm, U., Ibl, N., and Frei, A. M., *Electrochim. Acta* **11**, 421 (1966).
B10. Brown, O. R., and Thirsk, H. R., *Electrochim. Acta* **10**, 383 (1965).
B11a. Brunner, E., *Z. Phys. Chem.* **47**, 56 (1904).
B11b. Bucur, R. V., Mecea, V., and Bartes, A., *Electrochim. Acta* **22**, 499 (1977).
C1a. Campergue, D., and Cognet, G., *C. R. Acal. Sci. Paris* **275**, 619 (1972).
C1b. Cano, J., and Böhm, U., *Chem. Eng. Sci.* **32**, 213 (1977).
C1c. Carbin, D. C., and Gabe, D. R., *Electrochim. Acta* **19**, 645 (1974).
C1d. Carbin, D. C., and Gabe, D. R., *Electrochim. Acta* **19**, 653 (1974).
C1e. Carbin, D. C., and Gabe, D. R., *J. Appl. Electrochem.* **5**, 129 (1975).
C1f. Carbin, D. C., and Gabe, D. R., *J. Appl. Electrochem.* **5**, 137 (1975).
C2. Chao, B. T., and Cheema, L. S., *Int. J. Heat Mass Transfer* **11**, 131 (1968).
C3. Chilton, C. H., and Colburn, A. P., *Ind. Eng. Chem.* **26**, 1183 (1934).
C4. Chin, D. T., *J. Electrochem. Soc.* **118**, 1764 (1971).
C5a. Chin, D. T., *AIChE J.* **20**, 245 (1974).
C5b. Chin, D. T., *J. Electrochem. Soc.* **122**, 643 (1975).
C6a. Chin, D. T., and Litt, M., *J. Fluid Mech.* **54**, 613 (1972).
C6b. Chin, D. T., Visnawathan, K., and Gutowski, R., *J. Electrochem. Soc.* **124**, 713 (1977).
C7. Cobo, O. A., Marchiano, S. L., and Arvía, A. J., *Electrochim. Acta* **17**, 503 (1972).
C8. Coeuret, F., and Vergnes, F., *C. R. Acad. Sci. Paris* **273**, 580 (1971).
C9. Coeuret, F., Le Goff, P., and Vergnes, F., *Proc. Int. Symp. Fluidization, Eindhoven, 1967,* **1**, 537 (1967).
C10. Coeuret, F., Vergnes, F., and Le Goff, P., *Houille Blanche,* 275 (1969).
C11a. Cognet, G., *J. Mec.* **10**, 65 (1971).
C11b. Cognet, G., and Daguenet, M., *C. R. Acad. Sci. Paris* **270** 142 (1970).
C12a. Cole, A. F. W., and Gordon, A. R., *J. Phys. Chem.* **40**, 733 (1936).
C12b. Colello, R. G., and Springer, G. S., *Int. J. Heat Mass Transfer* **9**, 1391 (1966).
C13. Cornet, I., and Kappesser, R., *Trans. Inst. Chem. Eng.* **47**, T194 (1969).
D1. Daguenet, M., *C. R. Acad. Sci. Paris* **262**, 706 (1966).
D2. Daguenet, M., *Int. J. Heat Mass Transfer* **11**, 1581 (1968).
D3a. Daguenet, M., and Aïmeur, F., *J. Chim. Phys.* **69**, 605 (1972).
D3b. Daguenet, M., and Robert, J., *C. R. Acad. Sci. Paris* **262**, 1125 (1966).
D3c. Daguenet, M., and Robert, J., *J. Chim. Phys.* **64**, 395 (1967).
D3d. Daguenet, M., and Robert, J., *J. Chim. Phys.* **65**, 1668 (1968).

D3e. Daguenet, M., and Schuhmann, D., *C. R. Acad. Sci. Paris* **260**, 2811 (1965).
D4. Daguenet, M., and Schuhmann, D., *C. R. Acad. Sci. Paris* **260**, 4731 (1965).
D5. Daguenet, M., Aouanouk, F., and Cognet, G., *C. R. Acad. Sci. Paris* **271**, 328 (1970).
D6a. Daguenet, M., Meklati, M., and Cognet, G., *C. R. Acad. Sci. Paris* **272**, 1355 (1971).
D6b. Daum, P. H., and Enke, C. G., *Anal. Chem.* **41**, 653 (1969).
D7. Davis, R. E., Horvath, G. L., and Tobias, C. W., *Electrochim. Acta* **12**, 287 (1967).
D8. Dawson, D. A., and Trass, O., *Int. J. Heat Mass Transfer* **15**, 1317 (1972).
D9. Deissler, R. G., N.A.C.A. Report 1210, p. 69 (1955).
D10. de Leeuw den Bouter, J. A., de Munnik, B., and Heertjes, P. M., *Chem. Eng. Sci.* **23**, 1185 (1968).
D11a. Deslouis, C., and Keddam, M., *Int. J. Heat Mass Transfer* **16**, 1763 (1973).
D11b. Deslouis, C., Epelboin, I., Tribollet, B., and Viet, L., *Electrochim. Acta* **20**, 909 (1975).
D11c. Deslouis, C., Epelboin, I., Gabrielli, C., and Tribollet, B., *J. Electroanal. Chem.* **82**, 251 (1977).
D11d. Despic, A. R., Mitrovic, M. V., Nikolic, B. Z., and Cijovic, S. D., *J. Electroanal. Chem.* **60**, 141 (1975).
D11e. Despic, A. R., Konjovic, M. N., and Mitrovic, M., *J. Appl. Electrochem.* **7**, 545 (1977).
D12. Devanathan, M. A., and Gurusvami, V., *Soviet Electrochem.* **7**, 1746 (1971); **7**, 1755 (1971).
D13. Devanathan, M. A., and Gurusvami, V., *Soviet Electrochem.* **9**, 896 (1973); **9**, 1205 (1973).
D14a. de Voodg, P., Ingenieur (M.Sc.) Report, Laboratory for Physical Technology, Technical University Delft, 1961.
D14b. Dikusar, G. K., and Dikusar, A. I., *Soviet Electrochem.* **10**, 774 (1974).
D15. Dimopoulos, H. G., and Hanratty, T. J., *J. Fluid Mech.* **33**, 303 (1968).
D16a. Dobry, R., and Finn, R. K., *Ind. Eng. Chem.* **48**, 1540 (1956).
D16b. Dreesen, E. W., Heindrich, A., and Vielstich, W., *Ber. Bunsenges. Phys. Chem.* **79**, 12 (1975).
E1. Eisenberg, M., Tobias, C. W., and Wilke, C. R., *J. Electrochem. Soc.* **101**, 306 (1954).
E2. Eisenberg, M., Tobias, C. W., and Wilke, C. R., *J. Electrochem. Soc.* **102**, 415 (1955).
E3. Eisenberg, M., Tobias, C. W., and Wilke, C. R., *J. Electrochem. Soc.* **103**, 413 (1956).
E4. Ellison, B. T., and Cornet, I., *J. Electrochem. Soc.* **118**, 68 (1971).
E5. Emelyanenko, G. A., and Baibarova, E. Ya., *Ukr. Khim. Zh.* **25**, 727 (1959).
E6. Emery, C. A., and Hintermann, H. E., *Electrochim. Acta* **13**, 127 (1968).
E7. Eucken, A., *Z. Elektrochem.* **38**, 341 (1932).
F1. Fahidy, T. Z., *Can. J. Chem. Eng.* **51**, 521 (1973).
F2. Fenech, E. J., UCRL-9079; Ph.D. thesis, University of California, Berkeley, April 1960.
F3. Fenech, E. J., and Tobias, C. W., *Electrochim. Acta* **2**, 311 (1960).
F4. Fouad, M. G., and Ahmed, A. M., *Electrochim. Acta* **14**, 651 (1969).
F5. Fouad, M. G., and Gouda, T., *Electrochim. Acta* **9**, 1071 (1964).
F6. Fouad, M. G., and Ibl, N., *Electrochim. Acta* **3**, 233 (1960).
F7a. Fouad, M. G., and Sedahmed, G. H., *Electrochim. Acta* **19**, 861 (1974).
F7b. Fouad, M. G., and Sedahmed, G. H., *Electrochim. Acta* **20**, 615 (1975).
F7c. Fouad, M. G., and Zatout, A. A., *Electrochim. Acta* **14**, 909 (1969).
F8. Fouad, M. G., Sedahmed, G. H., and Elabd, H. A., *Electrochim. Acta* **18**, 279 (1973).
F9. Friend, W. L., and Metzner, A. B., *AIChE J.* **4**, 393 (1958).
G1a. Gabe, D. R., *J. Appl. Electrochem.* **4**, 91 (1947).
G1b. Gabe, D. R., and Robinson, D. J., *Electrochem. Acta* **17**, 1121 (1972).
G1c. Gabe, D. R., and Robinson, D. J., *Electrochim. Acta* **17**, 1129 (1972).

G1d. Galloway, T. R., and Seid, D. M., *Lett. Heat Mass Transfer* **2**, 247 (1975).
G2. Gibert, H., and Angelino, H., *Chem. Eng. Sci.* **28**, 855 (1973).
G3. Gibert, H., and Angelino, H., *Int. J. Heat Mass Transfer* **17**, 625 (1974).
G4a. Gibert, H., Couderc, J. P., and Angelino, H., *Chem. Eng. Sci.* **27**, 45 (1972).
G4b. Gileadi, E., Kirowa-Eisner, E., and Penciner, J., "Interfacial Electrochemistry." Addison-Wesley, Reading, 1975.
G5. Gordon, S. L., Newman, J. S., and Tobias, C. W., *Ber. Bunsenges. Phys. Chem.* **70**, 414 (1966).
G6. Goren, S. L., and Mani, R. V. S., *AIChE J.* **14**, 57 (1968).
G7. Grassmann, P., Ibl, N., and Trueb, J., *Chem. Ing. Tech.* **33**, 529 (1961).
G8. Gubbins, K. E., and Walker, R. D., Jr., *J. Electrochem. Soc.* **112**, 469 (1965).
H1. Hamann, C. H., Schöner, H., and Vielstich, W., *Ber. Bunsenges. Phys. Chem.* **77**, 484 (1973).
H2. Hanratty, T. J., *Phys. Fluids Suppl.* S126 (1967).
H3. Hickman, R. G., Ph.D. thesis, Univ. of California, Berkeley, December 1963.
H4. Hicks, R. E., and Mandersloot, W. G. B., *Chem. Eng. Sci.* **23**, 1201 (1968).
H5. Holmes, J. T., Wilke, C. R., and Olander, D. R., *J. Phys. Chem.* **67**, 1469 (1963).
H6a. Homsy, R. V., and Newman, J., *AIChE J.* **19**, 929 (1973).
H6b. Homsy, R. V., and Newman, J., *J. Electrochem. Soc.* **121**, 1448 (1974).
H6c. Howland, B., Springer, G. S., and Hill, M. G., *J. Fluid Mech.* **24**, 697 (1966).
H7. Hsueh, L., UCRL-18597; Ph.D. thesis, University of California, Berkeley, December 1968.
H8. Hsueh, L., and Newman, J., *Electrochim. Acta* **12**, 417 (1967).
H9. Hsueh, L., and Newman, J., *Electrochim. Acta* **12**, 429 (1967).
H10. Hsueh, L., and Newman, J., *Ind. Eng. Chem. Fund.* **10**, 615 (1971).
H11a. Hubbard, D. W., and Lightfoot, E. N., *Ind. Eng. Chem. Fundam.* **5**, 370 (1966).
H11b. Hutin, D., and Coeuret, F., *J. Appl. Electrochem.* **7**, 463 (1977).
I1. Ibl, N., Proc. 8th Reunion C.I.T.C.E., Madrid, 1956, 174 (1956).
I2. Ibl, N., *Adv. Electrochem. Electrochem. Eng.* **2**, 49 (1962).
I3. Ibl, N., *Chem. Ing. Tech.* **35**, 353 (1963).
I4. Ibl, N., *Proc. Int. Conf. Protection against Corrosion by Metal Finishing (Surface 66)*, Basel, 1966, 49. (1966).
I5a. Ibl, N., and Braun, V., *Chimia* **21**, 395 (1967).
I5b. Ibl, N., and Müller, R., *Z. Elektrochem.* **59**, 671 (1955).
I6. Ibl, N., and Müller, R. H., *J. Electrochem. Soc.* **105**, 346 and 536 (1959).
I7. Ibl, N., and Schadegg, K., *J. Electrochem. Soc.* **114**, 54 (1967).
I8. Ibl, N., Adam, E., Venczel, J., and Schalch, E., *Chem. Ing. Tech.* **43**, 202 (1971).
I9a. Ibl, N., Javet, P., and Stahel, F., *Electrochim. Acta* **17**, 733 (1972).
I9b. Ibl, N., Bindschedler, D., and Schenk, H. J., *Oberfläche-Surface* **16**, 278 (1975).
I10. Iribarne, A., Gosman, A. D., and Spalding, D. B., *Int. J. Heat Mass Transfer* **10**, 1661 (1967).
I11a. Iribarne, A. Palade de, Marchiano, S. L., and Arvía, A. J., *Electrochim. Acta* **15**, 1827 (1970).
I11b. Ismail, M. I., Al-Taweel, A. M., and El-Abd, M. Z., *J. Appl. Electrochem.* **4**, 347 (1974).
J1. Jagannadharaju, G. J. V., and Rao, C. V., *Indian J. Technol.* **3**, 201 (1965).
J2. Jahn, D., *Proc. Congr. Interfinish, 8th, Basel, 1972*, p. 146.
J3. Janssen, L. J. J., and Hoogland, J. G., *Electrochim. Acta* **15**, 1013 (1970).
J4a. Janssen, L. J. J., and Hoogland, J. G., *Electrochim. Acta* **18**, 543 (1973).
J4b. Jansson, R. E. W., and Ashworth, G. A., *Electrochim. Acta* **22**, 1301 (1977).

J5a. Javet, P., Ibl, N., and Hintermann, H. E., *Electrochim. Acta* **12**, 781 (1967).
J5b. Jennings, D., Kuhn, A. T., Stepanek, J. B., and Whitehead, R., *Electrochim. Acta* **20**, 903 (1975).
J6. Jolls, K. R., and Hanratty, T. J., *AIChE J.* **15**, 199 (1969).
J7. Jottrand, R., and Grunchard, F., *Proc. Symp. Interaction Fluids Particles, London, 1962*, 211 (1962).
K1a. Kadija, I. V., and Nakic, V. M., *J. Electroanal. Chem.* **34**, 15 (1972).
K1b. Kadija, I. V., and Nakic, V. M., *J. Electroanal. Chem.* **35**, 177 (1972).
K1c. Kappesser, R., Cornet, I., and Greif, R., *J. Electrochem. Soc.* **118**, 1957 (1971).
K2a. Karabelas, A. J., and Hanratty, T. J., *J. Fluid Mech.* **34**, 159 (1968).
K2b. Karabelas, A. J., Wegner, T. H., and Hanratty, T. J., *Chem. Eng. Sci.* **26**, 1521 (1971).
K2c. Kimla, A., and Strafelda, F., *Coll. Czech. Chem. Comm.* **32**, 56 (1967).
K3. King, C. V., and Cathcart, W. H., *J. Am. Chem. Soc.* **59**, 63 (1937).
K4a. King, C. V., and Schack, M., *J. Am. Chem. Soc.* **57**, 1212 (1935).
K4b. Kirowa-Eisner, E., and Gileadi, E., *J. Electrochem. Soc.* **123**, 22 (1976).
K5a. Kishinevskii, M. K., Kornienko, T. S., and Guber, Yu. E., *Soviet Electrochem.* **8**, 617 (1972).
K5b. Kishinevskii, M. K., Kornienko, T. S., and Loginov, A. V., *Soviet Electrochem.* **13**, 20 (1977).
K6. Krishna, M. S., D.Sc. thesis, Andhra University, Waltair, India, 1962.
K7. Krishna, M. S., *Indian J. Technol.* **10**, 163 (1972).
K8. Krishna, M. S., and Jagannadharaju, G. J. V., *Indian J. Technol.* **3**, 263 (1965).
K9. Krishna, M. S., and Jagannadharaju, G. J. V., *Indian J. Technol.* **4**, 97 (1966).
K10. Krishna, M. S., and Jagannadharaju, G. J. V., *Indian J. Technol.* **5**, 171 (1967).
K11. Krishna, M. S., Jagannadharaju, G. J. V., and Rao, C. V., *Indian J. Technol.* **4**, 8 (1966).
K12. Krishna, M. S., Jagannadharaju, G. J. V., and Rao, C. V., *Periodica Polytech.* **11**, 95 (1967).
K13. Krishna, M. S., Ramaraju, C. V., Jagannadharaju, G. J. V., and Rao, C. V., *Indian J. Technol.* **6**, 50 (1968).
K14. Kurvyakova, L. M., and Pomosov, A. V., *Soviet Electrochem.* **4**, 156 (1968).
L1a. Landau, U., LBL-2702; Ph.D. thesis, University of California, Berkeley, January 1976.
L1b. Landau, U., and Tobias, C. W., "Mass Transport and Current Distribution in Channel Type Electrolyzers in the Laminar and Turbulent Flow Regimes," Ext. Abstr., No. 266, Electrochemical Society Meeting, Washington D.C., May 1976, 663.
L1c. Le Goff, P., Vergnes, F., Coeuret, F., and Bordet, J., *Ind. Eng. Chem.* **61**, 8 (1969).
L1d. Leitz, F. B., and Marincic, L., *J. Appl. Electrochem.* **7**, 473 (1977).
L2. LeLan, A., and Angelino, H., *Chem. Eng. Sci.* **29**, 907 (1974).
L3. LeLan, A., and Angelino, H., *Chem. Eng. Sci.* **29**, 1557 (1974).
L4. LeLan, A., Gibert, H., and Angelino, H., *Chem. Eng. Sci.* **27**, 1979 (1972).
L5a. Lengas, T. E., Ph.D. thesis, University of Edinburgh, 1969.
L5b. Lévêque, J., *Ann. Mines* **13**, 201, 305, 381 (1928).
L6. Levich, B., *Acta Physicochim. URSS* **17**, 257 (1942); **19**, 117 (1944).
L7. Levich, B., *Discuss Faraday Soc.* **1**, 37 (1947).
L8. Levich, V. G., "Physicochemical Hydrodynamics." Prentice-Hall, Englewood Cliffs, New Jersey, 1962.
L9. Lin, C. S., Denton, E. B., Gaskill, H. S., and Putnam, G. L., *Ind. Eng. Chem.* **43**, 2136 (1951).
L10a. Lin, C. S., Moulton, R. W., and Putnam, G. L., *Ind. Eng. Chem.* **45**, 636 (1953).
L10b. Lloyd, J. R., and Moran, W. R., *J. Heat Transfer* **96**, 443 (1975).

M1. Mahato, B. K., and Shemilt, L. W., *Chem. Eng. Sci.* **23**, 183 (1968).
M2a. Makrides, A. C., and Hackerman, N., *J. Electrochem. Soc.* **105**, 156 (1958).
M2b. Mandelbaum, J. A., and Böhm, U., *Chem. Eng. Sci.* **28**, 569 (1973).
M3. Marchiano, S. L., and Arvia, A. J., *Electrochim. Acta* **14**, 741 (1969).
M4a. Matsuda, H., *J. Electroanal. Chem.* **15**, 109, 325 (1967); **21**, 433 (1969); **22**, 413 (1969); **25**, 461 (1970); **35**, 77 (1972); **44**, 199 (1973); **52**, 421 (1974).
M4b. Matsuda, H., *J. Electroanal. Chem.* **38**, 159 (1972).
M4c. Matsuda, H., and Yamada, J., *J. Electroanal. Chem.* **30**, 261 (1971).
M4d. Matsuda, H., and Yamada, J., *J. Electroanal. Chem.* **30**, 271 (1971).
M4e. Mattson, E., and Bockris, J. O'M., *Trans. Faraday Soc.* **55**, 586 (1959).
M5. Meklati, M., and Daguenet, M., *C. R. Acad. Sci. Paris* **272**, 2027 (1971).
M6a. Meklati, M., Daguenet, M., and Cognet, G., *C. R. Acad. Sci. Paris* **274**, 1373 (1972).
M6b. Meyer, R. E., Banta, C. M., Lantz, P. M., and Posey, F. A., *J. Electroana. Chem.* **30**, 345, 359 (1971).
M7. Mitchell, J. E., and Hanratty, T. J., *J. Fluid Mech.* **26**, 199 (1966).
M8. Mixon, F. O., Chon, W. Y., and Beatty, K. O., *Chem. Eng. Prog. Symp.* **56, 30**, 75 (1960).
M9. Mizushina, T., *Adv. Heat Transfer* **7**, 87 (1971).
M10. Mizushina, T., Ito, R., Hiraoka, S., Ibusuki, A., and Sakaguchi, I., *J. Chem. Eng. Jpn.* **1**, 89 (1969).
M11. Mizushina, T., Ito, R., Ueda, H., Tsubata, S., and Hayashi, H., *J. Chem. Eng. Jpn.* **3**, 34 (1970).
M12. Mizushina, T., Ogino, F., Oka, Y., and Fukuda, H., *Int. J. Heat Mass Transfer* **14**, 1705 (1971).
M13a. Mizushina, T., Ueda, H., Okada, N., and Uegaki, M., *J. Chem. Eng. Jpn.* **4**, 17 (1971).
M13b. Mohanta, S., and Fahidy, T. Z., *Electrochim. Acta* **19**, 771, 835 (1974).
M13c. Mohanta, S., and Fahidy, T. Z., *Electrochim. Acta* **21**, 149 (1976).
M13d. Mohanta, S., and Fahidy, T. Z., *Electrochim. Acta* **21**, 143 (1976).
M13e. Mohanta, S., and Fahidy, T. Z., *J. Appl. Electrochem.* **7**, 235 (1977).
M13f. Mohr, Jr., C. M., and Newman, J., *Electrochim. Acta* **18**, 761 (1973).
M13g. Mohr, Jr., C. M., and Newman, J., *J. Electrochem. Soc.* **122**, 928 (1975).
M13h. Mohr, Jr., C. M., and Newman, J., *J. Electrochem. Soc.* **123**, 1687 (1976).
M13i. Mollet, L., Dumargue, P., Daguenet, M., and Bodiot, D., *Electrochim. Acta* **19**, 841 (1974).
M14. Muller, R. H., *Adv. Electrochem. Electrochem. Eng.* **9**, 281 (1973).
N1a. Nadebaum, P. R., and Fahidy, T. A., *Can. J. Chem. Eng.* **53**, 259 (1975).
N1b. Naybour, R. D., *J. Electrochem. Soc.* **116**, 520 (1969).
N2. Nernst, W., *Z. Phys. Chem.* **47**, 52 (1904).
N3. Newman, J., *Ind. Eng. Chem. Fundam.* **5**, 525 (1966).
N4. Newman, J., *Adv. Electrochem. Electrochem. Eng.* **5**, 87 (1967).
N5. Newman, J., *Int. J. Heat Mass Transfer* **10**, 983 (1967).
N6. Newman, J., *Ind. Eng. Chem.* **60**, 12 (1968).
N7. Newman, J., *Electroanal. Chem.* (A. J. Bard, Ed.) **6**, 187 (1973).
N8a. Newman, J., " Electrochemical Systems," Prentice-Hall, Englewood Cliffs, New Jersey, 1973.
N8b. Newman, J., and Hsueh, L., *Ind. Eng. Chem. Fundam.* **9**, 677 (1970).
N8c. Newman, J., *Electrochim. Acta* **22**, 903 (1977).
N8d. Newman, J. S., and Tiedemann, W., *Adv. Electrochem. Electrochem. Eng.* **11**, 353 (1978).
N9a. Newson, J. D., and Riddiford, A. C., *J. Electrochem. Soc.* **108**, 695 (1961).
N9b. Nisancioglu, K., and Newman, J., *J. Electrochem. Soc.* **120**, 1339, 1356 (1973).

N9c. Nisancioglu, K., and Newman, J., *J. Electrochem. Soc.* **121**, 242 (1974).
N10. Noordsij, P., and Rotte, J. W., *Chem. Eng. Sci.* **22**, 1475 (1967).
O1. O'Brien, R. N., and Mukherjee, L. M., *J. Electrochem. Soc.* **111**, 1358 (1964); **112**, 1253 (1965).
O2. O'Brien, R. N., and Rosenfield, C., *J. Phys. Chem.* **67**, 643 (1963).
O3. Okada, S., Yoshizawa, S., Hine, F., and Asada, K., *J. Electrochem. Soc. Jpn.* **27**, E69 (1959).
P1. Parrish, W. R., and Newman, J., *J. Electrochem. Soc.* **117**, 43 (1970).
P2a. Paterson, J. A., Greif, R., and Cornet I., *Int. J. Heat Mass Transfer* **16**, 1017 (1973).
P2b. Patrick, M. A., and Wragg, A. A., *Int. J. Heat Mass Transfer* **18**, 1397 (1975).
P3a. Pickett, D. J., and Ong, K. L., *Electrochim. Acta* **19**, 875 (1974).
P3b. Pickett, D. J., and Stanmore, B. R., *J. Appl. Electrochem.* **2**, 151 (1972).
P3c. Pickett, D. J., and Stanmore, B. R., *J. Appl. Electrochem.* **5**, 95 (1975).
P3d. Pini, G. C., and De Anna, P. L., *Electrochim. Acta* **22**, 1423 (1977).
P4a. Piontelli, R., Mazza, B., and Pedeferri, P., *J. Electrochem. Soc.* **114**, 1267 (1967).
P4b. Pleskov, Yu. V., and Filinovskii, V. Yu., "The Rotating Disc Electrode." Consultants Bureau, New York, 1976.
P5. Podestá, J. J., Paus, G. F., and Arvia, A. J., *Electrochim. Acta* **19**, 583 (1974).
P6. Py, B., *Int. J. Heat Mass Transfer* **16**, 129 (1973).
R1. Raju, C. V. R., Jagannadharaju, G. J. V., and Rao, C. V. Venkata, *Indian J. Technol.* **5**, 171 (1967).
R2. Raju, C. V. R., Sastry, A. R., and Jagannadharaju, G. J. V., *Indian J. Technol.* **7**, 35 (1969).
R3. Ranz, W. E., *AIChE J.* **4**, 338 (1958).
R4. Rao, K. S., and Sharma, R. S., *Indian J. Technol.* **10**, 166 (1972).
R5. Rao, K. S., Jagannadharaju, G. J. V., and Rao, C. V., *Indian J. Technol.* **3**, 38 (1965).
R6. Rao, K. S., Jagannadharaju, G. J. V., and Rao, C. V., *Indian J. Technol.* **6**, 46 (1968).
R7. Rao, K. S., Raju, C. V. R., and Jagannadharaju, G. J. V., *Indian J. Technol.* **6**, 129 (1968).
R8. Ravoo, E., *Ingenieur*, **83**, 11 (1971).
R9. Ravoo, E., Diss., Technical University Twente, Netherlands, July 1971, 101–119.
R10. Ravoo, E., Rotte, J. W., and Sevenstern, F. W., *Chem. Eng. Sci.* **25**, 1637 (1970).
R11. Regner, A., and Rousar, I., *Coll. Trav. Chim. Tcnecoslov.* **25**, 1132 (1960).
R12. Reiss, L. P., and Hanratty, T. J., *AIChE J.* **8**, 245 (1962).
R13. Reiss, L. P., and Hanratty, T. J., *AIChE J.* **9**, 154 (1963).
R14. Riley, N., *J. Fluid Mech.* **17**, 97 (1964).
R15. Robinson, D. J., *Electrochim. Acta* **17**, 791 (1972).
R16. Robinson, D. J., and Gabe, D. R., *Trans. Inst. Metal Finishing* **48**, 35 (1970).
R17. Ross, T. K., and Badhwer, R. K., *Corros. Sci.* **5**, 29 (1965).
R18. Ross, T. K., and Wragg, A. A., *Electrochim. Acta* **10**, 1093 (1965).
R19. Ross, T. K., Campbell, D. U., and Wragg, A. A., *Trans. Inst. Chem. Eng.* **45**, T401 (1967).
R20. Rotte, J. W., Tummers, G. L. J., and Dekker, J. L., *Chem. Eng. Sci.* **24**, 1009 (1969).
R21a. Rousar, I., Hostomsky, J., Cezner, V., and Stverak, B., *J. Electrochem. Soc.* **118**, 881 (1971).
R21b. Rousar, I., Kacin, J., Smirous, F., and Cezner, V., *Electrochim. Acta* **20**, 295 (1975).
S1. Sakiadis, B. C., *AIChE J.* **7**, 26 (1961); **7**, 221 (1961); **7**, 467 (1961).
S2. Satyanarayana, A., Krishna, M. S., Jagannadharaju, G. J. V., and Rao, C. V., *Indian J. Technol.* **6**, 42 (1968).
S3. Schalch, E., and Ibl, N., *Electrochim. Acta* **20**, 435 (1975).

S4a. Scheller, F., Müller, S., Landsberg, R., and Spitzer, H. J., *J. Electroanal. Chem.* **19**, 187 (1968).
S4b. Scheller, F., Landsberg, R., and Müller, S., *J. Electroanal. Chem.* **20**, 375 (1969).
S5. Schmidt, H. G., Diss. ETH Zürich No. 3095 (1961).
S6. Schütz, G., *Int. J. Heat Mass Transfer* **6**, 873 (1963).
S7. Schütz, G., *Int. J. Heat Mass Transfer* **7**, 1077 (1964).
S8. Selman, J. R., UCRL-20557; Ph.D. thesis, University of California, Berkeley.
S9a. Selman, J. R., and Newman J., *J. Electrochem. Soc.* **118**, 1070 (1971).
S9b. Selman, J. R., and Newman, J., UCRL-20322, January (1971).
S10. Selman, J. R., and Tobias, C. W., *J. Electroanal. Chem.* **65**, 67 (1975).
S11a. Shaw, D. A., and Hanratty, T. J., *AIChE J.* **23**, 28 (1977).
S11b. Shemilt, L. W., and Sedahmed, G. H., *J. Appl. Electrochem.* **6**, 471 (1976).
S11c. Sih, P. H., UCRL-20509; Ph.D. thesis, University of California, Berkeley, April 1971.
S12a. Sioda, R. E., *Electrochim. Acta* **13**, 375 (1968).
S12b. Sioda, R. E., *Electrochim. Acta* **16**, 1569 (1971).
S12c. Sioda, R. E., *J. Appl. Electrochem.* **5**, 221 (1975).
S12d. Sioda, R. E., *Electrochim. Acta* **22**, 439 (1977).
S13. Sirkar, K. K., and Hanratty, T. J., *Ind. Eng. Chem. Fundam.* **8**, 189 (1969).
S14. Siver, Yu. G., and Kabanov, B. I., *J. Phys. Chem. USSR* **22**, 53 (1948).
S15a. Smith, A. F. J., and Wragg, A. A., *J. Appl. Electrochem.* **4**, 219 (1974).
S15b. Smith, R. N., and Greif, R., *Int. J. Heat Mass Transfer* **18**, 1249 (1975).
S16. Smyrl, W. H., and Newman, J., *J. Electrochem. Soc.* **119**, 212 (1972).
S17a. Sohr, R., and Müller, L., *Electrochim. Acta* **20**, 451 (1975).
S17b. Sohr, R., Müller, L., and Landsberg, R., *J. Electroanal. Chem.* **50**, 55 (1974).
S17c. Soliman, M., and Chambré, P. L., *Int. J. Heat Mass Transfer* **11**, 1311 (1968).
S18. Son, J. S., and Hanratty, T. J., *AIChE J.* **13**, 689 (1967).
S19. Son, J. S., and Hanratty, T. J., *J. Fluid Mech.* **35**, 353 (1969).
S20a. Spalding, D. B., *Int. J. Heat Mass Transfer* **7**, 743 (1964).
S20b. Spiro, M., *Electrochim. Acta* **9**, 1531 (1964).
S20c. Storck, A., and Coeuret, F., *Electrochim. Acta* **22**, 1155 (1977).
S21. Subramaniyan, V., Krishna, M. S., and Adivarahan, P., *Indian J. Technol.* **4**, 353 (1966).
S22. Sutey, A. M., and Knudsen, J. G., *Ind. Eng. Chem. Fundam.* **6**, 132 (1967).
T1a. Tanaka, N., and Tamamushi, R., *Electrochim. Acta* **9**, 963 (1964).
T1b. Taylor, J. L., and Hanratty, T. J., *Electrochim. Acta* **19**, 529 (1974).
T2. Tobias, C. W., and Hickman, R. G., *Z. Phys. Chem. (Leipzig)* **229**, 145 (1965).
T3. Tobias, C. W., Eisenberg, M., and Wilke, C. R., *J. Electrochem. Soc.* **99**, 359C (1952).
T4a. Tokuda, K., Bruckenstein, S., and Miller, B., *J. Electrochem. Soc.* **122**, 1316 (1975).
T4b. Tvarusko, A., *J. Electrochem. Soc.* **120**, 87 (1973).
T4c. Tvarusko, A., *J. Electrochem. Soc.* **123**, 49 (1976).
V1. Van Shaw, P., and Hanratty, T. J., *AIChE J.* **10**, 475 (1964).
V2. Van Shaw, P., Reiss, L. P., and Hanratty, T. J., *AIChE J.* **9**, 362 (1963).
V3. Venczel, J., Diss. ETH, 3019, Zurich (1961).
V4. Vetter, K. J., "Electrochemical Kinetics, Theoretical and Experimental Aspects." Academic Press, New York, 1967.
V5. Vieth, W. R., Porter, J. H., and Sherwood, T. K., *Ind. Eng. Chem. Fundam.* **2**, 1 (1963).
V6. Vogtländer, P. H., and Bakker, C. A. P., *Chem. Eng. Sci.* **18**, 583 (1963).
W1. Wagner, C., *Trans. Am. Electrochem. Soc.* **95**, 61 (1949).
W2. Weder, E., *Chem. Ing. Tech.* **39**, 914 (1967).

W3a. Weder, E., *Waerme Stoffuebertrag.* **1**, 10 (1968).
W3b. White, R., and Newman, J., *J. Electroanal. Chem.* **82**, 173 (1977).
W4. Wilke, C. R., Eisenberg, M., and Tobias, C. W., *J. Electrochem. Soc.* **100**, 513 (1953).
W5. Wilke, C. R., Tobias, C. W., and Eisenberg, M., *Chem. Eng. Prog.* **49**, 663 (1953).
W6. Wragg, A. A., *Electrochim. Acta* **13**, 2159 (1968).
W7. Wragg, A. A., *Int. J. Heat Mass Transfer* **11**, 979 (1968).
W8a. Wragg, A. A., *Electrochim. Acta* **16**, 373 (1971).
W8b. Wragg, A. A., *Electrochim. Acta* **20**, 917 (1975).
W9. Wragg, A. A., and Loomba, R. P., *Int. J. Heat Mass Transfer* **13**, 439 (1970).
W10. Wragg, A. A., and Nasiruddin, A. K., *Electrochim. Acta* **18**, 619 (1973).
W11. Wragg, A. A., and Patrick, M. A., *Electrochim. Acta* **19**, 929 (1974).
W12. Wragg, A. A., and Ross, T. K., *Electrochim. Acta* **12**, 1421 (1967).
W13a. Wragg, A. A., and Ross, T. K., *Electrochim. Acta* **13**, 2192 (1968).
W13b. Wragg, A. A., Serafinidis, P., and Einarsson, A., *Int. J. Heat Mass Transfer* **11**, 1287 (1968).
W13c. Wragg, A. A., Einarsson, A., and Dawson, J. L., *Electrochim. Acta* **19**, 503 (1974).
W14. Wranglen, G., and Nilsson, O., *Electrochim. Acta* **7**, 121 (1962).

AUTHOR INDEX

Numbers in parentheses are reference numbers and indicate that an author's work is referred to although his name is not cited in the text. Numbers in italics show the page on which the complete reference is listed.

A

Acosta, R. E., 256, 282(A1a), 298(A1a), *310*
Adam, E., 275(I8), 276(I8), 303(I8), *313*
Adamiak, Z. P., 102(N3), 104, *122*
Addams, J. N., 43, 44, *51*
Adivarahan, P., 283(S21), 285,(S21), *317*
Adorjan, L. A., *123*
Agar, J. N., 215(A2), 217, 218, *310*
Ahmed, A. M., 266, 291, *312*
Aïmeur, F., 263, *310, 311*
Albery, W. J., 261(A3b), *310*
Aldis, D., 183(F2), *206*
Alexander, S. M., 140, *205*
Allen, J. W., 150, 155(K9), *207*
Al-Sheikh, J. N., *51*
Al-Taweel, A. M., 299(A3c, A3d, I11b), *310, 313*
Alves, G. E., 2, *51*
Anderson, R. J., 5, 8, 14, 31, *51*
Angelino, H., 263, 273, 275, 288, 300(G3), 301(L3), 302(L2, L3, L4), *313, 314*
Angell, D. H., 227(A3e), *310*
Aoki, Y., 171(A3, A4, A5), 183(A3, A4, A5), *205*
Aouanouk, F., 285(D5), 294(D5), 295(D5), *312*
Appel, P. W., 277, 307(A3f), *310*
Appleby, J. E., 104(B4), *120*
Arora, S. C. D., 85(K7), 102(K7), 103(K7), *121*
Arruda, P. J., 7, *51*

Arvia, A. J., 219(A6a), 234, 235, 236, 251, 256, 260(B2, B3), 261(I11a), 265, 272, 280(C7), 286(B2, B3), 287(I11a), 289(M3), 295(A4), 299(P5), *310, 311, 313, 315, 316*
Asada, K., 274(O3), 280(O3), 283(O3), *316*
Ashton, M. D., 64, *120*
Ashworth, G. A., 259, 284(J4b), 295(A6b), *311, 314*
Atherton, R. W., 31, *51*
Ayers, P., *123*
Aziz, K., 5, 8, *51*

B

Babayev, D. A., 171, *205*
Badhwer, R. K., 248(R17), *316*
Baibarova, E. Ya., 250(E5), *312*
Baker, O., 5, 6, 50, *51*
Bakker, C. A. P., 263, 276(V6), 288(V6), *318*
Balasubramanian, M., 63(T5), *122*
Ball, D. F., 58(B2), 60(B1), 61(B1), 102(B2), 104(B2), *120*
Bankoff, S. G., 38, 42, *51*
Banta, C. M., 256(M6b), *315*
Barlow, J. F., 155, *205*
Barnartt, S., 223(B1), *311*
Bartes, A., 280(B11b), *311*
Batey, E. H., 190(W6), *208*
Bauer, W. J., 137, *205*
Bazán, J. C., 234(A5), 235(A5), 251(A5), 256(A5), 260(B2, B3), 286(B2, B3), *311*

Bazilevich, S. V., 60, *120*
Beach, K. W., 216(B4a), *311*
Beale, C. V., 104(B4), *120*
Beatty, K. O., 275(M8), *315*
Beightler, C. S., 172(W5), *208*
Bellman, R., 187(B4), *205*
Bending, M. J., 137, 143, 150, 155, 156(B5), 157, 168, *205*
Bentwich, M., 11, *51*
Berge, C., 135, *205*
Berger, F. P., 296(B4b), 297(B4b), *311*
Berquin, Y. F., 58(B5), *120*
Beveridge, G. S. G., 77(H5), *121*
Bhrany, U. N., 59(B6), 60, *120*
Bickel, T. C., 140(B7), 171, 172, 175(B7), 181, 183, *205*
Bijwaard, G., 42(Z1), *53*
Bindschedler, D., 249(I9b), *313*
Bird, D. W., 140(A1), *205*
Bird, R. B., *31*, *51*, 217(B4c), 286(B4c), *311*
Birkett, M. D., 260, *311*
Blaedel, W. J., *311*
Blunt, J. C., 192(H3), *207*
Bockris, J. O'M., 225, 226, *311*, *315*
Bode, H., 306(B6, B7), *311*
Bodiot, D., 285(M13i), *315*
Boëffard, A. J. L. P. M., 241, *311*
Böhm, U., 231(B9), 240(B9), 267, 291(B9), 307(M2b), 308(C1b), *311*, *315*
Bombled, J. P., 60(P1), *122*
Bonansinga, P., 171, 199, *205*
Bordet, J., 278(L1c), 279(L1c), *314*
Bowden, F. P., 215(A2), 217, *310*
Boyer, H. M., 198, *205*
Brameller, A., 144(B10), 150, 162, 171, 172, 184(B10), 185(B10), *206*
Branscome, D., 154(G7), *207*
Brauer, H., 8, *51*
Braun, V., 231(I5a), *313*
Brayton, R. K., 168, *206*
Brociner, R. E., 62, *120*
Brodkey, R. S., *51*
Bromley, L. A., 32, *52*
Brown, K. M., 152, 158, *205*
Brown, O. R., 225, *311*
Brown, R. A. S., 5(B6), *51*
Browning, J. E., 56(B8), *120*
Broyden, C. G., 152, 157, 158, *206*
Bruckenstein, S., 274(T4a), *317*
Brunemann, H., 45, *51*

Brunner, E., 217, *311*
Bryson, A. E., 174, *206*
Bucur, R. V., 280(B11b), *311*
Burghardt, O., 57(R2), 63(R2), 102(R2), 104(R2), *122*
Butensky, M., 63(B9), 106, 107, 108, *120*
Butterfield, P., 104(B4), *120*
Byers, C. H., 11, *51*

C

Cahn, D. S., 57(C1), 104, *120*, *122*
Cahn, J. W., 68(H6), *121*
Campbell, D. U., 287(R19), *316*
Campergue, D., 282(C1a), *311*
Cano, J., 308(C1b), *311*
Capes, C. E., 58(S11), 72(C3, C4), 73, 79, 80(C6), 81(C5), 85, 86, 87, 89, 90(C6), 101, 106, 115(C7, S11), 116, *120*, *121*, *122*, *123*
Carbin, D. C., 260, 278, 279, 286(C1c), 308(C1c, C1e), *311*
Carnahan, B., 143, 150, 151, 200, *206*
Carnahan, N. F., 137(C3), 143, 150, 152(C3), 153, 160, 161, 164, *206*
Carrozza, J. S. W., 234(A5), 235(A5), 251(A5), 256(A5), 272, 295(A4), *310*, *311*
Catanzaro, G. V., 137(N1), 159, 175(N1), 192, *208*
Cathcart, W. H., 217(K3), 272(K3), *314*
Cavanagh, P. E., 61(C8), *120*
Cembrowicz, R. G., 135(C4), 137(C4), 171, 172, 175(C4), 184, *206*
Cezner, V., 256(R21a), 257(R21a), 282(R21a), 304(R21b), *316*
Chaddock, J. B., 45, *51*
Chambré, P. L., 241(S17c), 242(S17c), 244, *317*
Chancellor, V. E., 144(B10), 150, 162(B10), 171(B10), 172(B10), 184(B10), 185(B10), *206*
Chandrasekhar, S., 91(C9), *120*
Chang, A., 166, *206*
Chao, B. T., 241(C2), *311*
Chari, K. S., 58(C10), 61(C10), 62(C10), *120*
Charles, C. O. A., 143, 150, 155, 156, 157, *209*
Charles, M. E., 5, 6, 8, 10, *51*
Cheema, L. S., 241(C2), *311*
Cheesman, A. P., 171, 175(C6), *206*
Chen, T. C., 183, *206*

Cheng, C. T., 26(W7), 53
Cheng, D. C. H., 64(C11), 65, 120
Cheng, W. B., 145(C7), 146, 161, 169, 171, 172, 175(C7), 176, 178, 179, 180, 205, 206
Chilton, C. H., 139, 208, 270, 271, 297, 298, 311
Chin, D. T., 259, 272, 280(C4), 283(C5b, C6b), 293(C4), 311
Chon, W. Y., 275(M8), 315
Christensen, J. H., 137(C3), 143, 150, 152(C3), 153, 160, 161, 164, 206
Cichy, P. T., 5, 9, 10, 14, 51
Cijovic, S. D., 261(D11d), 312
Clark, W. C., 68, 121
Cleveland, R. G., 8(D4), 27(D4), 51
Cobo, O. A., 280(C7), 311
Coeuret, F., 260, 278, 279(L1c), 286(C9, C10), 287(C8), 297(S20c), 308(C9, C10), 311, 313, 314, 317
Cognet, G., 281(C11a), 282(C1a), 285(D5), 286(C11a), 294(D5, D6a, M6a), 295(D5), 311, 312, 315
Cohen, E., 104(R1), 122
Colburn, A. P., 270, 271, 297, 298, 311
Cole, A. F. W., 233, 234, 256, 257, 311
Colello, R. G., 254(C12b), 311
Collier, J. G., 44, 51
Collins, A. G., 137(C10), 143, 150, 206
Collins, S. B., 27, 51
Conway-Jones, J. M., 66, 71(N2), 81(N2), 101(N2), 106, 122
Cooper, A. R., 75(C12), 120
Cornell, D., 138(K3), 207
Cornet, I., 219(E4), 272, 281(C13), 285(P2a), 293(E4), 295(K1c), 311, 312, 314, 316
Corney, J. D., 60(C13), 121
Couderc, J. P., 263(G4a), 313
Cousins, L. B., 14, 51
Cross, H., 143, 150, 154, 155(C11), 159, 160, 163(C11), 206

D

Daguenet, M., 219(D2), 233(D1, D3e, D4), 254, 257(D2), 263(A3a), 272, 280(D2), 285(M13i), 293(D2), 294(D5, D6a, M6a), 295, 310, 311, 312, 315
Dajani, J. S., 171, 172, 206

Danckwerts, P. V., 79, 80(C6), 81(C5), 85, 86, 87, 89, 90(C6), 101, 106, 120
Daniel, P. T., 135(D2), 137(D2), 150, 154(D2), 155(D2), 163(D2), 192, 206
Daum, P. H., 227(D6b), 312
Davidson, J. F., 2, 52
Davis, R. E., 236, 256, 312
Dawson, D. A., 270, 272, 297(D8), 312
Dawson, J. L., 287(W13c), 318
Dawson, P. R., 58(B2), 102(B2), 104(B2), 120
Deb, A. K., 137(D3), 184, 206
Debbas, S., 64(D1), 72(D1), 121
DeGance, A. E., 31, 51
Deissler, R. G., 271, 272, 312
Dekker, J. L., 259(R20), 283(R20), 316
de Leeuw den Bouter, J. A., 265, 289(D10), 290(D10), 312
de Munnik, B., 265(D10), 289(D10), 290(D10), 312
Dengler, C. E., 43, 44, 51
Denton, E. B., 219(L9), 236(L9), 255(L9), 256(L9), 282(L9), 296(L9), 315
Denton, W. H., 14(C8), 51
Deo, N., 132, 133(D4), 135, 148(D4), 206
Dergarabedian, P., 34, 51
Deslouis, C., 274(D11c), 293(D11a, D11b), 299(D11c), 312
Despic, A. R., 259, 261(D11d), 312
Devanathan, M. A., 260(D12, D13), 312
de Voogd, P., 219(D14a), 312
Dickinson, T., 227(A3e), 310
Dijkstra, E. W., 166, 206
Dikusar, A. I., 285(D14b), 312
Dikusar, G. K., 285(D14b), 312
Dillingham, J. H., 154(D6), 162(D6), 206
Dimopoulos, H. G., 262(D15), 312
Distefano, G. P., 192(D7), 206
Dobry, 263, 287(D16a), 312
Donachie, R. P., 139(D8), 140(D8), 143, 150, 152, 171, 173, 174, 206
Downing, D. M., 141(M10), 208
Dreesen, E. W., 295(D16b), 296(D16b), 312
Duff, J. S., 168, 206
Duffin, R. J., 168, 206
Duffy, G. J., 80(L1), 81(L1), 83(L1), 101(L1), 121
Dukler, A. E., 8, 23, 27, 51, 52, 53
Dumargue, P., 285(M13i), 315
Dunn, J. S. C., 5(G3), 51
Dupont, T., 193, 195, 197, 198, 208

E

Eaton, L. E., 75(C12), *120*
Edgar, T. F., 140(B7), 171(B7), 172(B7), 175(B7), 181(B7), 183(B7), *205*
Einarsson, A., 260(W13b), 276(W13b), 287(W13c), *318*
Eisenberg, M., 215(E1), 217(T3), 218(T3), 219(E1, W5), 221(E1), 223(E2), 231(W4, W5), 236(E3), 256(W4, W5), 272, 283(W4, W5), 294(E1), *312, 317, 318*
Elabd, H. A., 305(F8), *312*
El-Abd, M. Z., 299(I11b), *313*
Elenbaas, J. R., 138(K3), *207*
Ellison, B. T., 219(E4), 272, 293(E4), *312*
Emelyanenko, G. A., 250(E5), *312*
Emery, C. A., 280(E6), *312*
Engelleitner, W. H., 57(E1, E2), 58(E2), *121*
Enger, T., 136, *206*
English, A., 57(E3), 59(E3), *121*
Enke, C. G., 227(D6b), *312*
Epelboin, I., 274(D11c), 293(D11b), 299(D11c), *312*
Epp, R., 135(E2), 136, 137(E2), 140(E2), 143, 144, 150, 157, 162(E2), 163, 164, 165, *206*
Estermann, I., 34, *53*
Etchells, A. W., 5(E1), 6, 7, *51, 52*
Eucken, A., 217, *312*

F

Fahidy, T. A., 259, 283(N1a), *315*
Fahidy, T. Z., 212(M13b, M13c), 218, 281(M13e), 286(M13d), 292(M13d), 298(N1a), *312, 315*
Fair, G. M., 137(F1), 154, *206*
Fan, L. T., 183, *206*
Fanning, J. T., 137(F9), *206*
Farley, R., 64(A2), *120*
Farnand, J. R., 56(F1), 115(F1), *121*
Fenech, E. J., 219(F3), 234, 235(F2), 238, 240, 245, 251, 257, 266, 267(F3), 268, 290(F2, F3), *312*
Feng, C. C., 136, *206*
Fietz, T. R., 163(F3), *206*
Filinovskii, V. Yu., 261(P4b), *316*
Finn, R. K., 263, 287(D16a), *312*
Firth, C. V., 76, *121*
Fitton, J. T., 58(B2), 102(B2), 104(B2), *120*
Flanigan, O., 138(F4), 140(F4), 171, 172, 174, 175(F4), 199, *206*
Fletcher, R., 184, *206*
Ford, L. H., 58(F3), 62, 113(F3), 115, *121*
Fouad, M. G., 257, 266, 283(F5), 291(F5, F6, F7c), 305(F7a, F7b, F8), *312*
Fowler, A. G., 135(E2), 136, 137(E2), 140(E2), 143, 144, 150, 157, 162(E2), 163, 164, 165, *206*
Frank, H., 138(R5), 171(R5), 172(R5), 175(R5), 178(R5), 185(R5), 188(R5), 189(R5), 190(R5), 199(R5), *208*
Frei, A. M., 231(B9), 240(B9), 267(B9), 291(B9), *311*
Friend, W. L., 255, *312*
Fuerstenau, D. W., 76(S5), 77, 78(K5, S5), 79, 80, 81(S2, S5), 83(S5), 85(S4), 90, 91(K6, S1), 93, 99, 100(S3), 101(K5), 102(S4), 104, *121, 122, 123*
Fujisawa, T., 168, 169, *206*
Fukuda, H., 297(M12), *315*
Fulkerson, D. R., 160(F8), *206*
Furman, N. H., *314*

G

Gabe, D. R., 260, 272, 278, 279, 281, 286(C1c), 294(G1a, R16), 308(C1c, C1e), *311, 313, 316*
Gabrielli, C., 274(D11c), 299(D11c), *312*
Gadish, Y., 171(K2), 172(K2), 175(K2), 181(K2), *207*
Galloway, T. R., 278(G1d), 298(G1d), *313*
Garrett, K. H., 57(G1), 62(G1), 75(G1), 77(G1), 78(G1), *121*
Gaskill, H. S., 219(L9), 236(L9), 255(L9), 256(L9), 282(L9), 296(L9), *315*
Gaudin, A. M., 117, *123*
Gay, B., 143, 155, 158, 162, 163(G1), 201, *207*
Gemmell, R. S., 171(D1), 172(D1), *206*
Gerlt, J. L., 150, 154(G3), 171, 172, 184(G3), *207*
Geyer, J. C., 137(F1), 154, *206*
Gibert, H., 263, 273, 275(L4), 288, 300(G3), 302(L4), *313, 314*
Gileadi, E., 223(G4b), 280(K4b), *313, 314*
Glastonbury, J. R., 80(L1), 81(L1), 83(L1), 101(L1), *121*
Glenn, N. L., 140(A1), *205*

Goacher, P. S., 191(G4), 192, *207*
Goda, T., 172, *207*
Goksel, M. A., *123*
Goldstein, R. P., 173, *207*
Gordon, A. R., 233, 234, 256, 257, *311*
Gordon, S. L., 236, 256, 293(G5), *313*
Goren, S. L., 273, 300(G6), *313*
Gosman, A. D., 260(I10), 261(I10), 269(I10), 287(I10), 296(I10), 297(I10), *313*
Gouda, T., 257, 283(F5), 291(F5), *312*
Govier, G. W., 5, 6, 8, 10(C3), *51, 52*
Grassmann, P., 263, 287(G7), *313*
Graves, Q. B., 154(G7), *207*
Greaves, M. J., 57(E3), 59(E3), *121*
Greif, R., 285(P2a, S15b), 295(K1c), *314, 316, 317*
Grossman, L. M., 44, 45, *52*
Grunchard, F., 278, 308(J7), *314*
Gubbins, K. E., 236, *313*
Guber, Yu. E., 294(K5a), 295(K5a), 304(K5a), *314*
Guerrieri, S. A., 43, 44, *51*
Gupta, P. K., 145, *207*
Gurusvami, V., 260(D12, D13), *312*
Gustavson, F. G., 167, 168(B11), *206, 207*
Gutowski, R., 259(C6b), 283(C6b), *311*

H

Hackerman, N., 272(M2a), *315*
Hac-Taylor, N. S., 8, 14(H1), *51*
Haddix, G. F., 150, 154(G3), 171, 172, 184(G3), *207*
Hadley, H. G., 172, *207*
Hahn, G. T., 85(H1), *121*
Hakala, R. W., 65(H2), *121*
Hamam, Y., 144(B10), 150, 162(B10), 171(B10), 172(B10), 184(B10), 185(B10), *206*
Hamann, C. H., 272, 293(H1), *313*
Han, C. D., 105(H3), 112, *121*
Hannah, K. W., 190(W6), *208*
Hanratty, T. J., 212(V2), 231(T1b), 236(V2), 257, 261, 262, 263(K2a), 264, 269(V2), 270, 277(K2b), 283, 289(T1b), 295(V2), 296(S19, V1), 297(S11a), 307(J6, K2b), *312, 313, 314, 315, 316, 317*
Harary, F., 135, *207*.
Harrington, J. J., 135(C4), 137(C4), 171, 172, 175(C4), 184, *206*

Hau, K.-F. F.-L., 296(B4b), 297(B4b), *311*
Haughey, D. P., 77(H5), *121*
Hayashi, H., 298(M11), *315*
Heady, R. B., 68(H6), *121*
Heath, M. J., 192(H3), *207*
Heertjes, P. M., 265(D10), 289(D10), 290(D10), *312*
Heindrich, A., 295(D16b), 296(D16b), *312*
Hendry, J. E., 146(G8), *207*
Herrmann, W., 74(S8), 75(S8), *122*
Hewitt, G. F., 8, 14, *51*
Hickman, R. G., 213, 235(H3), 237, 238, 241, 245, 246, 256, 268, 282(T2), 292(T2), *313, 317*
Hicks, G. C., *123*
Hicks, R. E., 277, 307(H4), *313*
Hignett, T. P., 57(H7), *121*
Hill, M. G., 254(H6c), *313*
Himmelblau, D. M., 140(B7), 160(L4), 171(B7), 172(B7, H4), 175, 181(B7), 183(B7, H4), *205, 207*
Hine, F., 274(O3), 280(O3), 283(O3), *316*
Hintermann, H. E., 249(J5a), *314*
Hinze, J. O., 26, *51*
Hiraoka, S., 301(M10), *315*
Hitchman, M. L., 261(A3b), *310*
Ho, Y.-C., 174, *206*
Hoag, L. N., 154(H5), *207*
Hodgson, G. W., 5(C3), 6(C3, R6), 10(C3, R6), *51, 52*
Hoffman, E., 70(P8), 71(P8), *122*
Holliday, D. V., 190(W6), *208*
Holmes, J. T., 275, 302(H5), *313*
Homsy, R. V., 254(H6b), 274, 280(H6a), *313*
Hoogland, J. G., 219(J3), 276, 303(J3), 304(J3), 305(J4a), *314*
Horvath, G. L., 236(D7), 256(D7), *312*
Hostomsky, J., 256(R21a), 257(R21a), 282(R21a), *316*
Houghton, G., 34, *53*
Howard, D. D., 137(S2), 143, 145, 150, 151(S2), 173, 198, *208*
Howland, B., 254(H6c), *313*
Hsing, H. Y., 159, 160, *207*
Hsueh, L., 220, 229(H9), 232(H10), 233(H8), 234(H7, H8), 250(H7), 251, 254(N8b), 280(H9), *313, 316*
Hubbard, D. W., 270, 297(H11a), *313*
Hughes, R. R., 146(G8), *207*
Hughmark, G. A., 23, 27, *52*

Hutchison, H. P., 137, 143, 150, 155, 156, 157, 168, *205*
Hutin, D., 278, *313*
Hyman, D., 63(B9), 106, 107, 108, *120*
Hyman, S. I., 163(H7), *207*

I

Ibl, N., 218, 219(I5b, I6), 222(I2, I4), 231(B9, I5a), 240(B9), 247, 248, 249, 250, 263(G7), 266, 267(B9), 275, 276, 287(G7), 291(B9, F6), 303(I8, S3), *311, 312, 313, 314, 317*
Ibusuki, A., 301(M10), *315*
Illig, H. J., 57, *121*
Ilmoni, P. A., 75, *122*
Ingels, D. M., 137(I1), 150, 154(I1), *207*
Iribarne, A., 260(I10), 261, 269, 287(I10, I11a), 296(I10), 297(I10), *313*
Isaacson, E., 151(I2), 192(I2), *207*
Ismail, M. I., 299(A3c, A3d, I11b), *310, 313*
Israel, R., 113(I2), 114, *121*
Ito, R., 298(M11), 301(M10), *315*

J

Jacoby, S. L. S., 140(J1), 171, 184, *207*
Jagannadharaju, G. J. V., 273(R1, R2, R5), 274(K8, S2), 276(R7), 277(K12), 278(K11), 284(K13), 288(R6), 290(K8), 298(R2), 299(R1, R2, R5), 301(K8, S2), 306(K10, K12, R7), 308(J1, K9, K11), *313, 314, 316, 317*
Jahn, D., 218, *313*
Janssen, L. J. J., 219(J3), 276, 303(J3), 304(J3), 305(J4a), *314*
Jansson, R. E. W., 259, 284(J4b), 295(A6b), *311, 314*
Javet, P., 248(I9a), 249(I9a, J5a), *313, 314*
Jennings, D., 305(J5b), *314*
Jeppson, R. W., 137(J2), 140(J2), 143, 150, 156, 157, *207*
Jipping, M. J., 57(R2), 63(R2), 102(R2), 104(R2), *122*
Johnson, A. P., 63(N1), 118(N1), *122*
Johnson, R. L., 137(C10), 143, 150, *206*
Johnson, R. T., 59(B6), 60(B6), *120*
Jolls, K. R., 277, 307(J6), *314*
Jones, H. A., 104(J1), *121*
Jones, R. I., 163(H7), *207*

Jottrand, R., 278, 308(J7), *314*

K

Kabanov, B. I., 218(S14), 280(S14), *317*
Kacin, J., 304(R21b), *316*
Kadija, I. V., *314*
Kaiser, R., 56(M6), 63(M4), 64(M5), 73(M4), 76(M6), 80(M4), 113(M4), 114(M6), 115(M6), *121*
Kally, E., 171, 172, 175(K1), 181, 184(K1), *207*
Kanetkar, V. V., 85, 100, 101, *121*
Kappesser, R., 281(C13), 295(K1c), *311, 314*
Kapur, P. C., 76, 77, 78(K5), 80, 81, 82, 84, 85(K7), 87, 88, 89(K3), 90, 91, 92(K4), 93, 94, 95, 96, 97, 101, 102(K7), 103, *121, 123*
Karabelas, A. J., 263(K2a), 277, 307(K2b), *314*
Karayeva, E. M., 171, *205*
Karmeli, D., 171, 172, 175(K2), 181(K2), *207*
Kater, T., 57(R2), 63(R2), 102(R2), 104(R2), *122*
Katz, D. L., 138(K3), *207*
Kawashima, Y., 115, 116, *121, 123*
Kayatz, K., 60, *121*
Keddam, M., 293(D11a), *312*
Kehat, E., 25, *52*
Keller, H. B., 151(I2), 192(I2), *207*
Kermiche, F., 263(A3a), *310*
Kern, D. Q., 31, *52*
Kern, R., 200(K4, K5, K6), *207*
Khangaonkar, P. R., 57(M7), *121*
Khanna, P., 171(S8), 175(S8), 184(S8), *208*
Kihlstedt, P. G., 104, *121*
Kimla, A., 281(K2c), *314*
King, C. J., 11, *51*
King, C. V., 217, 272(K3, K4a), *314*
King, I. P., 149, 163, *207*
Kirowa-Eisner, E., 223(G4b), 280(K4b), *313, 314*
Kishinevskii, M. K., 294(K5a), 295, 296(K5b), 304(K5a, K5b), *314*
Klatt, H., 60(K11), 61(K11), *121*
Kleitman, D. J., 138(R5), 171(R5), 172(R5), 175(R5), 178(R5), 185(R5), 188(R5), 189(R5), 190(R5), 199(R5), *208*
Kniebes, D. V., 150, 154(K8), *207*
Knights, I. A., 150, 155(K9), *207*

AUTHOR INDEX

Knudsen, J. G., 27, 32, 34, *51*, 295(S22), *317*
Kobayashi, R., 138(K3), *207*
Kocova, S., 64(K12), 65, *121*
Kolthoff, I. M., *314*
Konjovic, M. N., 259(D11e), *312*
Kornienko, T. S., 294(K5a), 295(K5a), 296(K5b), 304(K5a, K5b), *314*
Kortmann, H. A., 57(R2), 63(R2), 102(R2), 104(R2), *122*
Koumoutsos, N., 42, *52*
Kramer, W. E., 104, *121*
Krishna, M. S., 274, 277(K12), 278, 283(S21), 284(K13), 285(S21), 290(K8), 301(K8, S2), 306(K6, K10, K12), 308(K7, K9, K11), *314, 317*
Kron, G., 162(K10), *207*
Krupp, H., 65(K14), *121*
Kuh, S. E., 168(F7), 169(F7), *206*
Kuhn, A. T., 260, 305(J5b), *311, 314*
Kumar, V., 171(S8), 175(S8), 184(S8), *208*
Kurvyakova, L. M., 249(K14), *314*

L

Lacey, P. M. C., 44(C7), *51*
Ladesma, V. L., 26(W7), *53*
Lam, C. F., 136, 137(L1, L2), 143, 150, 152, 153, 157, 171, *207*
Lamb, D. E., 14, *52*
Landau, U., 234(L1a), 235, 256, 270, 282(L1b), 296(L1b), 297(L1b), *314*
Landsberg, R., 284(S4a, S4b), *317*
Lang, C., 104(R1), *122*
Langmuir, I., 34, *52*
Lantz, P. M., 256(M6b), *315*
Larson, R. E., 171, 172, 175(L3, W10), *207, 209*
Ledet, W. P., 160(L4), *207*
Le Goff, P., 260(C9, C10), 278(C9, C10), 279, 286(C9, C10), 308(C9, C10), *311, 314*
Leiter, G. G., 63(N1), 118(N1), *122*
Leitz, F. B., 297(L1d), *314*
LeLan, A., 275, 301(L3), 302(L2, L3, L4), *314*
Lemieux, P. F., 137(L5), 143, 150, 152, *207*
Lengas, T. E., 270, 279, *314*
Leppert, G., 38, 42, *52*
Letan, R., 25, *52*
Levenspiel, O., 20, 24, *52*
Lévêque, J., 217, 234, 236, 241, 255, 260, 261, 262, 268, 269, 272, *314*
Levich, B., 218, 234, 274, 280(L6), *314*
Levich, V. G., 248, 271, *315*
Levy, S., 42, *52*
Liang, T., 171, 172, 175(L6, Y1), *207, 209*
Lightfoot, E. N., 31(B4), *51*, 217(B4c), 270, 286(B4c), 297(H11a), *311, 313*
Lilliheht, L. U., 8, *51*
Lin, C. S., 219(L9, L10a), 236, 255, 256, 271, 282(L9), 296(L9, L10a), *315*
Lin, T. D., 168, *207*
Linkson, P. B., 80, 81(L1), 83(L1), 101, *121*
Litt, M., 272, *311*
Liu, K. T. H., 137(L8), 143, 150, 155, *207*
Lloyd, J. R., 290(L10b), *315*
Lockhart, R. W., 7, 8, 21, 23, 27, 31, 50, *52*
Loginov, A. V., 296(K5b), 304(K5b), *314*
Loomba, R. P., 290(W9), *318*
Loui, D. S., 137(B3), *205*
Luther, H. A., 151(C1), *206*
Lynch, W. O., 137(T2), 171(T2), 183(T2), *208*

M

Maa, J. R., 32, 33, 34, 35, *52*
Macavei, G., 60, *121*
Macbeth, R. V., 41, *52*
McCall, N. J., 171, 175(M3), *207*
McCamy, I. W., *123*
McIlhinney, A. E., *123*
Madigan, D. C., 59(M2), 76(M2), 102(M2), *121*
Magiros, P. G., 23, *52*
Mah, R. S. H., 135, 136, 141(M10), 143, 146(M2), 150, 161(M1), 163, 165, 166, 168, 171, 172, 173, 176, 178, 179, 180, *206, 207, 208*
Mahato, B. K., 248(M1), *315*
Makrides, A. C., 272(M2a), *315*
Mandelbaum, J. A., 307(M2b), *315*
Mandersloot, W. G. B., 277, 307(H4), *313*
Mani, R. V. S., 273, 300(G6), *313*
Marchiano, S. L., 219(A6a), 236(A6a), 261(I11a), 265, 280(C7), 287(I11a), 289(M3), *311, 313, 315*
Marincic, L., 297(L1d), *314*
Markland, E., 155, *205*
Martch, H. B., 171, 175(M3), *207*

Martin, D. W., 150, *208*
Martinelli, R. C., 7, 8, 21, 23, 27, 31, 43, 50, *52*
Masliyah, J. H., 192, *208*
Mason, G., 68, *121*
Matsuda, H., 254, 280(M4b), 281(M4c, M4d), *315*
Matsuda, N., 171(M6, M7, M8), 175(M6, M7, M8), 183(M6, M7, M8), *208*
Mattson, E., 225, 226, *315*
Mazza, B., *316*
Mecea, V., 280(B11b), *311*
Meissner, H. P., 56(M6), 63(M4), 64(M5), 73(M4), 76(M6), 80(M4), 113(M4), 114, 115(M6), *121*
Meklati, M., 263(A3a), 294(D6a, M6a), *310, 312, 315*
Metzner, A. B., 255, *312*
Meyer, R. E., 256, *315*
Meyers, S., 171(K2), 172(K2), 175(K2), 181(K2), *207*
Michaels, A. S., 56(M6), 63(M4), 64(M5), 73(M4), 76(M6), 80(M4), 113(M4), 114(M6), 115(M6), *121*
Middleton, P., 143, 155, 158, 162, 163(G1), 201, *207*
Miller, B., 274(T4a), *317*
Misra, V. N., 57(M7), *121*
Mitchell, J. E., 261(M7), *315*
Mitrovic, M., 259(D11e), *312*
Mitrovic, M. V., 261(D11d), *312*
Mixon, F. O., 26, *52*, 275(M8), *315*
Mizushina, T., 218, 297(M12), 298(M11, M13a), 301(M10), *315*
Mohanta, S., 212(M13b, M13c), 281(M13e), 286(M13d), 292(M13d), *315*
Mohr, C. M., Jr., 259, 281, 284(M13g), 292(M13h), 293(M13h), 304(M13f), *315*
Moissis, R., 42(K2), *52*
Mollet, L., 285(M13i), *315*
Moran, W. R., 290(L10b), *315*
Morlok, E. D., 171(D1), 172(D1), *206*
Morrow, N. R., 70(M8), *122*
Moulton, R. W., 219(L10a), 271(L10a), 296(L10a), *315*
Müller, L., 227(S17a, S17b), *317*
Müller, R., 219(I5b), *313*
Müller, R. H., 219(I6), *313*
Müller, S., 284(S4a, S4b), *317*
Mukherjee, L. M., 267, *316*

Muller, R. H., 216(B4a), 218, 256(A1a), 282(A1a), 298(A1a), *310, 311, 315*
Murtagh, B. A., 171, 172, 175(M9), 176(M9), 177, 179(M9), *208*
Myron, T. L., 59(B6), 60(B6), *120*

N

Nabavian, K., 32, *52*
Nadebaum, P. R., 259, 283(N1a), 298(N1a), *315*
Nahavandi, A. N., 137(N1), 159, 175(N1), 192, *208*
Nakajima, S., 137(N2), 171, 175(N2), 184(N2), *208*
Nakic, V. M., *314*
Nash, J. H., 63(N1), 118, *122*
Nasiruddin, A. K., 265, 291(W10), 306(W10), *318*
Naybour, R. D., 247(N1b), *315*
Nelson, D. B., 8, 31, 43, *52*
Nelson, J. C., 57(V1), *122*
Nernst, W., 214, 215, 217, 224, 233, 310, *315*
Neufville, R., 199, *208*
Newitt, D. M., 66, 71(N2), 81(N2), 101(N2), 106, *122*
Newman, J., 212(N8a), 216, 223(N8a), 224, 229(H9), 231, 232(H10, N5, N6, S9a), 233(H8, N8a), 234(H8), 236(S16), 246, 247(W3b), 251, 254, 257(S9a), 259, 261, 274, 277, 280(H6a, H9, N3), 281, 282(N6), 284(M13g), 292(M13h), 293(M13h), 304(M13f), 307(A3f), *310, 313, 315, 316, 317, 318*
Newman, J. S., 236(G5), 256(G5), 276, 293(G5), *313, 316*
Newson, J. D., *316*
Nicklin, D. J., 2, *52*.
Nicol, S. K., 102(N3), 104, *122*
Nikolic, B. Z., 261(D11d), *312*
Nilsson, O., 286(W14), *318*
Nisancioglu, K., 254(N9b, N9c), *316*
Noordsij, P., 236, 273, 285(N10), 300(N10), *316*
Norton, M. M., *123*

O

O'Brien, R. N., 219(O2), 267, *316*
O'Callaghan, R. T., 171, 198, *208*

O'Connor, G. E., 11, 21, 22, 34, 47, *52*
O'Connor, T. F., 137(T2), 171(T2), 183(T2), *208*
Ogino, F., 297(M12), *315*
Ogura, Y., 172, *207*
Ohtusuki, T., 168(F7), 169(F7), *206*
Oka, Y., 297(M12), *315*
Okada, N., 298(M13a), *315*
Okada, S., 104(W2), *123*, 274, 280(O3), 283(O3), *316*
Olander, D. R., 275(H5), 302(H5), *313*
Olson, J. H., 12, 26, 28, 35, 36, *52*
Ong, K. L., 260, *316*
Orcutt, J. C., 26(M8), *52*
Ore, O., 135, *208*
Ostermaier, J., 25, *52*
Ouchiyama, N., 81(O1), 98, 99, *122*, *123*

P

Palade de, A., 261(I11a), 287(I11a), *313*
Papadakis, M., 60(P1), *122*
Parrish, W. R., 246, *316*
Paterson, J. A., 285(P2a), *316*
Patrick, M. A., 290(P2b), 291(W11), *316*, *318*
Paus, G. F., 299(P5), *316*
Pavlica, R. T., 12, 26, 28, *52*
Peck, W. C., 56(P2), *122*
Pedeferri, P., *316*
Pelezarski, E. A., 59(B6), 60(B6), *120*
Penciner, J., 223(G4b), *313*
Perry, R. H., 139, *208*
Pestowski, T., 150, 198, *208*
Peters, G., 150, *208*
Pickett, D. J., 260, 282(P3b), 297(P3b), *316*
Pietsch, W., 59(P3), 60(P3), 61(P3), 67, 68, 69(P6), 70(P8), 71(P8), 72(P5, P7), 73, 117, *122*
Pilpel, N., 57(P9), 64(K12), 65, *121*, *122*
Piontelli, R., *316*
Pitts, C. C., 38, 42, *52*
Pleskov, Yu. V., 261(P4b), *316*
Podestá, J. J., 219(A6a), 236(A6a), 299(P5), *311*, *316*
Poettmann, F. H., 138(K3), *207*
Pomosov, A. V., 249(K14), *314*
Porter, J. H., 270(V5), 271(V5), *317*
Porter, J. W., 26(W7), *53*
Posey, F. A., 256(M6b), *315*
Powell, M. J. D., 184, *206*

Powers, J. E., 137(I1), 150, 154(I1), *207*
Preece, P. E., 143, 162, *207*
Puddington, I. E., 56(F1), 58(S11), 115(F1, S11), *121*, *122*, *123*
Pujol, L., 44, 45, *52*
Pulling, D. J., 44(C7), *51*
Pulvermacher, B., 96, 98, *122*, *123*
Putnam, G. L., 219(L9, L10a), 236(L9), 255(L9), 256(L9), 271(L10a), 282(L9), 296(L9, L10a), *315*
Py, B., 263(P6), *316*

R

Rachford, H. H., 193, 195, 197, 198, *208*
Radford, B. A., 5(G3), *51*
Rafal, M., 146(M2), 173, *207*
Raju, C. V. R., 273(R1), 276(R7), 298(R2), 299(R1, R2), 306(R7), *316*
Ramabhadran, T. E., *123*
Raman, V., 126(R3), *208*
Ramaraju, C. V., 284(K13), *314*
Ranz, W. E., 263, *316*
Rao, C. V., 273(R1, R5), 274(S2), 277(K12), 278(K11), 284(K13), 288(R6), 299(R1, R5), 301(S2), 306(K12), 308(J1, K11), *313*, *314*, *316*, *317*
Rao, K. S., 259, 273, 276, 288(R6), 298(R4), 299(R5), 306(R7), *316*
Ravoo, E., 264, 267, 275, 289(R10), 291(R10), 303(R8), *316*
Records, F. A., 57(G1), 62(G1), 75(G1), 77(G1), 78(G1), *121*
Reddy, A. K. N., *311*
Reeves, C. M., *206*
Regner, A., 275, 302(R11), *316*
Reid, J. K., 168, *206*
Reiss, L. P., 212(V2), 236(V2), 261(R12, R13, V2), 262, 269(V2), 295(V2), *316*, *317*
Riddiford, A. C., *316*
Ridgion, J. M., 104(R1), *122*
Riley, N., 241(R14), *316*
Robert, J., *312*
Robinson, D. J., 268, 272, 281, 292(R15), 294(R16), *313*, *316*
Rogers, R. W., 16, *52*
Rohsenow, W. M., 42, *52*
Roorda, H. J., 57(R2), 63(R2), 102(R2), 104, *122*

Rosenbaum, D. M., 138(R5), 171(R5),
 172(R5), 175(R5), 178(R5), 185(R5),
 188(R5), 189(R5), 190(R5), 199(R5), *208*
Rosenfield, C., 219(O2), *316*
Rosenhan, A. K., 137, 150, 159(R4), 198, *208*
Ross, T. K., 234, 248(R17), 256, 268,
 282(R18, W13a), 287(R19), 292(W12),
 316, 318
Rothfarb, R., 138(R5), 171, 172, 175(R5), 178,
 185(R5), 188, 189, 190, 199, *208*
Rotte, J. W., 236, 259, 264(R10), 273,
 283(R20), 285(N10), 289(R10), 291(R10),
 300(N10), *316*
Rousar, I., 219(R21a), 223, 256, 257, 275,
 282(R21a), 302(R11), 304(R21b), *316*
Rubina, S. I., 113(V2), *123*
Ruckenstein, E., 42, *52*, 96, 98, *122, 123*
Rudd, D. F., 161(C9), *206*
Rumpf, H., 62, 63(R3), 64, 65, 66, 67, 68,
 69(P6), 70, 71(P8, R4, R5), 72(D1, P7),
 73, 74, 75(S8), 118, *121, 122, 123*
Russell, T. W. F., 5, 6, 8, 9(C5), 10, 14, 16,
 31, *51, 52*

S

Sakaguchi, I., 301(M10), *315*
Sakiadis, B. C., 258, *317*
Sani, R. L., *52*
Sarkar, A. K., 137(D3), 184, *206*
Sastry, A. R., 273(R2), 298(R2), 299(R2),
 316
Sastry, K. V. S., 76(S5), 77(S3), 78(S5), 79,
 80(S5), 81(S2, S5), 83(S5), 85(S4), 91(S1),
 99, 100(S3), 102(S4), 104, *122, 123*
Satyanarayana, A., 274(S2), 301(S2), *317*
Saunders, D. E., *51*
Schack, M., 217(K4a), 272(K4a), *314*
Schadegg, K., 250, *313*
Schalch, E., 275(I8), 276(I8), 303(I8, S3),
 313, 317
Scheller, F., 284(S4a, S4b), *317*
Schenk, H. J., 249(I9b), *313*
Schieffer, G. W., *311*
Schmidt, H. G., 254, 255, 258, 261, 267,
 270, 272, *317*
Schöner, H., 272(H1), 293(H1), *313*
Schrage, R. W., 32, *52*
Schrock, V. E., 44, 45, *52*
Schubert, H., 64, 69, 70(56), 71, 74, 75, *122,
 123*

Schütz, G., 264, 289(S6), 295(S7), 296(S7),
 317
Schuhmann, D., 233(D3e, D4), *312*
Scott, D. S., 5, 31, *52*
Sedahmed, G. H., 290(S11b), 305(F7a, F7b,
 F8), *312, 317*
Seid, D. M., 278(G1d), 298(G1d), *313*
Selman, J. R., 216, 220, 226, 231, 232(S9a),
 234(S8), 235(S8), 239(S8, S10), 242, 243,
 248, 249, 250(S8), 251, 257(S9a), *317*
Semin, V. S., 58(C10), 61(C10), 62(C10), *120*
Serafinidis, P., 260(W13b), 276(W13b), *318*
Sevenstern, F. W., 264(R10), 289(R10),
 291(R10), *316*
Shaake, J., 199(N3), *208*
Shamir, U., 137(S2), 143, 145, 150, 151(S2),
 171, 172, 173, 185, 186, 187, 198, 199(S1),
 208
Shapiro, S. S., 85(H1), *121*
Sharma, R. S., 259, 298(R4), *316*
Shaw, D. A., 270, 297(S11a), *317*
Shemilt, L. W., 248(M1), 290(S11b), *315, 317*
Shennan, J. V., 58(F3), 62, 113(F3), 115, *121*
Sherrington, P. J., 61(S10), 62, 78(S10),
 79(S10), 106, 107, 108, 109(S9, S10), 110,
 111, 112, *122*
Sherwood, T. K., 260, 270(V5), 271(V5), *317*
Shook, C. A., 192, *208*
Short, W. L., 5, *51*
Sideman, S., 11, 26, 27, 28, *51, 52*
Sih, P. H., 236, *317*
Silvestri, M., 41, *52*
Sinvhal, R. C., 57(M7), *121*
Sioda, R. E., 276, 307(S12d), *317*
Sirianni, A. F., 58(S11), 115(S11), *122, 123*
Sirkar, K. K., 270, *317*
Sivashankaran, V. S., 58(C10), 61(C10),
 62(C10), *120*
Siver, Yu. G., 218(S14), 280(S14), *317*
Smirous, F., 304(R21b), *316*
Smith, A. F. J., 264, 289(S15a), *317*
Smith, B., 199, *208*
Smith, H. M., 56(F1), 115(F1), *121*
Smith, R. N., 285(S15b), *317*
Smyrl, W. H., 236, *317*
Schr, 227(S17a, S17b), *317*
Soliman, M., 241(S17c), 242(S17c), 244, *317*
Somerville, G. F., 44, 45, *52*
Son, J. S., 262(S18), 263, 270, 296(S19), *317*
Spalding, D. B., 260(I10), 261(I10), 269(I10),
 271, 287(I10), 296(I10), 297, *313, 317*

AUTHOR INDEX

Sparks, B. D., 123
Sparrow, E. M., 7, 53
Spiro, M., 227(S20b), 317
Spitzer, H. J., 284(S4a), 317
Springer, G. S., 254(C12b, H6c), 311, 313
Spyridonos, A., 42(K2), 52
Stafford, J. H., 199(N3), 208
Stahel, F., 248(I9a), 249(I9a), 313
Stanfield, R. B., 173, 207
Stanley, G. M., 141(M10), 208
Stanmore, B. R., 282(P3b), 297(P3b), 316
Staub, F. W., 42, 52, 53
Stearns, D. E., 137(T2), 171(T2), 183(T2), 208
Steiglitz, K., 138(R5), 171(R5), 172(R5), 175(R5), 178(R5), 185(R5), 188(R5), 189(R5), 190(R5), 199(R5), 208
Stenning, A. H., 44, 45, 52
Stepanek, J. B., 305(J5b), 314
Stevenson, D. G., 57(G1), 62(G1), 75(G1), 77(G1), 78(G1), 121
Steward, D. V., 145(S4), 146, 208
Stewart, W. E., 31(B4), 51, 217(B4c), 286(B4c), 311
Stirling, H. T., 61, 122
Stoev, S. M., 61, 77(S13), 122
Stone, R. L., 102(S14), 104, 122
Stoner, M. A., 138(S5), 139, 140(S5, W12), 150, 166, 168, 171, 173, 192(W12), 193(W13), 195(W13), 196(W12), 198(W13), 209
Storck, A., 297(S20c), 317
Strafelda, F., 281(K2c), 314
Streeter, V. L., 140(W12), 190(S6), 192, 193(W13), 194, 195(W13), 196, 198, 209
Stuckey, A. T., 137(S7), 147(S7), 150, 154, 208
Stverak, B., 256(R21a), 257(R21a), 282(R21a), 316
Subbarao, S. V. B., 85(K7), 102(K7), 103(K7), 121
Subramaniyan, V., 283(S21), 285(S21), 317
Sullivan, G. A., 5(B6), 51
Sutey, A. M., 295(S22), 317
Sutherland, J. P., 115(C7), 116, 120
Swamee, P. K., 171, 175(S8), 184(S8), 208

T

Talty, R. D., 43, 44, 51
Tamamushi, R., 227(T1a), 317
Tanaka, N., 227(T1a), 317
Tanaka, T., 81(O1), 98, 99, 122, 123
Taïrjan, G., 76, 122
Tavallaee, A., 137(J2), 140(J2), 143, 150, 156, 157, 207
Taylor, J. L., 231(T1b), 257, 264, 283, 289(T1b), 317
Tewarson, R. P., 168, 208
Thirsk, H. R., 225, 311
Tiedemann, W., 276, 316
Tigerschoid, M., 75, 104(T2), 122
Tobias, C. W., 215(E1), 217, 218, 219(E1, F3, W5), 220, 221(E1), 223(E2), 231(W4, W5), 234, 236(D7, E3, G5), 238, 239(S10), 242, 243, 251(F3), 256(A1a, D7, G5, L1b), 257(F3), 266, 267(F3), 268, 272(E1), 282(A1a, L1b, T2), 283(W4, W5), 290(F3), 292(T2), 293(G5), 294(E1), 296(L1b), 297(L1b), 298(A1a), 310, 312, 313, 314, 317, 318
Tokuda, K., 274(T4a), 317
Tong, A. L., 137(T2), 171, 183, 208
Tong, L. S., 31, 53
Tonry, J. R., 57(T4), 122
Trass, O., 270, 272, 297(D8), 312
Tribollet, B., 274(D11c), 293(D11b), 299(D11c), 312
Trueb, J., 263(G7), 287(G7), 313
Tschudin, K., 34, 53
Tsubata, S., 298(M11), 315
Tsuchiya, O., 63(W1), 104(W2), 123
Tummers, G. L. J., 259(R20), 283(R20), 316
Turba, E., 73(R6), 122
Turner, G. A., 63(T5), 122
Tvarusko, A., 275, 287(T4b), 288(T4c), 289(T4b), 317

U

Ueda, H., 298(M11, M13a), 315
Uegaki, M., 298(M13a), 315
Ultman, J. S., 5(C5), 9(C5), 51

V

Valentin, F. H. H., 64(A2), 120
Van Shaw, P., 212(V2), 236, 261(V1, V2), 269, 295(V2), 296(V1), 317
Varga, R. S., 158, 208
Vary, J. H., 138(K3), 207

Venczel, J., 219(V3), 275, 276(I8), 303(I8, V3), *313*, *317*
Venkateswarlu, D., 113(I2), 114, *121*
Vergnes, F., 260(C9, C10), 278(C9, C10, L1c), 279(L1c), 286(C9, C10), 287(C8), 308(C9, C10), *311*, *314*
Vetter, K. J., 214(V4), 225(V4), *317*
Vielstich, W., 272(H1), 293(H1), 295(D16b), 296(D16b), *312*, *313*
Viet, L., 293(D11b), *312*
Vieth, W. R., 270(V5), 271, *317*
Violetta, D. C., 57(V1), *122*
Visnawathan, K., 259(C6b), 283(C6b), *311*
Vogtländer, P. H., 263, 276(V6), 288(V6), *318*
Volmer, M., 34, *53*
Voorduin, W. L., 137(B3), *205*
Voyles, C. F., 154(V2), *208*
Voyutski, S. S., 113(V2), *123*

W

Wada, M., 63(W1), 104, *123*
Wagner, C., 218, 283(W1), *318*
Waldman, L. A., 34, *53*
Walker, R. D., Jr., 236, *313*
Wallis, G. B., 8, *53*
Ward, W. J., 104(K13), *121*
Warga, J., 143, 150, 151(W1), 168, *208*
Watanatada, T., 137(W2), 171, 172, 175(W2), 184, *208*
Watson, D., 61, 77(S13), *122*
Weder, E., 264, 265, 266, 275, 289(W3a), 306(W2), *318*
Wegner, T. H., 277(K2b), 307(K2b), *314*
Wehner, J. F., 12, *53*
Weinaug, C. H., 138(K3), *207*
Weinberg, G., 154(H5), *207*
Westerberg, A. W., 146(G8), *207*
Weston Starratt, F., 78(W3), *123*
Westwater, J. W., 38, 42, *53*
Whitaker, D. R., 26(M8), *52*
White, R., 247, *318*
Whitehead, R., 305(J5b), *314*
Wicks, M., 8(D4), 23, 27(D4), *51*, *53*
Wilde, D. J., 172(W3, W5), 174(W4), *208*
Wilenitz, I., 112, *121*
Wilhelm, R. H., 12, *53*
Wilke, C. R., 26, *53*, 215(E1), 217(T3), 218(T3), 219(E1, W5), 221(E1), 223(E2), 231, 236(E3), 256(W4, W5), 272(E1), 275(H5), 283(W4, W5), 294(E1), 302(H5), *312*, *313*, *317*, *318*
Wilke, R., 154(V2), *208*
Wilkes, J. O., 143, 150, 151, 200, *206*
Wilkinson, J. F., 190(W6), *208*
Williams, G. N., 143, 150, 155, *208*
Williams, J. D., 150, 198, *208*
Willoughby, R. A., 168(B11), *206*
Wilson, G. G., 150, 154(K8), *207*
Wolfe, P., 153, 154, 158, *209*
Wolla. M. L., 136, 137(L1, L2), 143, 150, 152, 153, 157, 171, *207*
Wong, P. J., 171, 172, 175(L3, W10), *207*, *209*
Wood, D. J., 143, 150, 155, 156, 157, *209*
Wragg, A. A., 234, 256, 260, 264, 265, 267, 268, 276(W13b), 282(R18, W13a), 287(R19, W13c), 289(S15a, W7), 290(P2b, W6, W9), 291(W10, W11), 292(W8a, W12), 306(W10), *316*, *317*, *318*
Wranglen, G., 286(W14), *318*
Wright, R. M., 44, 45, *53*
Wu, I-P., 171(Y1), 175(Y1), *209*
Wylie, E. B., 140(W12), 190(S6), 192(S6, W12), 193(W13), 194, 195, 196, 198, *209*

Y

Yalcindag, C., 144(B10), 150, 162(B10), 171(B10), 172(B10), 184(B10), 185(B10), *206*
Yamada, J., 254(M4c), 281(M4c, M4d), *315*
Yang, K-P., 171, 175(Y1), *209*
Yao. K. M., 137(Y2), *209*
Yoshizawa, S., 274(O3), 280(O3), 283(O3), *316*
Young, P. A., 104(B4), *120*
Young, W. E., 104(K13), *121*
Yow, W., 193(Y3), 195, *209*
Yu, H. S., 7, *53*
Yusa, M., 117, *123*

Z

Zaionchkovskii, A. D., 113(V2), *123*
Zatout, A. A., 291(F7c), *312*
Zimmer, H. I., 138(Z1), 171, 172, 175(Z1), *209*
Zuber, N., 42, *53*

SUBJECT INDEX

A

Abrasion transfer, in balling and granulation, 81
Absolute rate expressions, in heat transfer with phase change, 34
Agglomerate growth, regions of, 81–84
Agglomerates
  bonding mechanisms in, 62–63
  deformation of, 74
  tensile strength of, 63–66
Agglomeration
  inadvertent, 117–118
  spherical, 56, 115–117
  spontaneous, 117

B

Balancing heads and flows, in pipeline network problems, 154–155
Ball growth region, in balling and granulation, 83–84
Balling, defined, 56
Balling and granulation, 55–120
  abrasion transfer in, 81
  attractive forces in absence of material bridges, 73–74
  ball growth region in, 83–84
  bonding liquid and additives in, 100–105
  capillary bonds in, 61–71
  compaction in, 75–77
  crushing and layering in, 78–80, 86–90
  empirical kinetic models in, 99–100
  fertilizer granulation in, 105–112
  growth mechanisms in, 77–81
  inadvertent agglomeration in, 117–118
  kinetics of, 84–100

miscellaneous topics in, 112–118
nonrandom coalescence kinetics and, 93–99
nucleation in, 56
random coalescence kinetics in, 90–93
snowballing in, 78–79, 85–86
solid bridges in, 72–73
Balling and granulation equipment, 57–62
  drums and disks in, 58–60
  miscellaneous devices in, 61–62
Bentonite additive, in iron ore balling, 102–105
Bonding liquid and additives
  in balling and granulation, 100–104
  bentonite additive and, 102–105
  water content of, 100–102
Bonds, in balling and granulation, 66–71
Broyden's method, in pipeline network problems, 152–153
Burn-out, defined, 29

C

Capillary bonds, in balling and granulation, 66–71
Cathodic reactions, limiting current and, 215
Clausius-Clapeyron equation, 33
Coalescence, in balling and granulation, 80–81, 106–109
Column row reordering, in pipeline network problems, 162–166
Compaction, in balling and granulation, 75–77
Condensation
  absolute rate of, 44
  in heat transfer with phase change, 47–48

SUBJECT INDEX

Copper deposition
 limiting current in, 241
 surface roughness and, 247
Crushing and layering, in balling and granulation, 79–80, 86–90
Cycle selection algorithms, in pipeline network problems, 163–164
Cyclic networks, in pipeline network design, 183–185

D

Diakoptics, in pipeline network problems, 162
Diffusion layer, limiting current and, 213–215
Diffusivity, effective, 216, 233–235
Disk pelletizer, 58
Drums and disks, in balling and granulation, 58–61
Dry pelletization, 113–115

E

Electrochemical mass-transfer studies, model reactions used in, 219–223
Electrochemical systems, in mass-transfer studies, 219–223
Electrode boundary conditions, in limiting-current measurements, 227–228
Electrode potential measurement, in limiting-current measurements, 223–227
Electrolytic gas evolution, in heat or mass transfer, 275–276
Elongated electrodes, potential distribution at, 243–247
Erdey-Gruz-Volmer-Butler equation, in limiting-current measurements, 225

F

Fertilizer granulation, 56–57, 105–112
 in coalescence mode, 106–109
 in snowballing mode, 109–112
Fertilizer granulation loop, 105, 109
 in snowballing mode, 111–112

Finite difference methods, in pipeline network design, 195–197
Flow configurations
 basic model equations for, 11–19
 in liquid-liquid systems, 10
 parameter evaluation in, 19–28
 Regime I, 19–22
 Regime II, 22–25
 Regime IV, 25–28
Flow networks, description and characterization of, 127–136
Fluid-fluid systems, see Liquid-liquid systems
Forced convection, two-phase, 39–40
Funicular bonds, in balling and granulation, 70–71

G

Gas-liquid flow patterns, 2–6
Gas-liquid systems
 heat transfer in, 12–13, 18–19
 parameter evaluation in, 21–22, 27–28
Generalized secant methods, in pipeline network problems, 152–154
Granulation, defined, 56
 see also Balling and granulation
Granulation loop, in fertilizer granulation, 105–109, 111–112
Granules, growth of, 77–81
Graphs and digraphs
 matrix representation of, 130–136
 in pipeline network problems, 127–130
Grinding equilibrium, defined, 117

H

Hardy-Cross method, in pipeline network problems, 154–155, 159
Heat transfer
 liquid-deficient, 40–41
 transition zones in, 40–41
 in tubular-fluid-fluid systems, 1–49
 as two-phase flow, 9
Heat transfer coefficient
 gas-phase wall, 22
 interfacial, 15, 27

SUBJECT INDEX 333

Heat-transfer studies
 Forms I and II for, 44–45
Heat transfer with phase change, 28–48
 absolute rate expressions in, 34–35
 basic model equations in, 37–41
 model behavior in, 46–48
 nucleate boiling and, 38, 41–44
 parameter evaluation in, 41–46
 rate of phase change and, 31–37
 two-phase forced convection and, 39–40, 43–46
 vaporization phenomena in, 28–31
Heat transfer without phase change, 9–28
 basic model equations in, 11–19
 parameter evaluation in, 19–28
Hertz-Knudsen equation, 32–33
Hydrodynamics, two-phase, 6–9

I

Initializing procedures, in pipeline network problems, 156–157
Iron ore
 agglomeration of, 56–57
 pelletization of, 56–57, 104–105
Iron ore balling, bentonite additive in, 102–105
Iterative methods, in pipeline network problems, 157–159

K

Kinetics
 of balling and granulation, 84–100
 nonrandom coalescence, 93–99
 random coalescence, 90–93

L

Laminar flows
 with complications, 259–263
 mass-transfer rate in, 254–259
Laminar-laminar flow, 11
Layering, in balling and granulation, 78–79
Limiting current
 approach to, 228–229
 cathode reactions and, 215
 conditions for valid measurement and interpretation of, 252–253
 diffusion layer and, 213–215
 gradual approach to, 244
 migration and, 216
 and potential distribution at elongated electrodes, 243–247
Limiting-current density, defined, 213
Limiting-current mass transfer, applications of, 253–279
Limiting-current measurement
 approach to limiting current in, 228–229
 correlations established by, 280–308
 counterelectrode in, 227
 diffusion coefficients in, 233–235
 electrochemical reactions used in, 219–223
 electrode boundary conditions and, 227–228
 electrode potential and, 223–227
 forced convection in, 241–243
 free convections in, 235–241
 historical survey of, 217–218
 in mass-transfer measurements, 211–308
 migration effects in, 231–233
 overpotential in, 223–225, 245–247
 particulate electrodes and, 276–279
 redox reaction in, 222
 surface overpotential in, 224, 245–247
 surface roughness and, 247–251
 transition times in, 239
 unsteady-state effects in, 235–243
Limiting-current plateau, defined, 229–230
Linearization method, in pipeline network problems, 155–156
Liquid-liquid flow patterns, 4–6
Liquid-liquid systems
 flow configurations for, 10
 heat transfer in, 12, 17–18
 parameter evaluation for, 19–20, 25–27
Liquid suspension, spherical agglomeration in, 115–117
Loop-defining algorithm, 164

M

Mass transfer
 electrolytic gas evolution in, 275–276
 free and mixed convection in, 263–268
 oscillating flows in, 273–274

particulate electrodes in, 276–279
stirred cells and, 274–275
in turbulent forced convection, 268–272
two-phase, 9
Mass-transfer coefficient, 216
Mass-transfer measurements, *see also*
  Limiting-current measurements
  basic theory in, 213–216
  diffusion coefficients and, 233–235
  effective diffusivities in, 233–235
  interpretation of results in, 229–251
  limiting-current technique in, 211–308
Mass-transfer rate
  in laminar flows, 254–259
  measurement of, 215–216
  surface overpotential and, 212, 224, 247
Mass-transfer studies, heat transfer problems and, 9
  see also Mass-transfer measurements
Microroughness, in mass-transfer limited deposition, 248
Migration effects, limiting current and, 216, 231–233
Molecular interchange process, 31–37

N

Natural gas pipelines, offshore, 188–190
  see also Pipeline network problems
Nernst diffusion layer thickness, 214
Nernst potential, in limiting-current measurements, 224
Newton-Raphson method, in pipeline network problems, 149–152, 157
Nucleate boiling, 38–39
  in heat transfer with phase change, 38, 41–44
Nucleation, in balling and granulation, 56, 78, 82
Nuclei growth region, in balling and granulation, 82
Nusselt number, 264

O

Onion skinning, in balling and granulation, 78

Optimal pipeline routing, in pipeline network design, 185–188
Oscillating flows, in mass transfer, 273–274

P

Parameter evaluation
  in heat transfer with phase change, 41–46
  in heat transfer without phase change, 19–28
Particulate electrodes, in limiting-current mass-transfer measurements, 276–279
Peclet number, 20, 26
Pelletization
  dry, 63, 113–115
  spontaneous, 73
  wet or green, 56
Pelletizer, disk, 58
Pendular bonds, 66–69
Penetration theory, net rate expression for, 35–37
Phase change
  heat transfer and, 9–48
  inside horizontal and vertical tubes, 29–30
  rate of, 31–37
Pipeline network design, 125–205
  applications of, 170–190
  column row reordering and cycle selection in, 162–166
  computational performance and, 197–198
  computer-aided, 199–200
  cyclic networks in, 183–185
  discrete merging in, 179–180
  finite difference methods in, 195–197
  for large networks, 160–168
  long-distance gas transmission in, 181–183
  for networks with regulators and other nonlinear elements, 168–170
  offshore natural gas gathering networks in, 188–190
  optimal pipeline routing in, 185–188
  optimization methods in, 172–185
  parameter estimation in, 174
  pressure-relieving piping networks and, 175–180
  sensitivity analysis in, 173–174

sparse computation techniques in, 166–168
steady-state network simulation programs in, 198
synthesis in, 185–190
test problems in, 200–203
water distribution networks in, 180–181
Pipeline network elements
  modeling of, 136–140
  pipe sections in, 136–138
Pipeline network problems
  admissible specification sets in, 144–146
  alternative formulations in, 140–144
  balancing flows in, 155
  balancing heads in, 154–155
  Broyden's method in, 152–153
  column row reordering and cycle selection in, 162–166
  diakoptics in, 162
  electrical circuits compared with, 146–148
  formulation of in steady state, 143–144
  generalized methods in, 152–154
  graphs and digraphs in, 127–130
  Hardy-Cross method in, 154–155, 159
  implicit finite difference methods for, 195–197
  initializing procedures in, 156–157
  iterative methods for, 150, 157–159
  in larger networks, 160–168
  linearization method in, 155–156
  method of characteristics for, 193–195
  methods of solution in steady state, 148–170
  Newton-Raphson method in, 149–153, 157
  numerical methods in, 148–160
  problem specifications in, 144–146
  pumps and compressors in, 139–140
  steady-state, 127–148
  tearing in, 160–162
  test problems and, 200–203
  transient network flow equations in, 192–198
  transient solution methods in, 159–160
  valves and regulators in, 138–139
  Wolfe's method in, 153
Pipeline networks
  equations of state for, 190
  governing equations for, 190–192
  transient and compressible flows in, 190–198
  transient flow solution methods for, 192–198
  volume of traffic in, 126
Plating baths, 249
  see also Limiting-current mass transfer; Limiting-current measurements
Powders, dry pelletization of, 63
Prandtl number, 22
Pressure bonds, in balling and granulation, 70–71

R

Rayleigh number, 267
Redox reaction, in limiting-current measurements, 222
Regime I flows, model equations for, 11–13
Regime II flows
  model equations for, 13–16
  parameter evaluation for, 22–24
Regime IV flows, model equations for, 16–19
Reynolds number, 2, 19, 22–23
Rotating-disk electrode, 228, 261

S

Saturated vapor phase, in heat transfer with phase change, 47
Sensitivity analysis, in pipeline network design, 173–174
Sensitivity matrix, computation of, 173–174
Separated flows, defined, 11
Snowballing mode, granulation in, 109–112
Snowballing kinetics, in balling and granulation, 85–86
Solution-phase theory, 106–108
Sparse computation technique, in pipeline network design, 166–168
Stirred cells, in mass transfer, 274
Surface overpotential
  current distribution and, 245–247
  in limiting-current measurements, 224
Surface roughness
  in limiting-current measurements, 247–251
  in turbulent mass transfer, 272

T

Tearing, in pipeline network problems, 160–162

Transition region, in balling and granulation, 83

Tubular fluid-fluid systems, heat transfer in, 1–49

Tubular systems
 flow patterns in, 2–6
 pressure drop in, 8–9

Turbulent flow, temperature profile in, 20

Turbulent forced convection, in mass transfer, 268–272

Turbulent mass transfer
 expression for, 269–272
 surface roughness and, 272

Turbulent-turbulent flows, 11

Two-phase heat transfer devices, design of, 6–9

V

Valves and regulators, in pipeline network problems, 138–139

Van der Waals forces, 63, 73

Vaporization, absolute rate of, 44

Vapor phase, saturated, 47

W

Wolfe's method, in pipeline network problems, 153–154

TP
145
A4
v.10
1978

NOV 13 1978